The Book of

10,000

Incredible

Facts

Publications International, Ltd.

Scripture quotations from *The Holy Bible, King James Version*

Copyright © 2023 Publications International, Ltd. All rights reserved. This book may not be reproduced or quoted in whole or in part by any means whatsoever without written permission from:

Louis Weber, CEO
Publications International, Ltd.
8140 Lehigh Avenue
Morton Grove, IL 60053

ISBN: 978-1-63938-418-1

Manufactured in China.

8 7 6 5 4 3 2 1

Let's get social!

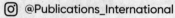 @Publications_International

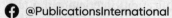 @PublicationsInternational

www.pilbooks.com

Contents

✳ ✳ ✳ ✳

Chapter 1

Animals

36 Freaky Facts About Animal Mating

1. The funnel-web spider knocks his mate unconscious with pheromones before mating.

2. The male *Argyrodes zonatus* spider secretes a drug that intoxicates the female, which is good because otherwise she would devour him.

3. Harlequin bass and hamlet fish take turns being male and female, including releasing sperm and eggs during mating.

4. Male North American fireflies flash their light every 5.8 seconds while females flash every 2 seconds, so there isn't any confusion.

5. An albatross can spend weeks courting, and their relationships can last for decades.

6. However, the actual act of mating lasts less than a minute.

7. Fruit flies perform an elaborate seven-step dance routine before mating.

8. No copulation takes place unless each part is completed perfectly.

9. Male mites mate with their sisters before they are born.

10. After birth, the females rush off in search of food and their brothers are left to die.

11. Mayflies live for one day, during which they do nothing but mate.

12. The rattlesnake has two penises.

13. The penis of the echidnas has four heads.

14. A pig's penis is shaped like a corkscrew.

15. The male swamp antechinus, a mouse-like marsupial in Australia, has sex until he dies, often from starvation.

16. Sometimes he's simply too weak after mating to escape predators.

17. It's no fun being a male honeybee. Those "lucky" enough to engage in a mating flight with a virgin queen usually die after their genitalia snap off inside her.

18. Flatworms have both male and female sex organs.

19. During mating, two worms will "fence" with their penises until one is pierced and impregnated.

20. During mating season, male frigate birds inflate their throat sacs while engaging in a wild dance.

21. The females usually hook up with the males possessing the largest, brightest sacs.

22. When it comes time to mate, male Galapagos giant tortoises rise on their legs and extend their necks.

23. The male with the longest neck gets the girl.

24. When a female red-sided garter snake awakens from hibernation, she releases a scent that attracts every male in the area.

25. The result: a huge, writhing "mating ball."

26. Male giraffes won't mate unless they know a female is in estrus.

27. To find out, they nudge a prospective mate's rump until she urinates—then taste her urine.

28. Percula clownfish live in families consisting of a mating male and female and several nonbreeding males.

29. If the female dies, the mating male becomes the female, and one of the nonbreeding males gets promoted to hubby.

30. Male dolphins have such strong libidos that they've been known

to mate with inanimate objects and even with other sea creatures, such as turtles.

31. When the male snowy owl wishes to arouse a female, he dances while swinging a dead lemming from his beak.

32. Amorous great gray slug couples hang from a rope of their slime as they twist around each other in the throes of slug passion.

33. In a Mediterranean species of cardinal fish, the male takes part in mouthbrooding—holding the fertilized eggs in his mouth until they're ready to hatch.

34. All shrimp form harems, usually consisting of one male and up to 10 females.

35. When the male leader dies, he may be replaced by a young female that can change her gender to take his place.

36. After the female emperor penguin lays an egg, the male protects it in his brood pouch—a roll of skin and feathers between his legs that drops over the egg—until it hatches.

34 Elephant Facts

1. The Asian elephant is endangered, while the African elephant is vulnerable.

2. A female elephant is pregnant for 22 months—almost two years.

3. Depending on the weather, an Asian elephant can guzzle 30 to 50 gallons of water every day.

4. The largest known specimen of the African savanna elephant is on display at the Smithsonian's National Museum of Natural History in Washington, D.C.

5. It stood 13 feet tall and weighed 22,000 pounds when it was alive.

6. Elephants never forget distant watering holes, other elephants, and humans they've encountered, even after many years.

7. Having a "fat day"? Forget it. The average male Asian elephant weighs between 10,000 and 12,000 pounds.

8. African savanna elephants are currently most common in Kenya, Tanzania, Botswana, Zimbabwe, Namibia, and South Africa.

9. Elephants supplement the sodium in their food by visiting nearby mineral licks.

10. The closest relatives to the elephant are the hyraxes (small chunky mammals that resemble fat gophers), dugongs and manatees, and aardvarks.

11. Several physical characteristics distinguish Asian and African elephants.

12. Asian elephants are generally smaller, with shorter tusks.

13. They also have two domed bulges on their foreheads, rounded backs, less wrinkly skin, and their trunks have a finger-like projection at the tip.

14. African forest elephants live in central and western Africa, with the largest populations found in Gabon and the Republic of the Congo.

15. African elephants speak their own special language.

16. Communication takes place using rumbles, moans, and growls.

17. These low-frequency sounds can travel a mile or more.

18. An elephant's trunk comprises 150,000 different muscle fibers.

19. Ever wonder why elephants have trunks? It's because an elephant's neck is so short that it wouldn't be able to reach the ground.

20. The trunk allows the elephant to eat from the ground as well as the treetops.

21. Both female and male African elephants have tusks.

22. But only male Asian elephants have tusks.

23. An elephant's trunk weighs about 400 pounds.

24. Between 1979 and 1989, Africa's elephant population went from 1.3 million to 750,000, as a result of ivory poaching.

25. About 415,000 African elephants remained as of 2016 (the last time we have good numbers).

26. Elephants greatly enjoy baths, but in the meantime the red-billed oxpecker (a relative to the starling) picks ticks and other parasites off the elephant's skin.

27. Those big, floppy ears aren't just for decoration—they help the huge animals cool off.

28. A group of elephants is called a herd.

29. When two herds join forces (which happens during migration) it's called a clan.

30. Elephants were first seen in Europe in 280 B.C., when an army of 25,000 men and 20 elephants crossed from North Africa to Italy.

31. Adult elephants can't jump.

32. Elephants use mud as sunscreen.

33. Under the right conditions, an elephant can smell water from approximately three miles away.

34. Elephants sometimes "hug" by wrapping their trunks together as a way of greeting one another.

16 Odd Ostrich Facts

1. Contrary to popular belief, ostriches don't bury their heads in the sand.

2. Ostriches can't fly, but they can run at speeds of 45 miles per hour for up to 30 minutes.

3. The African black ostrich (*Struthio camelus domesticus*) is farmed for meat, leather, and feathers.

4. Ostriches have the best feed-to-weight ratio gain of any farmed land animal in the world and produce the strongest commercially available leather.

5. The ostrich's eyes are about the size of billiard balls.

6. An ostrich's brain is smaller than either of its eyeballs. This may explain why it tends to run in circles.

7. The ostrich's intestines are 46 feet long—about twice as long as those of a human.

8. Ostriches in captivity have been known to swallow coins, bicycle valves, alarm clocks, and even small bottles.

9. The ostrich is the largest living bird, standing six to ten feet tall and weighing as much as 340 pounds.

10. This great bird has only two toes on each foot; all other birds have three or four.

11. Ostriches kick forward, not backward, because that's the direction in which their knees bend.

12. Their powerful, long legs can deliver lethal kicks.

13. Although ostrich eggs are the largest of any bird, they are the smallest eggs relative to the size of the adult bird.

14. A three-pound egg is only about 1 percent as heavy as the ostrich hen.

15. Ostriches never need to drink water.

16. Some of it they make internally, and the rest is derived from their food.

51 Rodent Facts

1. Rodents are the largest group of mammals, constituting nearly half of the class Mammalia's species.

2. The world's 4,000 known rodent species are divided into three sub-orders: squirrel-like rodents (Sciuromorpha), mouse-like rodents (Myomorpha), and porcupine-like rodents (Hystricomorpha).

3. Rodents include not only "true" rats and mice, but also such diverse groups as beavers, marmots, pocket gophers, and chinchillas.

4. Chinchillas have the thickest fur of all land animals with more than 50 hairs growing from a single follicle.

5. Like other rodents, a chinchilla's teeth never stop growing.

6. Chinchillas are native to the Andes Mountains in Chile.

7. Their thick fur allows them to survive harsh winds and plunging temperatures at elevations of 12,000 feet.

8. Some people keep chinchillas as pets. Only a few thousand survive in the wild.

9. Mice can get by with almost no water; they get most of the moisture they need from their food.

10. Mice become sexually mature at six to ten weeks and can breed year-round.

11. Female mice average six to ten litters annually.

12. If a pair of mice started breeding on January 1, they could have as many as 31,000 descendants by the end of the year.

13. Measuring over two feet tall, four feet long, and weighing an average of 100 pounds, the South American capybara is the largest rodent in the world.

14. Capybaras love the water and can remain submerged for up to five minutes.

15. Rats can jump three feet straight up and four feet outward from a standing position.

16. Rodents spread myriad diseases, including bubonic plague, leptospirosis, tularemia, salmonellosis, murine typhus, and hantavirus.

17. The bubonic plague that killed millions throughout Europe in the mid-1300s was predominantly caused by fleas that rats carried.

18. Thousands of Americans are bitten by rats each year.

19. Rats constantly gnaw anything softer than their teeth, including bricks, wood, and aluminum sheeting.

20. The average lifespan of a rat is less than three years.

21. But one pair of rats can produce 2,000 offspring in a year.

22. Rats use their tails to regulate their temperature, communicate, and balance.

23. Some rats can swim as far as half a mile in open water, dive through water-plumbing traps, travel in sewer lines against strong currents, and stay underwater for as long as three minutes.

24. A single rat can produce 25,000 droppings in a year.

25. Rats can enter a building through a hole just half an inch.

26. A rat can fall 50 feet without injury.

27. Rats can jump 36 inches vertically and 48 inches horizontally.

28. Rats are intelligent and have excellent memories.

29. Once rats learn a route, they never forget it.

30. Rats are color blind and cannot vomit or burp.

31. Beavers are the largest rodents in the world after the South American capybara.

32. Transparent eyelids allow beavers to see while swimming underwater.

33. Beavers can stay submerged in water for up to 15 minutes.

34. Using their webbed feet for speed and their flat tail as a rudder, they can swim as fast as five miles per hour.

35. A single beaver can fell an aspen tree with a six-inch diameter in about 20 minutes.

36. It then gnaws the tree into logs of a more manageable size and drags these logs back to the river.

37. Beavers live in colonies of six to eight and build dams and the lodges they live in.

38. Porcupines look a little like hedgehogs, but they are not related.

39. In fact, hedgehogs aren't rodents at all.

40. The average porcupine has 30,000 quills. And each is a sharp reminder that if you poke at a porcupine, it will poke you back!

41. Baby porcupines are born with soft quills that don't harden until a few days after birth— a fact for which mother porcupines are supremely grateful.

42. Porcupines can float.

43. The Baluchistan pygmy jerboa is the world's smallest rodent at only one-and-a-half inches long.

44. Pocket gophers, named for their large, fur-lined cheek pockets, can close their lips behind their protruding, chisel-like front teeth.

45. This allows them to excavate soil without ingesting it.

46. Voles have blunt noses, small furry ears, dense brown fur, and a tail with no fur.

47. Voles typically live about three to six months.

48. Marmots are giant land squirrels.

49. Flying squirrels glide through the air using parachute-like wings.

50. A squirrel can smell a nut buried under a foot of snow.

51. The naked mole-rat, a burrowing rodent native to East Africa, is almost completely hairless with wrinkly skin.

23 Fun Fish Facts

1. There are more than 30,000 different species of fish.

2. Three major types of fish include jawless (e.g. lamprey eels), cartilaginous (e.g. sharks), and bony (e.g. blue marlin).

3. While fossils show that the earliest fish were jawless, the only remaining survivors in the jawless fish group are hagfish and lampreys.

4. Hagfish feed on dead fish at the ocean bottom, using their tongues to rasp at food with a pair of brushes covered in horn-like teeth.

5. Hagfish are sometimes called slime hags because of the large amounts of mucus they produce.

6. One hagfish can fill a two-gallon bucket with slime in a matter of minutes.

7. All lampreys start life as freshwater larvae, filtering particles from the bottom of riverbeds before developing teeth as adults.

8. The heaviest bony fish is the ocean sunfish, which can weigh a whopping 5,000 pounds.

9. The ocean sunfish produces about as many eggs at one time as there are people in the United States.

10. Of all fish, seahorses are the slowest.

11. Some seahorses swim less than five feet per hour.

12. Australian leafy sea dragons mask themselves as seaweed.

13. The green creatures have long, tattered ribbons of skin resembling seaweed fronds growing from their bodies.

14. Scientists can figure out how old a fish is by counting growth rings on its scales or its otoliths (ear bones).

15. The orange roughy lives more than 100 years.

16. The Mariana snailfish (*Pseudoliparis swirei*) thrives at depths up to 26,200 feet below the surface.

17. These deep-ocean dwellers were discovered along the Mariana Trench.

18. Some species can swim backwards, but usually don't. Those that can are mostly eels.

19. An electric eel can produce 600 volts of electricity in a single jolt.

20. Whale sharks are the longest fish in the sea, with some growing over 40 feet long.

21. At just nine millimeters long, the dwarf goby fish is the world's smallest fish.

22. Sailfish can swim as fast as 68 miles per hour.

23. Goldfish only have a memory of three seconds.

22 Facts About Bats

1. Bats are the only mammals that can truly fly.

2. A quarter of all mammals are bats.

3. There are over 1,400 species of bats worldwide.

4. A single brown bat can eat up to 1,200 mosquito-size insects in one hour.

5. Because bats eat so many insects, which have exoskeletons made of a shiny material called chitin, some bat poop sparkles.

6. Bats are nocturnal, mostly because it's easier to hunt bugs and stay out of the way of predators when it's dark.

7. Bats use echolocation to navigate in the dark: They send out beeps and listen for variations in the echoes that bounce back at them.

8. Inside those drafty caves they like so much, bats keep warm by folding their wings around them, trapping air against their bodies for instant insulation.

9. Vampire bats don't suck blood. They puncture their prey's skin with sharp incisors, then lap up the flowing blood.

10. Vampire bat saliva contains an anticoagulant that prevents blood from clotting too quickly.

11. There are three species of vampire bats, which only drink blood: the common vampire bat, the hairy-legged vampire bat, and the white-winged vampire bat.

12. Bats have only one pup a year: Most smallish mammals have way more offspring.

13. Over 300 species of fruit—including bananas, mangoes, and avocados—rely on bats for pollination.

14. The rare suckerfooted bat of Madagascar has small suction cups

on its hands, allowing it to cling to smooth surfaces as it glides through forests.

15. The *Anoura fistulata* nectar bat has a longer tongue relative to its body length than any other mammal.

16. When the bat is not using its tongue to reach inside flowers and get pollen, the appendage curls up in the bat's rib cage.

17. Bats clean themselves and each other meticulously by licking and scratching for hours.

18. The world's smallest mammal is Kitti's hog-nosed bat, also called the bumblebee bat. It weighs only as much as a dime.

19. Flying foxes are among the world's largest bats with wingspans of up to six feet.

20. Millions of bats have died in recent years from white-nose syndrome, named for the white fungus on bats' muzzle and wings.

21. Some species of bats can live more than 30 years.

22. Nearly 1.5 million bats living in North America's largest urban bat colony call the Congress Avenue Bridge in Austin, Texas, home.

51 Marvelous Mammal Facts

1. Mammals are warm-blooded animals that breathe air.

2. Zebras are black with white stripes (not white with black stripes) and have black skin.

3. A polar bear's fur is actually transparent rather than white; it merely appears white due to the way it reflects light.

4. Howler monkeys are the loudest land animals; their calls can be heard up to three miles away.

5. The Japanese macaque is that rare monkey that likes cold weather.

6. When temperatures in its mountain habitat drop below freezing, the macaque lounges in natural hot springs.

7. It's not water in a camel's hump.

8. The fat stored in a camel's hump allows the animal to trek across the desert for up to a month without food.

9. Pigs, light-colored horses, and walruses can get sunburned.

10. Measuring up to 110 feet long and weighing up to 419,000 pounds, the blue whale is the largest mammal (and animal) on Earth.

11. Most elephants weigh less than the tongue of a blue whale.

12. Just as humans favor their right or left hand, elephants favor their right or left tusk.

13. The echidna is a spiny anteater that can grow to three feet long.

14. The echidna and the platypus are the only egg-laying mammals.

15. The platypus is also one of the few venomous mammals.

16. The male platypus has a spur on his hind feet that can deliver venom.

17. The pangolin is the only mammal with scales.

18. It has no teeth, and uses its powerful claws to tear open termite and ant mounds.

19. The African honey badger's tough hide can resist penetration and most poisons, a great help when the badger's dinner includes puff adders or beehive honey.

20. Hedgehogs have 3,000 to 5,000 quills on their backs.

21. A mature ewe yields seven to ten pounds of shorn wool per year.

22. The sleepiest mammals are armadillos, sloths, and opossums. They spend 80 percent of their lives dozing.

23. The nine-banded armadillo is the only mammal known to always give birth to four identical young.

24. The tallest mammals are giraffes, towering up to 20 feet tall.

25. A giraffe can clean its ears with its 21-inch-long tongue.

26. The dark purple color of a giraffe's tongue protects it from getting sunburned.

27. Both male and female caribou grow antlers—the only deer species to do so.

28. The red deer on the island of Rhum (in Scotland) kill seabird chicks and gnaw the bones to get nutrients otherwise unavailable on the isle.

29. Rhinoceroses are odd-toed hoofed mammals, with three toes on each foot.

30. Horses and zebras have only one toe on each foot.

31. Even-toed hoofed mammals include pigs, deer, cattle, antelopes, giraffes, camels, and hippopotamuses.

32. The hippopotamus can open its mouth up to 3.3 feet wide—wider than any other land animal.

33. Hippos can run at speeds of up to 20 miles per hour.

34. The rabbit-size mouse deer of Asia has long upper incisors that make it look like a vampire.

35. But these timid creatures quickly flee when they encounter people.

36. Koalas aren't bears.

37. Koalas are marsupials, which means they raise their young in special pouches.

38. Eucalyptus leaves are poisonous to most animals, but koalas have special bacteria in their stomachs that break down dangerous oils in the leaves.

39. A koala gets almost all the liquid it needs from licking dew off tree leaves.

40. Some kangaroos can jump five times their body length in one jump.

41. Tiny shrews, sometimes only a few inches long, can kill prey twice their size.

42. They are able to do this partly because their saliva contains a paralyzing substance similar to cobra venom.

43. Cows and other ruminants have four digestive chambers.

44. An opossum will empty its anal glands when "playing dead" to help it smell like a rotting corpse.

45. Wolverines may look like small bears, but they're actually the largest species of weasel.

46. The Tasmanian devil can give birth to as many as 30 live young at one time.

47. But because a mama Tasmanian devil has only four teats, her babies compete fiercely for milk.

48. Eventually, Mom ends up eating many of her kids.

49. The most widespread meat-eating mammals in the world are red foxes.

50. Their natural range includes much of the Northern Hemisphere.

51. Mammals have better hearing than non-mammals thanks to special ear bones.

35 Amazing Ant Facts

1. Ants outnumber humans a million to one.

2. Their combined weight outweighs the combined weight of all the humans.

3. Ants are found on every continent except Antarctica.

4. Ants can lift 10 to 50 times their body weight.

5. Ants don't have ears.

6. They "hear" by feeling vibrations in the ground through their feet.

7. There are three kinds of ants in a colony: the queen, the female workers, and the males.

8. Depending on the species, a colony may have one queen or multiple queens.

9. Male ants' only job is to mate with the queen.

10. Males die soon afterwards.

11. Only a queen ant can lay eggs.

12. During a queen's 10- to 30-year reign, she may lay millions of eggs.

13. A queen's fertilized eggs become females; unfertilized eggs become males.

14. Most females are born sterile, consigned to be workers.

15. Ants pass through four life stages: egg, larvae, pupae, and adult.

16. The tiny ant eggs are sticky, allowing them to bond together for ease of care.

17. Young workers care for the queen and larvae, then graduate to nest duties such as engineering, digging, and sanitation.

18. Finally, they advance to the dangerous positions of security and foraging.

19. When foraging, ants leave a pheromone trail to show where they've been.

20. Only four species engage in agriculture: humans, termites, bark beetles, and ants. But ants were the first.

21. Leafcutter ants carefully cultivate subterranean fungus gardens by spraying their crops with self-produced antibiotics to ward off disease, then fertilizing them with their protease-laced anal secretions.

22. Leafcutter colonies can contain as many as two million ants.

23. A leafcutter colony can strip a citrus tree's leaves in less than a day.

24. Ants engage in livestock farming.

25. Ants domesticate and raise aphids, which they milk for honeydew.

26. Honeydew provides important nourishment for ants, which are incapable of chewing or swallowing solids.

27. Though blind, nomadic South American army ants fearlessly attack reptiles, birds, small mammals, and other insects (which they kill but don't eat) in their path.

28. There are over 12,000 ant species in the world.

29. The barbarous *Polyergus rufescens* species, or slave-maker ants, raid neighboring nests to steal their young.

30. No males of the species *Mycocepurus smithii* have yet been found.

31. The queen ant reproduces asexually, so all offspring are clones of the queen.

32. Carpenter ants build their homes, called galleries, within wood.

33. They prefer slightly moist wood.

34. Ants stretch when they wake up.

35. Ants also appear to yawn in a very human manner before taking up the tasks of the day.

11 Facts About Opossums

1. The rat-tailed opossum is the only marsupial native to North America.

2. Contrary to popular belief, opossums do not sleep hanging from trees by their tails.

3. Opossums possess a whopping 50 teeth—more than any other land-dwelling North American mammal.

4. The lowly opossum has opposable, thumblike digits on all four paws and a tail that can grasp food and tree branches.

5. A male opossum is called a jack.

6. A female opossum is called a jill.

7. After a gestation of just 12 to 13 days, females give birth to as many as 20 live young at a time.

8. Baby opossums, called joeys, are the size of jelly beans at birth.

9. After baby opossums leave the pouch (at between two and three months), the mother carries all of them on her back for the next month or so whenever the family leaves its den.

10. Opossums really do play dead to trick predators.

11. Opossums are immune to snake bites, bee stings, and other toxins.

25 Amphibian Facts

1. Amphibians live both on land and in water.

2. *Amphibian* comes from a Greek term meaning "double life."

3. Amphibians include frogs and toads, salamanders and newts, and wormlike caecilians.

4. The largest caecilians can grow more than five feet in length.

5. Axolotls are rare amphibians that live in Central American lakes.

6. All amphibians have gills; some only as larvae and others for their entire lives.

7. The largest amphibian in the world is the Chinese giant salamander, which can grow up to six feet long.

8. The world's smallest amphibian is the *Paedophryne amauensis*, a frog from Papua New Guinea measuring just 0.3 inch in length.

9. Amphibians are found everywhere in the world except Antarctica.

10. Some salamanders can regrow entire limbs and regenerate parts of major organs.

11. At about four weeks old, tadpoles get a bunch of very tiny teeth, which help them turn their food into mushy, oxygenated particles.

12. Frogs swallow their food whole.

13. The size of what they can eat is determined by the size of their mouth and their stomach.

14. While swallowing, a frog's eyeballs retreat into its head, applying pressure that helps push food down its throat.

15. Some frogs glow after eating fireflies.

16. When frogs aren't near water, they will often secrete mucus to keep their skin moist.

17. Frogs typically eat their old skin once it's been shed.

18. The Goliath frog of West Africa is the largest frog in the world.

19. When fully stretched out, it is often more than 2.5 feet long!

20. A small, blind cave salamander called the olm (also called "the human fish" for its humanlike skin tone) is the world's longest-lived amphibian.

21. It can live up to 100 years.

22. A frog's ear is connected to its lungs. When a frog's eardrum vibrates, its lungs do, too.

23. This special pressure system keeps frogs from hurting themselves when they blast loud mating calls.

24. Wood frogs can continuously freeze and thaw throughout the winter in order to hibernate under surface leaves.

25. Just one golden poison frog has enough toxin to kill ten people.

27 Surprising Facts About Spiders

1. Spiders are arachnids, not insects.

2. Unlike insects, spiders cannot fly.

3. After they hatch, spiderlings drift away on silk string.

4. This first flight is called "ballooning."

5. Some young spiderlings have been seen ballooning at altitudes of 10,000 feet.

6. Studies have shown that jumping spiders can solve simple 3-D puzzles.

7. Jumping spiders also learn the behavior patterns of other spiders in order to capture them.

8. The fisher, or raft, spider can walk on water.

9. When it detects prey (insects or tiny fish) under the surface, it can quickly dive to capture its dinner.

10. Male spiders are almost always smaller than the females and are often much more colorful.

11. It is estimated that up to 1 million spiders live in one acre of land.

12. In the tropics, that the number might be closer to 3 million.

13. Some spiders live underwater all of their lives. They surface to collect a bubble of air, which acts as an underwater lung.

14. Male spiders are unique among all animals in having a secondary copulatory organ.

15. While most spiders live for a year or two, certain tarantulas can live for 20 years or longer.

16. Spiders eat more insects than birds and bats eat combined.

17. Some species of spiders have evolved to mimic ants in appearance and pheromones, which they use to prey on the ants.

18. The body of a typical spider has approximately 600 silk glands.

19. Spider silk is more flexible than nylon, and in some cases stronger than steel and Kevlar.

20. All spiders secrete silk from their abdomens.

21. Tarantulas can also shoot silk from their feet.

22. Most spiders are nearsighted.

23. To compensate, they rely on their body hair to feel their way around and to detect nearby animals.

24. While most spiders have eight eyes, some have fewer, though always an even number.

25. The world's most venomous spider is the male Sydney funnel-web spider *Atrax robustus*.

26. The female is far less dangerous, being four-to-six times less effective than the male.

27. The black widow is North America's most venomous spider.

50 Animals by the Bunch

1. A congregation of alligators
2. A shrewdness of apes
3. A cete of badgers
4. A battery of barracudas
5. A kaleidoscope of butterflies
6. A wake of buzzards
7. A gang or obstinacy of buffalo
8. A quiver of cobras
9. A bask of crocodiles
10. A convocation of eagles
11. A parade of elephants
12. A cast of falcons
13. A business of ferrets
14. A charm of finches
15. A flamboyance of flamingoes
16. A leash or skulk of foxes
17. A gaggle of geese
18. A tower of giraffes
19. A troubling of goldfish
20. A bloat of hippopotamuses
21. A cackle of hyenas
22. A mess of iguanas
23. A husk of jackrabbits
24. A shadow of jaguars
25. A smack of jellyfish
26. A mob or troop of kangaroos
27. A conspiracy of lemurs
28. A leap of leopards
29. A barrel of monkeys
30. A murder of magpies
31. A passel of opossums
32. A family or romp of otters
33. A parliament of owls
34. A pandemonium of parrots
35. A covey of partridges
36. An ostentation of peacocks
37. A prickle of porcupines
38. A bevy of quails
39. A gaze of raccoons
40. A mischief of rats
41. A rhumba of rattlesnakes
42. An unkindness of ravens
43. A shiver of sharks
44. A stench of skunks
45. A dray or scurry of squirrels
46. A fever of stingrays
47. An ambush or streak of tigers
48. A knot of toads
49. A wisdom of wombats
50. A zeal of zebras

48 Remarkable Reptile Facts

1. Reptiles are cold-blooded animals covered in scales or bony plates. They include lizards, turtles, snakes, and crocodiles.

2. Lizards are the most common reptiles, with nearly 6,000 species around the world.

3. Chameleons change color with the help of special skin cells called chromatophores.

4. Chameleons use their color-changing ability mainly to show their mood, adjust their temperature, or communicate.

5. Chameleons use each eye independently, so they can look in two directions at once.

6. Leaf chameleons of Madagascar are the world's smallest reptile, measuring just over an inch from nose to tail.

7. The Komodo dragon is the world's biggest lizard.

8. An average male weighs 175 to 200 pounds and measures 8.5 feet long.

9. The shingleback lizard of Australia has a thick, short, rounded tail that's shaped just like its head.

10. This confuses predators, giving the slow, sleepy reptile a better chance to escape.

11. Horned lizards, often called horny toads, can squirt blood from their eyeballs to attack predators.

12. This only happens in extreme cases, but they can shoot it up to three feet.

13. The longest reptile is the reticulated python, measuring over 32 feet.

14. The world's smallest snake is the Martinique thread snake, which measures just over four inches long when fully grown.

15. Some snakes have more than 300 pairs of ribs.

16. Snakes have a taste sensor on the roof of their mouth.

17. They use their tongue to pick up scents in the air, then deliver them back to the sensor inside their mouth.

18. Snakes have no eyelids.

19. Several species of snakes, including the spitting cobra, fake death by flipping over on their backs when threatened.

20. The female boomslang, a type of African tree snake, looks so much like a tree branch that birds—its main prey—will land right on it.

21. Boas and pythons kill by squeezing, or constricting, their prey.

22. These constrictors wrap themselves around their prey and use their powerful muscles to suffocate them.

23. Vipers inject their prey with venom through their hollow, needle-pointed fangs.

24. Venomous snakes kill more than 100,000 people each year.

25. Female pythons keep their eggs warm by coiling on top of them and shivering their bodies to generate heat.

26. Tortoises live on land, while turtles live mostly in water.

27. The marine leatherback turtle can weigh as much as 1,800 pounds with an 8-foot shell.

28. Sea turtles can hold their breath for several hours.

29. Sea turtles absorb a lot of salt from the seawater in which they live.

30. They excrete excess salt from their eyes, so it often looks as though they are crying.

31. Giant tortoises can live more than 200 years.

32. The ribs and backbones of turtles and tortoises are connected to their shell.

33. Shells are made of solid plates called scutes, which are fused together.

34. The domed top is called the carapace.

35. Alligators lack chromosomes for sex determination.

36. Hot temperatures during incubation produce males, while cooler temperatures produce females.

37. Crocodiles bury their eggs in riverside nests.

38. The mother waits nearby for up to three months.

39. Crocs and gators are surprisingly fast on land, but lack agility.

40. If you're being chased by one, run in a zigzag line.

41. Unborn crocodiles make noises inside their eggs to alert the mother that it's time to hatch.

42. The mother then digs down to the nest and carries her babies to the water inside her mouth.

43. A mother crocodile can carry as many as 15 babies in her mouth at one time.

44. Female Nile crocodiles often gently roll the eggs in their mouth to crack the shell and help hatching babies emerge.

45. While crocodiles have large teeth and powerful jaws, they lift their young with incredible care.

46. As an alligator's teeth wear down, they are replaced with new ones.

47. As a result, the average alligator can go through 2,000 to 3,000 teeth over its lifetime.

48. Earth's largest crocodilian is the saltwater crocodile, which can grow as long as 23 feet and weigh as much as 2,200 pounds.

53 Stupendous Shark Facts

1. Sharks have been around for over 400 million years.

2. Sharks have no tongues.

3. Their taste buds are in their teeth.

4. Sharks have different types of teeth. Mako sharks have very pointed teeth.

5. White sharks have triangular, serrated teeth.

6. A sandbar shark will have around 35,000 teeth over the course of its lifetime.

7. The tiny cookiecutter shark, rarely seen by human eyes, has big lips and a belly that glows a pale blue-green color to help camouflage it from prey.

8. Its name comes from the small, cookie-shape bite marks it leaves.

9. Sharks have no bones. Their skeletons are made of cartilage.

10. Hammerhead sharks have 360-degree vision.

11. Greater hammerheads are known to group in large schools of 100 or more off the Island of Cacos near Mexico.

12. Shark skin feels like sandpaper because it's covered in tiny, pointed structures called placoid scales.

13. Sharks are known for their "sixth sense."

14. They can sense electromagnetic fields of other creatures and objects and temperature shifts thanks to special sensors called the Ampullae of Lorenzini.

15. A shark can be put into a catatonic state called "tonic immobility" when it's flipped onto its back or when the Ampullae of Lorenzini are appropriately stimulated.

16. Several types of sharks have demonstrated an affinity for being touched or for being put into a state of tonic immobility.

17. Humans are much more dangerous to sharks than sharks are to humans.

18. People kill as many as 100 million sharks every year, often when the sharks are accidentally caught in fishing nets.

19. Many other sharks are caught only for their fins, which are used in shark-fin soup.

20. Tiger sharks are often called the "garbage cans of the sea" because they will eat nearly anything.

21. Contents of their stomachs have revealed tires, baseballs, and license plates.

22. Shortfin makos are the fastest sharks.

23. They have been clocked at 36 miles per hour and have been estimated to swim up to 60 miles per hour.

24. They need this extreme speed to chase down their favorite food—the lightning-fast yellowfin tuna.

25. Scientists have identified more than 400 species of sharks.

26. Approximately 30 of those species are considered dangerous to humans.

27. Bull sharks, one of the most dangerous, aggressive shark species, have the highest testosterone levels of any animal in the world.

28. Bull sharks can survive in both saltwater and freshwater.

29. Whale sharks are the biggest fish in the ocean.

30. They can grow up to 40 feet and weigh more than 74,000 pounds.

31. Despite their large size, whale sharks are no danger to humans.

32. They feed on mostly plankton and small fish.

33. A whale shark's spot pattern is unique as a fingerprint.

34. Basking sharks are the world's second largest fish, growing as long as 32 feet and weighing more than five tons.

35. The pygmy ribbontail catshark is the smallest shark in the world, with a maximum length of seven inches.

36. Most sharks are solitary hunters, but some are quite social.

37. Blacktip reef sharks frequently hunt in packs, helping one another grab fish and crabs out of the coral.

38. The great white shark can go nearly three months without food.

39. Great white attacks are usually caused by the animals' curiosity about an unfamiliar object.

40. Lacking hands, the curious creatures "feel out" the new object with their teeth, usually in a gentle bite.

41. A person is 1,000 times more likely to be bitten by a dog than by a shark, and dogs kill more people every year than sharks do.

42. If you happen to be attacked by a shark, try to gouge its eyes and gills, its most sensitive areas.

43. Sharks are opportunistic feeders and generally don't pursue prey that puts up a fight in which they could be injured.

44. Gansbaii, South Africa, touts itself as the "Great White Capital of the World."

45. Its shores host the greatest concentration of great white sharks in any ocean.

46. An extinct relative of the great white, *Carcharodon megalodon* grew to 50 feet long.

47. They died out around a million years ago.

48. Not all sharks live in the ocean.

49. River sharks have been found in the rivers of South Asia, New Guinea, and Australia.

50. The number of pups in a litter varies by species.

51. A blue shark once gave birth to 135 pups in a single litter.

52. Female sharks have been known to use the sperm from multiple males when they reproduce—meaning that pups they give birth to at the same time may be just half-siblings.

53. The spiny dogfish shark can take two years to gestate before delivery—making it the longest gestation period of any vertebrate.

10 Monogamous Animals

1. **Prairie Voles:** The male of this rodent breed prefers to stay with the female he loses his virginity to and will even attack other females.

2. Bald Eagles: Like all raptors, including the golden eagle, hawk, and condor, bald eagles remain faithful until their mate dies.

3. Penguins: Penguins practice serial monogamy. This means they have several mates throughout their lives but only one at a time. Penguins usually switch it up each mating season.

4. Wolves: Monogamy makes it easier for alphas to display their strength and superiority over the other male members of the wolf pack. In a grey wolf pack, the male and female alpha mate for life, cementing their position as pack leaders.

5. Anglerfish: Like a parasite, the male bites into the female, fusing his mouth to her skin. Their bloodstreams merge together, and the male hangs there, slowly degenerating until he becomes nothing more than a source of sperm for the female.

6. Beavers: Eurasian beavers team up for life as a way to increase their chances of survival. By pairing up, couples can split their workload.

7. Black Vultures: Gossip gets around—just ask a vulture. If one of them gets caught mating with a bird that's not its partner, it gets harassed not only by its "spouse" but also by all the other vultures in the area.

8. Gibbon Apes: These monkeys make a close-knit family unit, with mom, dad, and babies even traveling together as a group.

9. Red-Backed Salamanders: Males of this species are the jealous type, physically and sexually harassing their female mates if they suspect infidelity or even see them associating with another male. Yikes.

10. Termites: Termites can mate for life (that's about five years) and raise a family together—but they often leave each other within the first two hours of mating if something better comes along.

32 Baby Animal Facts

1. Just like humans, all of the great apes have babies called infants.

2. Baby elephants often suck on their trunk for comfort.

3. A baby hare is called a leveret.

4. Baby camels are born without humps.

5. They can run within hours of birth.

6. Shark fetuses will kill and eat other fetuses by the score in order to increase their chances of making it to birth.

7. A baby porcupine is a porcupette.

8. Only about one out of a thousand baby sea turtles survives after hatching.

9. A baby alpaca, llama, vicuña, or guanaco is a cria.

10. While play fighting, sometimes male puppies let female puppies win on purpose.

11. A baby swan is a cygnet or a flapper.

12. Kangaroos, koalas, opossums, wallabies, and wombats have babies called joeys.

13. When a baby joey is too big to fit in the pouch, a koala mom will carry her baby on her back.

14. Baby goats are called kids. Momma goats are called nannies.

15. Baby pandas weigh about as much as a cup of tea.

16. A young eel is called an elver.

17. Baby cheetahs are born without spots and develop them later.

18. In addition to "piglet," baby hogs and boars can also correctly be referred to as shoats (newly weaned pigs) or a farrow (a collective term for a group of young pigs).

19. Baby cougars are born with spots but lose them as they age.

20. A baby goose is a gosling.

21. Fox kits are born blind, deaf, and toothless, but mature quickly.

22. The name for a baby oyster is a spat.

23. Okapi calves can walk just 30 minutes after birth.

24. But they don't poop until they are four to eight weeks old.

25. When a giraffe is ready to give birth after a 15 month gestation period, she'll often return to the place she was born.

26. Giraffes give birth standing up, so the baby (calf) falls from five or six feet to the ground.

27. Baby giraffes have little horns called "ossicones" just like adults, but they are folded down inside the womb.

28. They stand up straight a few hours after birth.

29. A baby pigeon is a squab or squeaker.

30. A baby echidna or platypus is called a puggle.

31. Baby fish are initially called larvae, then fry, and finally fingerlings before they mature into adult fish.

32. A newborn turkey chick has to be taught to eat, or it will starve.

15 Humdinger Hummingbird Facts

1. There are about 330 different species of hummingbirds.

2. Most live in Central and South America.

3. Due to their small body size and lack of insulation, hummingbirds lose body heat rapidly.

4. To meet their energy demands, they enter torpor (a state similar to hibernation), during which they lower their metabolic rate.

5. During torpor, the hummingbird drops its body temperature by 30°F to 40°F, and it lowers its heart rate from more than 1,200 beats per minute to as few as 50.

6. Excluding insects, hummingbirds have the fastest metabolism of any animal.

7. To provide energy for flying, hummingbirds must consume up to three times their body weight in food each day.

8. Unlike other birds, a hummingbird can rotate its wings in a circle.

9. A hummingbird can also hover in one spot; fly up, down, sideways, and even upside down (for short distances); and is the only bird that can fly backward.

10. The wing muscles of a hummingbird account for 25 to 30 percent of its total body weight, making it well adapted to flight.

11. However, the hummingbird has poorly developed feet and cannot walk.

12. Like bees, hummingbirds carry pollen from one plant to another while they are feeding, playing an important role in plant pollination.

13. Each hummingbird can visit between 1,000 and 2,000 blossoms every day.

14. The smallest bird on Earth is the bee hummingbird (*Calypte helenae*), native to Cuba.

15. It can comfortably perch on the eraser of a pencil.

31 Big Cat Facts

1. The "big cats" include the lion, tiger, jaguar, and leopard—all members of the genus *panthera*.

2. The snow leopard has been included in the genus *panthera* since 2008, but there is some controversy over moving them from their own genus, *uncia*.

3. Not all big cats have distinct larynx characteristics that allow them to roar.

4. Big cats are divided into two groups—those that roar, such as tigers and African lions, and those that purr.

5. Mountain lions purr, hiss, scream, and snarl, but they cannot roar.

6. While cheetahs can't roar, they can purr while inhaling and exhaling.

7. Cheetahs have more in common with their domesticated small cousins.

8. The cheetah is the world's fastest land animal.

9. It can reach speeds of 70 miles per hour in under 3 seconds.

10. Cheetahs can only maintain this pace for around 20 seconds.

11. Tigers are the most powerful cats in the world.

12. A Siberian tiger can grow up to 13 feet long and weigh as much as 660 pounds.

13. If you shaved a tiger, you'd discover striped skin under its fur.

14. A fully-grown male tiger can leap 33 feet in one jump.

15. Though they spend most of their time on land, tigers are one of the few cats that love water.

16. The white tiger is a rare variety of Bengal tiger with white fur and black stripes.

17. A tiger can tell if it's in another tiger's territory by sniffing trees marked with urine.

18. Leopards are the only big cats that regularly climb trees.

19. A leopard can easily drag prey twice its own body weight up to the branches of a tree.

20. Leopards eat everything from dung beetles to antelopes and young giraffes.

21. Black panthers are actually leopards with dark coats.

22. Jaguars are the largest cats in the Americas.

23. A jaguar's roar sounds more like a cough.

24. All lions live in groups called prides, usually consisting of one or two adult males, plus six to eight females and their cubs.

25. Because they are smaller, quicker, and more agile than males, female lions do the hunting.

26. While the lionesses are at work, the males patrol the area and protect the pride from predators.

27. Even though female lions do the hunting, male lions often get to eat first.

28. The African lion has the largest larynx and the loudest roar—it can be heard from five miles away.

29. The serval has the longest legs relative to its body of any cat.

30. A serval uses its large ears to listen for prey—then pounce.

31. Cheetahs will use their tails to help turn when they are running at full speed.

44 Marine Mammal Facts

1. Marine mammals are warm-blooded animals that live in or very near the ocean, breathe air, give live birth, and nurse their young.

2. Whales, dolphins, and porpoises are classified in the cetacean group of marine mammal.

3. There are over 70 different species of cetacean.

4. There are two suborders of whales—baleen whales (mysticetes) and toothed whales (odontocetes).

5. Baleen whales (called mysticeti) filter their food through huge comb-like plates made of flexible keratin.

6. Baleen whales include humpback whales, blue whales, North Atlantic right whales, and bowhead whales.

7. A blue whale can eat up to four tons of krill in a day.

8. The call of the blue whale registers an incredible 188 decibels, making it the loudest animal on Earth.

9. The blue whale is the largest animal that has ever lived (yes, including dinosaurs).

10. A newborn blue whale calf is almost as long as a school bus.

11. Male humpback whales sing complex songs in winter breeding areas that can last up to 20 minutes and be heard miles away.

12. Bowhead whales have massive skulls that can be over 16.5 feet long—or about 30 to 40 percent of their entire body length—that they use to break through ice.

13. When a whale surfaces after a dive, it exhales air from its lungs through its blowhole.

14. As the warmed air hits the colder ocean air, it condenses into droplets, making it appear that the whale is spouting water.

15. Unsurprisingly, toothed whales (called odonotoceti) have teeth that they use to sense, capture, and/or eat prey.

16. Whales in this category include narwhals, belugas, and sperm whales.

17. When hunting squid, a sperm whale may spend as much as an hour on a dive to depths of more than 3,000 feet, where the pressure is more than 1,400 pounds per square inch.

18. Young sperm whales often form coed schools that gradually split up as dominant males drive off smaller ones until just one male is left with a harem of up to 25 females.

19. Some sperm whales swim enough miles during their lifetime to circle Earth several times.

20. Most marine mammals produce urine that is saltier than seawater.

21. Beluga whales have a five-inch-thick layer of blubber and dorsal ridge that help them navigate through harsh, icy waters.

22. Beluga whales' complex communication repertoire of whistles, clicks, and chirps has prompted the nickname "canaries of the sea."

23. Whales can't swim backwards.

24. The killer whale, or orca, will actually beach itself to feast on baby seals. It then worms its way back to the water.

25. Don't let the "killer whale" name fool you. The orca is really a dolphin, the largest member of the dolphin family.

26. More than 30 species of dolphin inhabit every ocean in the world—even some rivers.

27. Each dolphin has a unique whistle it uses to identify itself.

28. Bottlenose dolphins surface two or three times per minute for air.

29. Right whale dolphins are the only members of the dolphin family without dorsal fins.

30. Seals, sea lions, and walruses are all grouped together in the scientific order Pinnipedia, which is Latin for "fin-footed."

31. All pinnipeds have four "fins" or flippers.

32. Pinnipeds spend most of their lives swimming and eating in water and come onto land or ice to give birth, rest, and molt.

33. Phocidae ("earless" seals) have tiny ear holes but no external ear flaps.

34. Walruses in the Pacific migrate with moving ice, climbing atop it to rest and give birth.

35. Manatees and dugongs are sirenians.

36. Like cetaceans, sirenians spend their whole lives in water.

37. Sirenians are the only group of marine mammal that eat no meat.

38. Polar bears and sea otters are marine fissipeds, "split-footed" members of the order Carnivora.

39. Although skilled swimmers, polar bears are the marine mammal least adapted to aquatic existence.

40. Polar bears rest, mate, give birth, and nurse their young on the ice.

41. They are particularly vulnerable to reductions in sea ice.

42. Polar bears use a nose-to-nose greeting when asking another bear for something.

43. Sea otters hold hands while sleeping to keep from drifting apart.

44. Sea otters have a pouch under their forearm to store food and their favorite rocks.

22 Facts to Buzz About

1. Bees and wasps have two pairs of wings.

2. Tiny hooks link each pair.

3. Bees have five eyes.

4. There are three small eyes on the top of a bee's head and two larger eyes in front.

5. Some bees spend their lives on their own, but most bumblebees and honeybees live in large colonies.

6. Bumblebee colonies may have up to 50 bees, while honeybee colonies may have many thousands.

7. Honeybee colonies contain three types of bee: the queen, drones, and worker bees.

8. Each honeybee colony has one queen that lays between 1,500 and 2,000 eggs a day.

9. A colony has a few hundred drones (male honeybees) that mate with the queen.

10. All worker bees are female, but can't lay eggs.

11. A large colony can have more than 50,000 workers who collect nectar to feed the hive.

12. Inside the hive, the bees build hexagonal cells from wax.

13. Hundreds of these cells together make the nest.

14. A queen bee might survive up to five years, but a worker bee only lives about 40 days.

15. In the course of its lifetime, a single worker bee will only make $1/12$th of a teaspoon of honey.

16. Honeybees must gather nectar from two million flowers to make one pound of honey.

17. Honeybees communicate by dancing.

18. Common wasps build their nests out of dead wood that they chew to make paper.

19. Wasps have hairy heads. The hairs sense information about the wasp's surroundings.

20. Honeybees die after they use their stinger.

21. Drones (males) don't have stingers.

22. Killer bees are among the most aggressive insects, and their stings can be fatal.

24 Bizarre Bird Behaviors

1. **Poacher:** The world's smallest owl at about five inches, the elf owl moves into abandoned gila woodpecker holes in cacti.

2. **Home invader:** The kea of New Zealand, the world's only cold-weather parrot, swings on car antennas, sleds down snowy roofs, and invades a lodge through its chimney and then trashes the joint in search of food.

3. **Neighborhood lookout:** Bright-beaked puffins adopt a low-profile walk to tell other puffins they are just passing through. The other puffins adopt a sentry pose to warn the tourists not to get any bright ideas.

4. **Airborne garbage disposal:** The gull-like sheathbill eats dead fish, other birds' eggs and babies, and even seal and bird droppings.

5. **Mugger:** Skuas are gull-like seabirds that chase other birds and force them to drop or cough up their food.

6. **Aussie storm chaser:** Huge, flightless emus run after rain clouds, hoping for water.

7. **Family planner:** Similar to the cockatoo, the galah of Australia raises a larger or smaller clutch of chicks, depending on food availability.

8. **Vermonter at heart:** The widespread sapsucker bores holes in trees, then slurps up the sweet sap.

9. **The Stuka:** These Arctic terns know little fear and will dive-bomb larger predators, often in squadrons.

10. **Deep-sea diver:** The common loon can dive more than 250 feet below the water's surface.

11. **Sponge dad:** A male sand grouse, living in Asian and African deserts, soaks himself in water, then flies back to the nest so his chicks can drink from his feathers.

12. **Captain Ahab:** The wetlands-dwelling anhinga spears fish with a long, sharp, slightly barbed beak that keeps dinner from sliding off.

13. **Detox dieter:** The stunning scarlet macaw eats clay from riverside deposits, which may help it process the toxic seeds it consumes.

14. **Busy mom:** A gray partridge hen can lay up to 20 eggs. She has to, because many partridge chicks don't survive.

15. **Lazy mom:** A paradise whydah hen lays her eggs in a finch nest. This fools the finch, which raises the chicks as its own.

16. **Fears nothing:** The two-and-a-half-foot-tall great gray owl has a wingspan of five feet and fiercely attacks anything that gets too close to its nest and owlets.

17. **Preventive measures:** The southern carmine bee-eater rubs a bee's "butt" against a tree branch to break off its stinger.

18. **Camouflage:** The Eurasian bittern nests among tall river reeds. When alarmed, it stretches its neck and either becomes motionless or, if there is a breeze, begins to sway in time with the reeds. Because of its plumage pattern—vertical dark-brown stripes on beige—the bittern becomes extremely difficult to see.

19. **Sanitary engineer:** A malleefowl lays eggs in a nest full of rotting vegetation. The decay gives off heat to keep the eggs warm; the male bird checks the temperature often and adjusts the pile as necessary.

20. **Sturdy swimmer:** American dipper birds, also known as "water duzels," use their strong wings to "fly" under and through water to catch prey.

21. **No-holds-barred:** The roadrunner jabs a rattlesnake with its sharp bill, shakes it, body-slams it, then administers a final peck in the head before devouring its prey headfirst.

22. **Crafty dad:** The male starling will line a nesting area with

vegetation that helps the baby birds resist the impact of bloodthirsty lice.

23. **Bone-breaker:** The lammergeier, a high-flying vulture, drops bones repeatedly to get at the tasty marrow.

24. **Adept with tools:** In the Galapagos, the woodpecker finch digs bug larvae out of wood with a stick, twig, or cactus spine.

22 Deadly Animal Facts

1. About 500 million cases of mosquito-borne diseases such as malaria, dengue, Zika, and West Nile virus occur each year, killing as many as 2.7 million.

2. Venomous snake attacks account for close to 100,000 deaths per year, making these reptiles the deadliest on the planet.

3. Depending on the species of snake and the severity of the bite, death can ensue in a matter of hours if no treatment is given.

4. Dogs infected by the rabies virus cause up to 35,000 human deaths per year.

5. Experts estimate that one-sixth of Earth's population is infected with some kind of roundworm.

6. The Ascaris roundworm leads to an infection called aschariasis that kills an estimated 4,500 people a year.

7. The scorpion is a highly compact killing machine that takes more than 3,000 human lives every year.

8. The most lethal is the fat-tailed scorpion of North Africa.

9. The tsetse fly transmits sleeping sickness (African trypanosomiasis).

10. Sleeping sickness threatens some 65 million people in sub-Saharan Africa, but deaths have recently declined to under 1,000 annually.

11. Tapeworms are responsible for infections that kill up to 1,000 people per year.

12. Humans can contract any one of several species of tapeworm from eating raw or undercooked meat.

13. Undercooked fish can also contain the parasite's eggs.

14. The Nile crocodile is the most dangerous crocodile to humans, followed by the crocs that inhabit Australia.

15. Crocodiles kill approximately 1,000 people each year.

16. We generally think of elephants as friendly animals, but don't let that image fool you—each year they crush some 500 humans.

17. Hippos are responsible for approximately 500 human deaths per year, and, like elephants, they charge, trample, and gore their victims to death.

18. Nearly a hundred swimmers are fatally stung every year by box jellyfish, which live in the tropical waters of Australia and Southeast Asia.

19. Cardiac arrest can occur if antivenin is not taken immediately.

20. Cape buffalo, found in Africa south of the Sahara, are responsible for an estimated 100 human fatalities each year.

21. Leopards kill about 30 people each year.

22. Lions kill approximately 25 humans each year.

28 Facts About Dogs & Cats

1. There are about 450 different breeds of domestic dogs.

2. Cats were first kept as pets in the Middle East about 4,000 years ago.

3. An average-size dog's mouth can exert up to 200 pounds of pressure per square inch.

4. Larger breeds can exert up to 400 pounds per square inch.

5. Just like human fingerprints, the ridged pattern on every cat's nose is unique.

6. The same is true of dogs' noses.

7. A breed of dog known as the Lundehune has six toes and can close its ears.

8. Calico cats are almost always female.

9. There are more than 100 different domestic cat breeds.

10. Dogs are the only mammals (other than humans) that have walked to the North and South Poles.

11. The largest breed of domesticated cat is the ragdoll.

12. Males can weigh up to 20 pounds, and females up to 15 pounds.

13. The world's smallest dog breed is the Chihuahua, named after the Mexican state.

14. The Manx cat originated in Isle of Man.

15. The Manx's lack of tail is caused by a gene mutation.

16. One of the largest domestic dog breeds, the St. Bernard was bred by monks at the St. Bernard hospice in the Swiss Alps.

17. It was a well-known mountain rescue dog.

18. Moscow, Russia, has a museum dedicated to cats.

19. Basset hounds can't swim well, but are quite adorable.

20. Three dogs survived the sinking of the *Titanic*: a Newfoundland, a Pomeranian, and a Pekingese.

21. In Paris, France, there are more dogs than people.

22. The basenji is an African hunting dog that washes itself like a cat and yodels and chortles, although it does not naturally know how to bark.

23. There are more than 500 million domestic cats in the world.

24. The Scottish Fold is the only cat breed whose ears fold down toward the front of the head.

25. Scottish Folds are born with straight ears, but they begin folding forward within three weeks.

26. Dogs demonstrate incredible problem-solving abilities, an understanding of basic arithmetic, and mastery of navigating complex social relationships.

27. A 2009 study found that the average dog can learn 165 words, which is on par with a two-year-old child.

28. Border collies are generally considered the smartest dog breed, followed by poodles and German shepherds.

44 Incredible Insect Facts

1. All insects have a body composed of three segments—head, thorax, and abdomen—encased in a hard exoskeleton.

2. Insects also have six jointed legs and a pair of antenna.

3. Insects make up the vast majority of the animal kingdom.

4. There are about one million insect species currently identified.

5. Scientists estimate there are 200 million living insects for every living human being on Earth.

6. The study of insects is called entomology.

7. Insects are a wildly diverse bunch that include ants, beetles, wasps and bees, true flies, butterflies and moths, crickets and grasshoppers, and bugs.

8. True flies have just a single pair of wings.

9. Instead of a rear pair of wings, true flies have two tiny clubs called halters.

10. These help steady the insect in flight.

11. Bluebottles are the first flies to show up at dead animals' bodies.

12. They breed in the rotting flesh, where their eggs hatch into whitish larvae called maggots.

13. Disease-carrying mosquitos are responsible for nearly half of all human deaths in history.

14. Male mosquitos drink nectar but females need blood to make eggs.

15. Unlike most flies that bite, both male and female tsetse flies suck blood.

16. A tsetse fly can drink up to three times its weight in blood at one time.

17. True bugs come from a particular group of insects known as hemipterans.

18. All bugs have a beak-shaped mouth they use to stab prey or plants and suck up the body fluids or sap.

19. Scientists have discovered more than 80,000 species of bug.

20. The assassin bug catches other insects and sucks out the juice from inside their bodies.

21. Cicadas are noisy bugs. Each species of cicada has its own song.

22. Male cicadas rapidly vibrate a pair of drum-like organs, called tymbals, to attract a mate.

23. Mayflies only live about one day as an adult.

24. Adult ladybugs can force themselves to bleed from their leg joints, releasing an oily yellow toxin with a strong smell that's poisonous to small birds and lizards.

25. A cockroach can regrow its wings, legs, and antennae, and can live without a head for up to a week.

26. Although they can withstand 10 times as much radiation as humans, cockroaches would not survive a nuclear blast.

27. Beetles make up nearly 50 percent of all insect species.

28. Whirligig beetles, which live in ponds and streams, have two sections of eyes: one part suited for underwater viewing and the other for ogling the atmosphere.

29. The Hercules beetle can grow 7.5 inches long.

30. About half that length is consumed by a threatening, sword-shape horn and a second smaller horn that curves back toward its head.

31. Fleas can jump up to 200 times their own length, but they can't fly.

32. Froghopper insects clamp onto plant stems and drain them of juice while excreting a whitish foam from their abdomens until they are covered in their own bubble bath.

33. The crafty antlion digs a sand pit by using its head as a shovel.

34. It then hides in the pit with only its giant mouth sticking out and waits for an ant to tumble into the pit.

35. Ounce for ounce, the jaws of the giant Fijian longhorn beetle are as powerful as those of a killer shark.

36. At close to 13 inches in length (20 inches with its legs stretched), the Borneo stick insect is the longest on Earth.

37. Also known as the bent twig insect, it can bend its body at an acute angle and stay that way for hours.

38. The female praying mantis (Chinese mantid) eats her paramours after mating.

39. She often devours his head as an appetizer while he is still fertilizing her eggs.

40. Caddisfly larvae construct little houses in ponds to live in until they reach their winged stage.

41. They use bits of leaves, small pebbles, and shells.

42. Earwigs, once widely (and wrongly) reputed to crawl inside human ears to lay their eggs, actually feed mostly on caterpillars, slugs, and already dead flesh.

43. Tree crickets sing in exact mathematic ratio to the air temperature.

44. Just count the number of chirps a cricket makes in 15 seconds, add 40, and the result will be the current temperature in degrees Fahrenheit.

22 Facts About Butterflies

1. Butterflies taste with their hind feet.

2. A butterfly passes through four lifecycle stages: egg, larva (caterpillars), pupa (chrysalis), and adult (butterfly).

3. Butterflies attach their eggs to leaves with a special glue.

4. Butterflies hatch from their eggs as caterpillars, or larva.

5. Caterpillars spend most of their time eating to store enough energy to transform into their adult shape.

6. Caterpillars shed their skin five times as they grow.

7. After two weeks, a caterpillar finds a suitable plant stem and grips onto it with its hind legs.

8. Once attached to the twig, the caterpillar will shed its skin again, this time to reveal a bright green protective shell called a chrysalis.

9. Inside the chrysalis, or pupa, the caterpillar completes its metamorphosis to transform itself from larva (caterpillar) to adult (butterfly).

10. This process takes about 10 days.

11. After an adult butterfly emerges from the chrysalis, it must stretch its wings and let them dry out before flying for the first time.

12. There are approximately 20,000 known species of butterflies.

13. Millions of monarch butterflies fly south each winter, traveling up to 2,800 miles from Canada and the eastern United States to Mexico and California.

14. A migrating monarch butterfly can flap its wings up to 2,000 times per minute.

15. The monarch butterfly is foul-tasting and poisonous due to the toxic chemicals it ingests when feeding on milkweed during the caterpillar stage.

16. The viceroy butterfly, which looks like a small monarch, is even more unpleasant-tasting.

17. Before monarch and queen butterflies become beautiful winged creatures, they're less-than-adorable caterpillars that often eat the eggs of their own species.

18. A group of caterpillars is called an army.

19. Butterflies have four wings.

20. Most butterflies rest with their wings folded together above their bodies.

21. Glasswing butterflies have see-through wings.

22. The Queen Alexandra's birdwing is the largest butterfly with a wingspan of up to 11 inches.

33 Parrot Facts

1. Parrots are a broad order of nearly 400 bird species.

2. Within the order, there are three families: Psittacidae ("true" parrots), Cacatuoides (cockatoos), and Strigopoidea (New Zealand parrots).

3. Both mandibles of a parrot's beak can articulate. This means that the upper mandible can move up, just as the lower mandible can move down.

4. This is useful for cracking open seeds and nuts and for climbing branches.

5. The zygodactyl feet of parrots have two forward facing, two backward facing toes.

6. Compared to humans' 10,000 taste buds, parrots only have about 300, mostly located on the roof of its mouth.

7. Even so, parrots have more taste buds compared to other bird species and show preferences for certain foods.

8. Cockatoos are unlike other "true" parrots in that they have a gallbladder and a moveable crest of feathers.

9. The black palm cockatoo is one of the most difficult birds to breed and raise in captivity.

10. Black palms lay one egg per clutch, and chicks often die around one year of age.

11. Cockatoos lack the typical feature composition of other parrots, meaning that their feathers cannot be green or blue.

12. In other parrots, the small air cavities of keratin cancel out red and yellow wavelengths, causing blue wavelengths to reflect back to the viewer's eye.

13. One of the most elusive birds in the world, the nocturnal night parrot was presumed extinct for about 100 years.

14. Another nocturnal parrot, the kakapo, is also the only flightless parrot.

15. It is endemic to New Zealand.

16. The omnivorous diet of the kea parrot consists mostly of roots, leaves, and berries, but can also encompass carrion and live petrel chicks.

17. Most species of parrots sleep standing with one foot up, but Southeast Asia's blue-crowned hanging parrots sleep upside down like bats.

18. Male and female eclectus parrots are so different that they were originally classified as separate species.

19. The hyacinth macaw is longer than any other parrot, and is the largest of the flying parrot species.

20. Unlike other macaws, the hyacinth macaw's facial patch is reduced to a small ring around its eye.

21. Patagonian conures, also known as burrowing parrots, tunnel into cliff faces and ravines when building their nests.

22. They are native to Argentina and Chile.

23. The small, green Carolina parakeet was North America's only native parrot species.

24. Now extinct, the last Carolina parakeet in captivity died in 1918 at the Cincinnati Zoo.

25. The echo parakeet is the last surviving native parrot in the Mascarenes (an island group in the Indian Ocean).

26. During the 1980s, less than 20 individuals were counted in the wild, a result of habitat loss and invasive species.

27. Through conservation efforts, the population has bounced back from the brink of extinction, reaching 750 individuals in 2019.

28. Although nine subspecies of lovebirds are found on the African continent, their closest relatives are thought to be the hanging parrots of Asia.

29. African gray parrots are born with black irises, which change to light yellow as they age.

30. Alex, a famous African gray, was the subject of a 30-year experiment by animal psychologist Irene Pepperberg.

31. Pepperberg bought Alex at a pet shop.

32. Alex had a vocabulary of over 100 words.

33. He could also identify different objects by their quantity, shape, size, and color.

12 Stinky Skunk Facts

1. All 11 skunk species have stinky spray housed in their anal glands.

2. While no animal's anal glands are remotely fragrant, skunks' pack an especially pungent stench because they use their spray as a defense mechanism.

3. Skunks have strong muscles surrounding the glands, which allow them to spray 16 feet or more on a good day.

4. A skunk doesn't want to stink up the place. It warns predators before it douses its target with *eau de skunk*.

5. A skunk will jump up and down, stomp its feet, hiss, and lift its tail in the air, all in the hope that the predator will realize that it's dealing with a skunk and go away.

6. A skunk only does what it does best when it feels it has no choice. Then it releases the nauseating mix of thiols (chemicals that contain super-stinky sulfur).

7. Skunks have enough "stink juice" stored up for five or six sprays.

8. After skunks empty their anal glands, it takes up to ten days to replenish the supply.

9. Being sprayed by a skunk is an extremely unpleasant experience. Besides the smell, the spray from a skunk can cause nausea and temporary blindness.

10. Bobcats, foxes, coyotes, and badgers usually only hunt skunk if they are really, really hungry.

11. Only the great horned owl makes skunk a regular snack—and the fact that the great horned owl barely has a sense of smell probably has a lot to do with it.

12. Should you find yourself on the receiving end of a skunk shower, your best deodorizer is alkaline hydrogen peroxide.

43 Facts About Horses

1. Horses are related to zebras, wild asses, and donkeys.

2. The height of a horse is measured in hands, from the ground to the top of the withers (the ridge between the shoulders at the base of the neck).

3. One hand equals four inches.

4. Horses were first domesticated about 6,000 years ago.

5. Draft horses such as the Clydesdale, Percheron, Belgian, and Shire were bred to carry and pull heavy loads.

6. The Shire horse is the world's largest breed.

7. Shire stallions average just over 17 hands (68 inches) in height and weigh as much as 2,200 pounds.

8. Clydesdales have served as a mascot of various beer brands, including Budweiser.

9. Belgian foals weigh about 125 pounds at birth.

10. Light horses are bred for speed, agility, endurance, and riding.

11. Light horse breeds include the Arabian, Morgan, and American quarter horses.

12. The Arabian is one of the world's oldest horse breeds.

13. Nearly every light horse breed today has Arabian bloodlines.

14. The American quarter horse is North America's most popular breed.

15. The Morgan breed originated with a stallion named Figure given to Justin Morgan of Vermont in the late 1700s.

16. Warmbloods are intermediate-weight horses created by crossing draft horses (cold bloods) and light horses (hot bloods).

17. Ponies are horses shorter than 14.2 hands (56.8 inches).

18. The Fell pony gets its name from the fells (mountains or hills) of northern England where the breed originated.

19. It has been a recognizable breed since Roman times in England.

20. Horses are born without teeth.

21. Friesian horses have a long mane and tail that are often wavy.

22. The American miniature is a height breed; these horses must measure no more than 34 inches from the last hairs of the mane (at the withers) to the ground.

23. American miniature foals measure from 16 to 21 inches in height at birth.

24. Unlike most other breeds, Shetland ponies (named for the Shetland Islands of Scotland) are measured in inches rather than hands.

25. Shetland ponies must not exceed 42 inches in height at the withers.

26. In the 1840s, Shetland ponies replaced women and children working in Britain's coal mines.

27. The King of England's Royal Mews (stables) use Cleveland Bays in ceremonial duties.

28. During the First Crusade (1096–99), so many horses died in

the desert that the Crusaders were reduced to carrying their supplies on the backs of goats and dogs.

29. Spotted horses have existed for millennia. Cave paintings dating to 25,000 years ago depict spotted horses.

30. Appaloosas, developed by the Nez Percé people in North America, have patterned coats and striped hooves.

31. The Akhal-Teke horse of Turkmenistan is noted for its distinctive metallic sheen.

32. The Gotland pony is a primitive breed native to the island of Gotland in Sweden, where ponies of this type have been documented as far back as the Stone Age.

33. Lipizzaner horses are famous for the difficult "airs above the ground" dressage movements.

34. The Norwegian fjord horse has a distinctive mane that is dark in the center and light on the outside.

35. Manes are often trimmed to accentuate the dark stripe.

36. The windswept barrier island of Assateague, off the coasts of Maryland and Virginia, is home to horses known as Assateague horses in Maryland and Chincoteague ponies in Virginia.

37. Though commonly called "wild," the feral horses of Assateague Island descend from domesticated horses brought to the island more than 300 years ago.

38. The Przewalski's horse, also known as the Mongolian wild horse or the takhi, is a rare and endangered horse native to the steppes of Central Asia.

39. The brumby is a free-roaming feral horse of Australia.

40. The Falabella miniature horse is one of the smallest horse breeds in the world, standing between 28 and 34 inches tall at the withers when fully-grown.

41. Despite its size, the Falabella is considered a miniature horse rather than a pony.

42. The Spanish introduced American mustangs to the Americas.

43. Christopher Columbus brought the first horses on his second journey to the New World.

23 Mammoth & Mastodon Facts

1. Mammoths and mastodons are ancient pachyderms, the distant cousins of modern elephants.

2. Unlike elephants today, which reside in Africa and South Asia, the remains of these ancient mammals are found predominantly in North America and North Asia.

3. Cave paintings and other archaeological evidence show that humans actively hunted mammoths about 30,000 years ago.

4. There is speculation about whether or not this led to the species' extinction.

5. In 1705, the Hudson River Valley village of Claverack, New York, discovered a large mastodon tooth.

6. This tooth discovery predated uncovering dinosaur fossils by over a century.

7. The tooth was sent to London as evidence of giants, like those described in the book of Genesis in the Bible.

8. The imagined giant creature was called *incognitum*.

9. Similar teeth were found later in South Carolina, where reportedly slaves pointed out that they closely resembled the teeth of African elephants.

10. George Cuvier was inspired by the mastodon's cone-shaped teeth when giving the ancient mammal its name.

11. *Mastos* and *odont* are Greek for "breast" and "tooth," respectively.

12. Mastodon teeth are suited for crushing twigs and other woody browse. They dwelled in forests.

13. Mammoth teeth are ridged. The flat surfaces were ideal for grazing and grinding tough grasses.

14. Mammoths ate about 400 pounds of grasses and sedges a day.

15. Mastodons were shorter in stature, with a flatter head and straight tusks.

16. Mammoth tusks grew in a curve. They had a distinctive knob at the top of their skulls.

17. The woolly mammoth is only one type of mammoth. It's the most well-known type because of its signature fur coat.

18. This woolly hair was just one of many of the mammoth's adaptations to living in colder climates.

19. Columbian mammoths were a North American species named after Christopher Columbus.

20. Columbian mammoths were bigger than modern African elephants. A full-grown adult weighed about ten tons.

21. The discovery of mammoth and mastodon remains helped scientists realize that species could go extinct.

22. Before the discovery of their remains, it was believed that the Earth was only 6,000 years old.

23. This finding meant that Earth was likely much, much older.

32 Facts That Bear Repeating

1. There are eight species of bear: the brown bear, polar bear, American black bear, Asiatic black bear, Andean bear, giant panda, sloth bear, and sun bear.

2. Named for its sluggish habits, the nocturnal sloth bear inhabits the tropical or subtropical forests of India and Sri Lanka.

3. Habitat loss and illegal hunting have led to the sloth bear's declining populations. It is classified as a vulnerable species.

4. It was once believed that the panda was a member of the raccoon family. Today, molecular analysis suggests that bears are the giant panda's closest relatives.

5. Giant pandas give birth to the smallest young of any mammal relative to its mother.

6. Pandas used to be carnivores, but committed to a bamboo diet approximately 3 million years ago.

7. Despite adaptations in its wrist bone, teeth, and jaws, the giant panda's digestive system is still unable to digest cellulose.

8. Because of China's success in conservation, the giant panda's status was changed from "endangered" to "vulnerable" in 2016.

9. The sun bear is the smallest species of bear, endemic to the forests of Southeast Asia.

10. According to legend, the crescent-shaped marking on its chest represents the rising sun.

11. The sun bear extracts insects from their nests using its flexible snout and long tongue.

12. It enjoys the taste of bees and termites, but will also eat fruit, honey, and small mammals.

13. The Andean bear, found only in the Andes Mountains, is the only bear species that lives south of the equator.

14. Andean bears are also called spectacled bears. They typically have spectacle-like beige markings across the face and upper chest.

15. The long, coarse neck and shoulder hair of the Asiatic black bear forms a modified mane.

16. Asiatic black bears are the most adept at bipedal walking of all the bear species.

17. American black bears are the most common and widely distributed bear species in North America.

18. The color of American black bears can vary, even in the same litter. For this reason, they are sometimes called cinnamon bears, blue-gray glacier bears, white bears, and brown bears. Actual brown bears are much larger.

19. American black bears spend their winter months in dens to avoid cold weather and low food supply.

20. Hibernation may last up to seven months.

21. No bears are native to Africa and Australia.

22. Polar bears are the largest species of bear.

23. Living in the snow and ice, polar bears have adapted black skin to help absorb heat, and translucent fur coats to provide camouflage and repel water.

24. Webbed paws allow polar bears to be powerful swimmers, but they are not fast enough to catch seals in the water.

25. Instead, they stalk and ambush their prey on the ice.

26. Polar bears eat more meat than any other bear species.

27. This is likely due to the lack of vegetation in their Arctic climates.

28. Mother polar bears have to protect their cubs from male polar bears because they sometimes kill the young of their own species.

29. Brown bears have a large geographic distribution, with around 200,000 individuals across North America, Europe, and Asia.

30. The brown bear's distinct shoulder hump is made entirely of muscle.

31. Brown bears have large brains relative to their body size and have been seen engaging in tool use.

32. A Syrian brown bear named Wojtek was drafted into the Polish Army as a private, and was later promoted to the rank of corporal. He retired to the Edinburgh Zoo.

28 Cambrian Explosion Facts

1. The Cambrian Explosion was a swift burst of new animal species during the Cambrian Period approximately 540 million years ago.

2. It's called an explosion because it all happened in a short amount of time relative to the geologic time scale: 20 million years.

3. What sparked this explosion is still debated, but a leading theory suggests that the amount of oxygen in the atmosphere finally reached levels that allowed for larger, more complex animals.

4. These oxygen emissions were a result of photosynthesizing cyanobacteria and algae.

5. A warming climate and rising sea levels also made the ocean much more hospitable.

6. Rising sea levels triggered erosion, allowing minerals like calcium and phosphorus to enter the seawater at unprecedented levels.

7. These minerals were the building blocks of hard body structures in Cambrian organisms.

8. Many different types of skeletons emerged during this time, including internal, external, multi-element, and single-element.

9. Many animal species independently evolved hard parts at about the same time.

10. Hard body parts like skeletons or calcium carbonate shells are able to fossilize more easily than soft tissues.

11. Because of this, the Cambrian fossil record is better preserved.

12. Lagerstätten deposits, like the Burgess Shale, are excellent fossil beds that even contain soft-bodied organisms.

13. Nearly all Cambrian animals were deposit feeders and lived near the ocean floor.

14. Fossilized worm burrows mark the existence of digging organisms. These are called trace fossils.

15. Trace fossils are traces of an animal instead of the animal itself. These can include footprints, trails, or burrows.

16. Colonies of archaeocyathids built up the reefs of Cambrian tropical and subtropical waters. They had characteristics of both sponges and corals.

17. Trilobites were the dominant species during the Cambrian Period. They are the most common fossil types of this period.

18. Trilobites are a kind of arthropod. Their closest living relatives are horseshoe crabs.

19. The exoskeleton of a trilobite was divided into three regions: the cephalon (head), thorax (body), and pygidium (tail).

20. The name trilobite comes not from their three main segments, but instead from their three longitudinal lobes.

21. Trilobites could roll up their external skeletons. This was probably for defensive purposes.

22. Brachiopods comprise the most common Cambrian fossil type after trilobites.

23. Wiwaxia were soft-bodied animals covered in scales and two rows of dorsal spines.

24. When hallucigenia was uncovered from the Burgess Shale, the odd appearance of the animal made Charles Doolittle Walcott think he was hallucinating.

25. Anomalocaris had stalked, compound eyes, a wide mouth, spiked arm-like appendages, and a fan-shaped tail.

26. Tamiscolaris was discovered in 2014.

27. It was about the same size as anomalocaris, but used the bristles on its front appendages to comb microorganisms out of the sea as a filter feeder.

28. The Cambrian Period's end saw a series of mass extinction events.

9 Facts & Fallacies About Sharks

1. **Sharks are vicious man-eaters.** False! People are not even on their preferred-food list. Every hunt poses a risk of injury to sharks, so they need to make every meal count. That's why they go for animals with a lot of high-calorie fat and blubber—they get more energy for less effort. Humans are usually too lean and bony to be worth the risk.

2. **Sharks are loners.** That depends on the shark. Some species, such as the great white, are rarely seen in the company of

other sharks. However, many other species aren't so antisocial. Blacktip reef sharks hunt in packs, working together to drive fish out of coral beds so every shark in the group gets a meal. Near Cocos Island off Costa Rica, hammerheads have been filmed cruising around in schools that consist of hundreds of sharks.

3. **You have a greater chance of being killed by a falling coconut than by a shark.** True! Falling coconuts kill close to 150 people every year. In comparison, sharks kill only five people per year on average. The International Shark Attack File estimates that the odds of a person being killed by a shark are approximately 1 in 264 million.

4. **Sharks have poor vision, and most attacks are cases of mistaken identity.** As popular as this belief is, it's wrong. Scientists have observed that sharks' behavior when they are hunting differs significantly from what most people report when a bite occurs. Sharks are extremely curious creatures and, since they don't have hands, they frequently explore their environment with the only things they have available—their mouths. Unfortunately for humans, a curious shark can do a lot of damage with a "test" nibble, especially if it's a big shark.

5. **Most shark attacks are not fatal.** True! There are approximately 70 unprovoked shark attacks around the world each year, and, on average, just 1 percent of those are fatal. What we usually call "attacks" are really only bites. Scientists report that an inquisitive shark that bites a surfboard (or an unlucky swimmer) shows far less aggression than when it is on the hunt and attacks fiercely and repeatedly.

6. **Shark cartilage is an effective treatment for cancer.** False! Anyone toting the benefits of shark cartilage as a nutritional supplement to cure cancer is selling snake oil. Multiple studies by Johns Hopkins University and other institutions have shown that shark cartilage has no benefit. This myth got started with the popular but incorrect notion that sharks don't get cancer.

7. **Most shark attacks occur in water less than six feet deep.** True! And the reason is obvious—that's where the majority of people are. It makes sense that most of the

interactions between humans and sharks happen where the concentration of people is greatest.

8. **Most sharks present no threat to humans.** True! There are more than 400 species of sharks, and approximately 80 percent of them are completely harmless to people. In fact, only four species are responsible for nearly 85 percent of unprovoked attacks: bull sharks, great white sharks, tiger sharks, and great hammerhead sharks.

9. **Sharks have to swim constantly or they drown.** There are a few species that need to keep moving, but most sharks can still get oxygen when they're "motionless." They just open their mouths to draw water in and over their gills.

14 Freaky Facts About Bedbugs

1. Bedbugs are oval-shaped wingless insects that actually do bite.

2. When bedbugs bite, proteins in their saliva prevent the wound from closing.

3. Bedbugs can consume as much as six times their weight in blood at one feeding.

4. When full of blood, bedbugs can swell as large as three times their normal size.

5. It takes a bedbug five minutes to drink its fill of blood.

6. A female bedbug can lay up to five eggs per day and 200 to 500 eggs in her lifetime.

7. Normally, a bedbug is brown.

8. After eating, however, its body appears dark red.

9. Bedbugs are nocturnal—they become active when humans are sleeping.

10. Bedbugs only eat once every seven to ten days.

11. Other names for bedbugs include "mahogany flats," "redcoats," and "chinches."

12. Bedbugs molt five times before reaching adulthood.

13. Bedbugs migrated from Europe to the U.S. during the 1600s.

14. Contrary to popular belief, the presence of bedbugs does not indicate a dirty house.

20 Mythical Creatures

1. **Dragon:** This legendary monster was thought to be a giant, winged, fire-breathing snake or lizard. Oddly, most ancient cultures that believed in dragons didn't know about dinosaurs.

2. **Griffin:** Originating in the Middle East around 2000 B.C., the griffin was said to have the body of a lion and the head of a bird (most often an eagle).

3. **Phoenix:** This ancient Egyptian bird was believed to live for 500 years or more before setting itself on fire. A new phoenix was supposed to have been born from the ashes of the previous bird.

4. **Chupacabra:** This vampiric creature, whose name means "goat sucker," has allegedly been spotted in the Americas from Maine to Chile. It has been described as a panther, dog, spined lizard, or large rodent that walks upright and smells ghastly.

5. **Yeti/Bigfoot/Sasquatch:** If these creatures existed, they would likely be the same animal, as they are all bearlike or apelike hominids that live in remote mountainous areas. (The Yeti has been reported in the Himalayas, while Bigfoot and Sasquatch are rumored to live in the northwestern U.S. and Canada.)

6. **Skunk Ape:** This hairy seven-foot critter weighs in at 300 pounds, smells like a garbage-covered skunk, inhabits the Florida Everglades, and is thought to be a relative of Bigfoot.

7. **Rukh/Roc:** Marco Polo returned from Madagascar claiming to have seen this enormous, horned bird of prey carry off elephants and other large creatures.

8. **Adjule:** It's reported to be a ghostly North African wolf dog but is likely just a Cape hunting dog, horned jackal, or ordinary wild dog.

9. **Vegetable Lamb of Tartary:** Mythical animal or vegetable? Eleventh-century travelers told tales of a Middle Eastern plant that grew sheep as fruit. Although the tales were false, the plant is real: It's the fern *Cibotium barometz*, which produces a tuft of woolly fiber.

10. **Jackalope:** Sometimes called a "warrior rabbit," the jack-alope is a legendary critter of the American West. Described as an aggressive, antlered rabbit, it appears to be related to the German wolperdinger and Swedish skvader.

11. **Andean Wolf/Hagenbeck's Wolf:** In 1927, a traveler to Buenos Aires bought a pelt belonging to what he was told was a mysterious wild dog from the Andes. This tale encouraged other crypto-enthusiasts to purchase skulls and pelts from the same market until DNA testing in 1995 revealed that the original sample came from a domestic dog.

12. **Fur-Bearing Trout (or Beaver Trout):** The rumor of a fur-covered fish dates back to Scottish visitors to the New World who regaled their countrymen with tales of "furried animals and fish" and photographs of pelt-wrapped trout. Occasional sightings of fish with saprolegnia (a fungal infection that causes a white, woolly growth) serve to perpetuate the tall tales.

13. **Gilled Antelope (or Gilled Deer):** This Cambodian deer or antelope is rumored to have gills on its neck or muzzle that enable it to breathe underwater. In reality, it's the rare Saola or Vu Quang ox, whose distinctive white facial markings only look like gills.

14. **Hodag:** In lumberjack circles, the hodag was believed to be a fetid-smelling, fanged, hairy lizard that rose from the ashes of cremated lumber oxen. In 1893, a prankster from Rhinelander, Wisconsin, led a successful "hunt" for the fearsome beast, which resulted in its capture and subsequent display at the county fair. It was later revealed that the hodag was actually a wooden statue covered in oxhide, but by then it had earned its place in Rhinelander lore.

15. **Sea Monk/Sea Bishop:** Reports and illustrations of strange fish that looked like clergymen were common from the 10th through 16th centuries, likely due to socioreligious struggles and

the idea that all land creatures had a nautical counterpart. Sea monks were likely angel sharks, often called "monkfish."

16. **Unicorn:** The unicorn's appearance varies by culture—in some it's a pure white horse and in others it's a bull or an ox-tailed deer—but the single horn in the middle of the forehead remains a constant. Unicorns have reportedly been seen by such luminaries as Genghis Khan and Julius Caesar, but it's likely that such reports were based on sightings of rhinoceroses, types of antelopes, or discarded narwhal horns.

17. **Cameleopard:** Thirteenth-century Romans described the cameleopard as the offspring of a camel and leopard, with a leopard's spots and horns on the top of its head. The legendary creature in question was actually a giraffe, whose modern name stems from the Arabic for "tallest" or "creature of grace."

18. **Basilisk/Cockatrice:** Pliny the Elder described this fearsome snake as having a golden crown, though others described it as having the head of a human or fowl. In fact, no one would have been able to describe it at all, because it was believed to be so terrifying that a glimpse would kill the viewer instantly.

19. **Mokele-mbembe:** Since 1776 in the Congo rainforests, there have been reports of this elephant-size creature with a long tail and a muscular neck. Its name translates as "one that stops the flow of rivers." Hopeful believers—who point out that the okapi was also thought to be mythical until the early 1900s—suggest that it may be a surviving sauropod dinosaur.

20. **Lake Ainslie Monster:** Lake Ainslie in Cape Breton, Nova Scotia, is frequently said to be home to a sea monster, usually described as having a snakelike head and long neck (similar in appearance to the Loch Ness Monster). Recently, these sightings have been attributed to "eel balls," a large group of eels that knot themselves together in clumps as large as six feet in diameter.

14 Narwhal Facts

1. A narwhal is a medium-sized, toothed whale found only in Arctic waters.

2. A typical narwhal averages somewhere between 11.5 and 16.4 feet, and weighs in at around 3,500 pounds.

3. Narwhals swim upside down! Researchers are still trying to figure out exactly why narwhals spend 80 percent of their time inverted, but they think it's because the animals send sonar signals underwater to detect prey.

4. Swimming upside down may direct the sonar beam downwards, where lunch is likely to be most abundant.

5. Narwhal enemies include polar bears and killer whales, but human poachers pose the biggest threat.

6. The most distinguishing characteristic of a narwhal is its long, unicornlike tusk.

7. The tusk points slightly downward, which is another reason narwhals may swim on their backs—while looking for food on the sea floor, they don't want to bust their horns on a rock.

8. Over 10 million tiny nerve endings are found on the surface of a narwhal tusk, making it an essential tool for sensory perception.

9. Narwhals lack a dorsal fin, which is unusual in underwater creatures of their kind.

10. Scientists believe the narwhal evolved without a dorsal fin as an adaptation to navigate beneath ice-covered waters.

11. Newborn narwhals are called "calves."

12. Calves are weaned after a year or so, and then they move on to a regular diet of fish, squid, and shrimp.

13. If you see a group of narwhals above water, rubbing their tusks together, they're "tusking."

14. This activity helps the narwhals clean their tusks—kind of like when humans brush their teeth.

9 Fallacies & Facts About Snakes

1. Fallacy: You can identify venomous snakes by their triangular heads. **Fact:** Many non-venomous snakes have triangular heads, and many venomous snakes don't. You would not enjoy testing this theory on a coral snake, whose head is not triangular. It's not the kind of test you can retake if you flunk. Boa constrictors and some species of water snakes have triangular heads, but they aren't venomous.

2. Fallacy: A coral snake is too little to cause much harm. **Fact:** Coral snakes are indeed small and lack long viper fangs, but their mouths can open wider than you might imagine—wide enough to grab an ankle or wrist. If they get ahold of you, they can inject an extremely potent venom.

3. Fallacy: A snake will not cross a hemp rope. **Fact:** Snakes couldn't care less about a rope or its formative material, and they will readily cross not only a rope but a live electrical wire.

4. Fallacy: Some snakes, including the common garter snake, protect their young by swallowing them temporarily in the face of danger. **Fact:** The maternal instinct just isn't that strong for a mother snake. If a snake has another snake in its mouth, the former is the diner and the latter is dinner.

5. Fallacy: When threatened, a hoop snake will grab its tail with its mouth, form a "hoop" with its body, and roll away. In another version of the myth, the snake forms a hoop in order to chase prey and people! **Fact:** There's actually no such thing as a hoop snake. But even if there were, unless the supposed snake were rolling itself downhill, it wouldn't necessarily go any faster than it would with its usual slither.

6. Fallacy: A snake must be coiled in order to strike. **Fact:** A snake can strike at half its length from any stable position. It can also swivel swiftly to bite anything that grabs it—even, on occasion, professional snake handlers. Anyone born with a "must grab snake" gene should consider the dangers.

7. Fallacy: The puff adder can kill you with its venom-laced breath. **Fact:** "Puff adder" refers to a number of snakes, from a common and dangerous African variety to the less aggressive hog-nosed snakes of North America. You can't defeat any of them with a breath mint, because they aren't in the habit of breathing on people, nor is their breath poisonous.

8. **Fallacy:** Snakes do more harm than good. **Fact:** How fond are you of rats and mice? Anyone who despises such varmints should love snakes, which dine on rodents and keep their numbers down.

9. **Fallacy:** Snakes travel in pairs to protect each other. **Fact:** Most snakes are solitary except during breeding season, when male snakes follow potential mates closely. Otherwise, snakes aren't particularly social and are clueless about the buddy system.

25 Rarest Animals

1. Vancouver Island marmot
2. Seychelles sheath-tailed bat
3. Javan rhinoceros
4. Hispid hare (Assam rabbit)
5. Northern hairy-nosed wombat
6. Tamaraw (dwarf water buffalo)
7. Amur leopard
8. Red wolf
9. Philippine crocodile
10. Yellow-tailed woolly monkey
11. Kouprey (Cambodian forest ox)
12. Riverine rabbit
13. Malabar large spotted civet
14. Saola (Vu Quang ox)
15. Tonkin snub-nosed monkey
16. Sumatran rhinoceros
17. Northern muriqui (woolly spider monkey)
18. Kakapu (owl parrot)
19. Hirola (Hunter's hartebeest)
20. Addax (Sahara antelope)
21. Gobi bear
22. Superagüi (or black-faced) lion tamarin
23. Vaquita (Gulf of California porpoise)
24. Mediterranean monk seal
25. African wild ass

12 Flying Fish Facts

1. To escape predators such as swordfish, tuna, and dolphins, flying fish extend their freakishly large pectoral fins as they approach the surface of the water.

2. Their velocity then launches them into the air.

3. A flying fish can glide through the air for 10 to 20 feet—farther if it has a decent tailwind.

4. It holds its outstretched pectoral fins steady and "sails" through the air, using much the same action as a flying squirrel.

5. If you'd like to catch sight of a member of the Exocoetidae family, you'll have to travel to the Atlantic, Indian, or Pacific oceans, where there are more than 50 species of flying fish.

6. Whiskers are not an indication that a flying fish is up there in years; actually, it's the opposite.

7. Young flying fish have long whiskers that sprout from the bottom jaw. These whiskers are often longer than the fish itself and disappear by adulthood.

8. To get a taste of flying fish, stop by a Japanese restaurant. The eggs of flying fish, called tobiko, are often used in sushi.

9. Attached to the eggs of the Atlantic flying fish are long, adhesive filaments that enable the eggs to affix to clumps of floating seaweed or debris for the gestation period. Without these filaments, the eggs (which are denser than water) would sink.

10. Beyond their useful pectoral fins, all flying fish have unevenly forked tails, with the lower lobe longer than the upper lobe.

11. Many species also have enlarged pelvic fins and are known as four-winged flying fish.

12. Flying fish can soar high enough that sailors often find them on the decks of their ships.

Chapter 2

Planet Earth

35 Amazing Earth Facts

1. Earth weighs approximately 5,940,000,000,000,000,000,000 metric tons.

2. The surface of Earth is approximately 70.9 percent water and 29.1 percent land.

3. Earth is the only planet not named after a god.

4. The deepest place on Earth is Challenger Deep in the Mariana Trench, with a depth of 35,840 feet below sea level.

5. Earth is not a perfect sphere; it's more of a squished ball.

6. Thanks to Earth's quirky shape, Mount Chimborazo, which rises only 20,702 feet from sea level, is about a mile and a half closer to the moon than Everest's peak (29,035 feet from sea level).

7. If you stood at the equator, you would weigh slightly less than if you stood at the North or South Pole.

8. The distance around the equator is 24,901 miles.

9. Earth is fatter in the middle near the equator where gravity pushes to create a bulge.

10. Planet Earth is in constant motion.

11. This movement causes mountains to rise, earthquakes to rumble, and volcanoes to spew out hot rock.

12. Made of solid iron and nickel, Earth's inner core is as hot as the surface of the sun—about 10,000°F.

13. It doesn't melt, though, because it's compressed by the weight of the planet all around it.

14. Surrounding the inner core is the outer core. This, too, is mostly iron and nickel, and very hot.

15. The outer core's rock is liquid because it's not under as much pressure as the inner core.

16. Around the outer core is the mantle. It is the thickest part of Earth, about 1,800 miles deep.

17. The mantle is solid rock, but so hot that it flows like a thick, gummy liquid.

18. Earth's thin outermost layer is the crust, which covers the mantle.

19. Veryovkina Cave in the country of Georgia is the world's deepest cave, plunging 7,257 feet.

20. Earth is the only planet in our solar system to have water in its three states of matter: liquid, solid, and gas.

21. The Caspian Sea, situated where southeastern Europe meets Asia, is the world's largest lake.

22. It has a surface area of 143,243 square miles.

23. Stretching for more than 270 miles, the Grand Canyon in Arizona is the world's largest canyon.

24. The Himalayas are the tallest and youngest mountains on Earth.

25. Today's continents were split apart from a super continent named Pangea.

26. Bouvet Island in the South Atlantic is Earth's most remote island.

27. It lies almost a thousand miles from the nearest land (Queen Maude Land, Antarctica).

28. Earth is thought to be about 4.54 billion years old.

29. Compared to geologic time, humans have lived on Earth for the blink of an eye.

30. Chicxulub Crater in the Yucatán Peninsula is the largest confirmed impact crater on Earth.

31. The asteroid that formed this 105-mile-diameter crater is widely credited with killing off the dinosaurs.

32. Earth is moving through the solar system at around 67,000 miles per hour.

33. If the sun suddenly stopped emitting light and warmth, Earth would get dark in about eight minutes—the length of time it takes for light to reach us once it escapes the sun—and would gradually become colder.

34. There are 195 countries in the world.

35. Russia is the world's largest country by area.

12 Blazing Hot Facts

1. The world's hottest temperature of 134°F was recorded on July 10, 1913, in Furnace Creek, Death Valley, California.

2. The Sahara is the world's largest hot desert.

3. Despite the popular saying, it's never hot enough to fry an egg on the sidewalk.

4. The pavement would need to hit at least 158°F for the egg to cook, and even blacktop only reaches 145°F.

5. Hot water should boil at 212°F (100°C), but the actual boiling point varies depending on atmospheric pressure, which changes according to elevation.

6. A typical lightning bolt is hotter than the surface of the sun.

7. The Chicago heat wave of 1995 killed 739 people, mostly poor elderly residents who couldn't afford air conditioning.

8. The Red Sea is the world's hottest. It's 87°F at its warmest spot.

9. In Death Valley, California, the average monthly temperature for July is 101°F.

10. Phoenix, Arizona, recorded 128 days with a temperature of 100°F or higher in 2018.

11. The city had 143 days with triple-digit temps in 1989.

12. Massive heat waves across North America were persistent in the 1930s during the Dust Bowl, killing at least 5,000 people in the United States alone.

34 Cold Hard Facts

1. Water is supposed to freeze at 32 degrees Fahrenheit (0 degrees Celsius), but scientists have found liquid water as cold as −40°F in clouds and even cooled water down to −42°F in the lab.

2. The world's lowest temperature of −128.5°F was recorded at Vostok Station, Antarctica, on July 21, 1983.

3. Icebergs are formed from glaciers on land and drift out to sea.

4. They are mostly made from freshwater, not saltwater.

5. Snowflakes are collections of ice crystals.

6. The way that ice crystals join together gives every snowflake a unique design.

7. One snowflake can contain as many as 100 ice crystals.

8. What shape a snowflake takes depends on the temperature and the amount of moisture in the cloud.

9. In 1988, two identical snowflakes collected from a Wisconsin storm were confirmed to be twins at an atmospheric research center in Colorado.

10. The Rainier Paradise Ranger Station in Washington State gets an average 676 inches of snow annually.

11. Earth's two ice sheets cover most of Greenland and Antarctica.

12. They make up more than 99 percent of the world's glacial ice.

13. Most North Atlantic icebergs calve (detach) from the Greenland ice sheets.

14. "Growlers" are mini-icebergs and "bergy bits" are big ice chunks.

15. Ninety percent of Greenland is covered by an ice cap and smaller glaciers.

16. The coldest temperature ever recorded in North America was −87.0°F on January 9, 1954, at the North Ice research station on Greenland.

17. Harbin, China, often called the "ice city," hosts the Harbin International Snow and Ice Festival each year.

18. Temperatures as low as −44°F have been recorded in Harbin.

19. The village of Oymyakon, Russia, is the coldest permanently occupied human settlement in the world (population 500).

20. During the winter, temperatures average −58°F and 21 hours per day are spent in darkness.

21. The all-time coldest temperature in Oymyakon, −98°F, was recorded in 2013.

22. International Falls, Minnesota, took another town to court over the title of coldest place in the continental United States.

23. Temperatures as low as −40°F have been recorded there.

24. The title of coldest place in the continental U.S. actually belongs to Stanley, Idaho, with a record cold temperature of −52.6°F.

25. Not all blizzards have falling snow.

26. Ground blizzards occur when no snow is falling.

27. Ulaanbaatar, Mongolia, is the coldest capital city.

28. The average annual temperature is 29.6°F in Ulaanbaatar, but temperatures range between 5°F and −40°F in winter months.

29. While it seems counterintuitive, Earth is actually closer to the sun during winter in the Northern Hemisphere.

30. When you hear thunder during a snowstorm, it's called thundersnow.

31. Lightning is harder to see in the winter and the snow sometimes dampens the thunderous sound.

32. Snow falls one to six feet per second.

33. More than 22 million tons of salt are used to de-ice U.S. roads each winter.

34. The average monthly temperature for January in Verkhoyansk, Russia, is −53°F.

36 Facts About North America

1. North America is the world's third largest landmass with an area of 9,449,078 square miles.

2. Only about eight percent of the world's population live in North America.

3. The United States is the world's third most populous country with an estimated 334,233,854 people.

4. North America has no landlocked countries.

5. Central America refers to the seven countries of North America between Mexico and South America: Belize, Costa Rica, El Salvador, Guatemala, Honduras, Nicaragua, and Panama.

6. Honduras is the only Central American country without an active volcano.

7. Mexico City, Mexico, is the highest North American capital with an elevation of 7,349 feet.

8. North America was first populated about 10,000 years ago when people moved across the Bering Sea between Siberia and Alaska.

9. Bermuda is the most densely populated country in North America, with 3,200 people per square mile.

10. Groups of mountain quail (*Oreortyx pictus*), native to western North America, often walk up to 5,000 feet in winter in the longest bird migration on foot.

11. Guatemala's national bird is the quetzal.

12. Canada is the world's second largest country with a total area of 3,855,102 square miles.

13. The world's biggest island, Greenland, is part of North America, although it belongs to the European country of Denmark.

14. Canada is the world's largest country that borders only one other country (United States).

15. Sinkholes called *cenotes* on Mexico's Yucatán Peninsula were sacred to the ancient Maya.

16. Nearly four million people of Maya descent still live in southern Mexico and Central America.

17. North America's longest mountain chain is the Rocky Mountains, which stretch 3,000 miles across the U.S. and Canada.

18. Grenada, nicknamed the Spice Island, is one of the world's leading sources of nutmeg and other spices.

19. The Panama Canal divides North and South America.

20. The world's longest shared border is the 3,987-mile boundary between the United States and Canada. That doesn't include Canada's 1,539-mile border with Alaska.

21. Approximately 90 percent of Canadians live within 100 miles of the U.S. border.

22. Alaska's Denali (Mount McKinley) is North America's highest point at 20,308 feet.

23. Grant's caribou (*Rangifer tarandus granti*) of Alaska and the Yukon Territory of North America travel up to 2,982 miles per year.

24. Lake Superior is North America's largest lake and one of the largest freshwater lakes on the planet.

25. Canada has up to three million lakes—more than all other countries combined.

26. Bracken Cave outside of San Antonio, Texas, is the world's

largest bat cave. An estimated 20 million Mexican free-tailed bats roost in the cave from March to October.

27. The Yosemite Falls are North America's highest waterfalls, dropping 2,425 feet.

28. Hawaii's Mauna Kea is the world's tallest mountain as measured from base to summit.

29. Its base begins 19,685 feet below sea level; total height estimates range from 32,696 to 33,474 feet.

30. The Canadian Arctic Archipelago consists of 36,563 islands.

31. Canada has the longest coastline of any country in the world by far.

32. Off the coast of Belize in the Caribbean Sea is the Great Blue Hole, a submarine sinkhole measuring more than 1,000 feet across and 400 feet deep.

33. Haiti and the Dominican Republic share the island of Hispaniola.

34. Cuba is the largest Caribbean island with a total area of 42,803 square miles.

35. The Mississippi River runs 2,202 miles from Minnesota down to the Gulf of Mexico.

36. An earthquake on December 16, 1811, caused parts of the Mississippi River to flow backward.

34 Bridge Facts

1. With a total length of 102 miles, the Danyang-Kunshan Grand Bridge, on the Jinghu High-Speed Railway in China, is the world's longest bridge.

2. There are no bridges that span the entire Amazon River.

3. There are more than 600,000 bridges in the United States.

4. China's Zhaozhou (or Anji) Bridge is the world's oldest stone segmental arch bridge.

5. Built around A.D. 600, it is still standing today.

6. The Rockville Bridge in Harrisburg, Pennsylvania, is the longest stone arch bridge in the world.

7. It was built in 1902 and measures 3,820 feet.

8. Fifteen major bridges span the Allegheny and Monongahela rivers as they flow together to become the Ohio River.

9. The Sydney Harbour Bridge in Australia can rise or fall up to seven inches depending on the temperature due to the steel expanding or contracting.

10. One of the longest steel-arch bridges in the world, it features six million rivets.

11. California's Golden Gate Bridge spans the narrow body of water between San Francisco Bay and the Pacific Ocean.

12. Despite its name, the famous suspension bridge is painted a reddish orange, not gold.

13. The bright color is called "international orange."

14. The main span of the Golden Gate Bridge is 4,200 feet long.

15. It was the longest bridge in the world from 1937 to 1964.

16. During the construction, 11 workers fell from the bridge and died, though a safety net saved the lives of 19 others.

17. The Charles Bridge, which spans the Vltava River, in Prague, Czech Republic, dates from 1357.

18. The stone bridge is lined with statues and has towers with gates at each end.

19. The nearly five-mile-long Mackinac Bridge links Michigan's Upper and Lower Peninsulas.

20. The world's longest cable suspension bridge is located in Turkey. It spans the Dardanelles Strait.

21. The main span of the 1915 Çanakkale Bridge is 6,637 feet, or 1.25 miles.

22. The total length of the bridge is 11,690 feet, or 2.21 miles.

23. Before Turkey's Çanakkale Bridge opened in 2022, Japan's Akashi-Kaikyo bridge held the record for world's longest cable suspension bridge.

24. The main span of the Akashi-Kaikyo bridge is 6,532 feet, or 1.24 miles, long.

25. The Brooklyn Bridge, designed by architect John Augustus Roebling, towers over New York City's East River.

26. It opened on May 24, 1883, linking what would become the boroughs of Brooklyn and Manhattan in New York City.

27. In its day, the Brooklyn Bridge was the longest suspension bridge in the world, with the main span measuring 1,595 feet long.

28. The Øresund Bridge, connecting Sweden with Denmark, is part overwater and part underwater tunnel.

29. The New River Gorge Bridge is one of the most photographed places in West Virginia.

30. The bridge in the Appalachian Mountains is one of the highest bridges open to vehicles in the world.

31. The New River Gorge Bridge cut the travel time across the gorge from an arduous 45 minutes on mountain roads to an easy minute across the bridge.

32. The world's longest steel truss-arch bridge is the Chongqing-Chaotianmen Bridge over the Yangtze River in China.

33. The main span is 1,811 feet long.

34. Oklahoma City honored its state bird, the scissor-tailed fly-catcher, with the design of the Skydance Bridge.

11 Facts About Giant's Causeway

1. Giant's Causeway is Northern Ireland's only UNESCO World Heritage Site.

2. Over 40,000 interlocking basalt columns make up this geological formation.

3. The columns of Giant's Causeway typically have five to seven irregular sides.

4. Most are hexagonal in shape.

5. The strange landscape was the result of an ancient volcanic fissure eruption during the Paleogene Period.

6. The tallest columns in Giant's Causeway are approximately 39 feet high.

7. The site also notes more mythological origins: Long ago, the Irish giant Fionn mac Cumhaill tore up the Antrim coastline during a battle with Scottish giant Benandonner. As he hurled the rocks into the water, he paved a route over the sea to Scotland.

8. Other popular examples of columnar jointing include Fingal's Cave in Scotland, Columbia River flood basalts in Oregon, Devils Postpile in California, and Devil's Tower in Wyoming.

9. German composer Felix Mendelssohn based his 1830 "Hebrides Overture" on the sound of waves flowing into Fingal's Cave.

10. Led Zeppelin's album cover for *Houses of the Holy* features Giant's Causeway.

11. Nicknames for several of the formations in this peculiar landscape include "the organ pipes," "the honeycomb," and "the chimneys."

43 Wild Weather Facts

1. All weather occurs within eight miles of Earth's surface.

2. All snowflakes have six sides or points.

3. The world's greatest 24-hour rainfall record was set in January 1966 on the French island of Réunion in the Indian Ocean.

4. A record 71.8 inches of rain fell.

5. Tropical storms rotate clockwise south of the equator and counterclockwise in the Northern Hemisphere.

6. A hurricane is given a rating on the Saffir-Simpson scale based on its sustained wind speed.

7. The Enhanced Fujita Scale (EF-Scale) is a system for classifying tornado intensity.

8. Waterspouts are tornadolike rotating columns of air over water.

9. A tropical storm must have winds of 74 miles per hour or greater to be classified a hurricane.

10. Tornadoes have been reported on all continents except Antarctica.

11. Contrary to popular belief, tornadoes can and do hit downtown areas of major cities.

12. The city center of St. Louis, Missouri, has been flattened four times since 1871.

13. All thunderstorms produce lightning, which is what makes the sound of thunder.

14. The world's heaviest hailstones weighed up to two pounds each and killed 92 people in Gopalganj district, Bangladesh, on April 14, 1986.

15. The United States has more tornadoes than any other country.

16. The Great Blizzard of 1888 blanketed some areas of the Atlantic coast with as much as 50 inches of snow.

17. Waterspouts can make sea creatures rain down from the sky.

18. The eye of a hurricane is calm, but the eye wall that surrounds it has the strongest, most violent winds.

19. Snowflakes formed in clouds take about an hour to reach the ground.

20. Hurricanes, tropical cyclones, and typhoons all describe the same type of storm—what they're called depends on where they form.

21. When lightning strikes, the heated air creates a sonic shock wave, which is the sound of thunder.

22. We see lightning before we hear thunder because light travels much faster than sound.

23. The top speed of a falling raindrop is 18 miles per hour.

24. Crickets make good thermometers—to get a rough estimate of the temperature, count the number of times a cricket chirps in 15 seconds and add 40.

25. Tornadoes usually spin counterclockwise in the Northern Hemisphere and clockwise in the Southern Hemisphere.

26. The winds of a tornado are the strongest on Earth. They may reach a speed of 300 miles per hour.

27. Most hurricanes stretch about 300 miles across.

28. The Tri-State Tornado was the deadliest tornado in U.S. history.

29. It traveled through Missouri, Illinois, and Indiana on March 18, 1925, killing 695 people.

30. Dirty snow melts faster than clean snow.

31. The Great Blizzard of 1888 convinced many people that passenger trains should run underground and helped to inspire cities like Boston and New York to build the first subway systems.

32. The fastest wind on Earth blew through the suburbs of Oklahoma City, Oklahoma, on May 3, 1999.

33. The 301-mile-per-hour gusts were recorded during an EF5 tornado that destroyed hundreds of homes.

34. The most active Atlantic hurricane season on record was in 2020.

35. The National Hurricane Center ran out of names and had to use Greek letters for the last several storms.

36. The Tri-State Tornado lasted 3.5 hours and traveled 219 miles—setting records for both duration and distance traveled.

37. Low barometric pressure generally indicates stormy weather, and high pressure signals calm, sunny skies.

38. The lowest barometric pressure ever recorded was 25.69 inches (870 mb) during Typhoon Tip in 1979.

39. The highest barometric pressure, 32.01 inches (1,083.8 mb), was measured in 1968 on a cold New Year's Eve in northern Siberia.

40. A tropical storm officially becomes a hurricane, cyclone, or typhoon when winds reach at least 74 miles per hour.

41. The deadliest storm on record was the Bhola Cyclone, which hit East Pakistan (now Bangladesh) on November 13, 1970, killing at least 500,000 people.

42. The Bhola Cyclone's winds were in excess of 120 miles per hour when it finally hit land.

43. It generated an astonishing storm surge of 12 to 20 feet, which flooded densely populated coastal areas.

50 South America Facts

1. South America is home to the Andes, the world's longest continental mountain range.

2. At 4,700 miles in length, the Andes span seven countries.

3. South America has 12 independent countries.

4. Two territories in South America are still administered by European powers: French Guiana (France) and the Falkland Islands (United Kingdom).

5. Easter Island, a Chilean territory, is some 2,300 miles from the coast of South America.

6. Brazil is the largest country in South America and in the Southern Hemisphere with a total area of 3,287,957 square miles.

7. Colombia has coastline on both the Pacific and Atlantic oceans.

8. The Galápagos, a volcanic chain of islands belonging to Ecuador, is home to the world's only marine iguanas.

9. Suriname is the smallest country in South America with a total area of 63,251 square miles.

10. Both North and South America are named after Italian explorer Amerigo Vespucci.

11. Guyana has up to 300 species of catfish.

12. Ecuador is the most densely populated country in South America with 135 people per square mile.

13. The Pantanal region of South America is the world's largest freshwater wetland.

14. It is almost ten times the size of the Florida Everglades.

15. The potato originates in South America.

16. The Amazon River runs through South America and is surrounded by the world's largest rain forest.

17. The world's highest gas pipeline tops out at over 16,077 feet in the Peruvian Andes.

18. The world's largest river basin is that drained by the Amazon, which covers about 2,720,000 square miles.

19. The Andean condor has a wingspan up to 10 feet 10 inches—one of the largest wingspans of any bird.

20. The Uyuni Salt Flats cover more than 4,000 square miles of Bolivia at an altitude of 12,500 feet.

21. Because of brine just below the surface, any crack in the salt soon repairs itself.

22. La Paz, Bolivia, is the world's highest administrative capital at 11,942 feet above sea level.

23. The Incas were the largest group of indigenous people in South America.

24. The Incan Empire lasted from 1438 until 1533.

25. Brazil's men's soccer team boasts the most World Cup wins with five titles.

26. Cerro Aconcagua, an Andean peak in Argentina, is South America's highest point at 22,834 feet above sea level.

27. Brazil shares a border with every other South American country except Chile and Ecuador.

28. Uruguayan cowboys are called "gauchos."

29. Spanning the border between Peru and Bolivia, Lake Titicaca is South America's largest lake at 3,200 square miles.

30. Argentina's president works in a pink building called the Casa Rosada ("Pink House").

31. The Itaipu Dam between Brazil and Paraguay is the world's second-largest hydroelectric facility.

32. It supplies three-quarters of the electricity used in Paraguay.

33. Venezuela's Angel Falls is the world's highest waterfall.

34. The main free-falling segment is almost 20 times higher than Niagara Falls.

35. Argentina and Chile share South America's largest island: Tierra del Fuego.

36. The southernmost city in the world is Ushuaia, located on the Argentinian part of the Tierra del Fuego.

37. Brazil is the most populous South American country with an estimated 214.3 million people.

38. Overlooking Rio de Janeiro, Brazil, stands a colossal statue of Jesus Christ called *Christ the Redeemer*, the world's largest Art Deco-style sculpture.

39. Suriname is the least populous South American country with an estimated 612,985 people.

40. Chile is a narrow strip on the southwestern edge of South America.

41. It offers 3,998 miles of South Pacific coastline.

42. Built by an Inca ruler between 1460 and 1470, Machu Picchu reveals the Inca's skills as stone masons.

43. The Amazon River carries more water than any other river on the planet.

44. The Atacama Desert in Chile is the world's driest non-polar desert.

45. Parts of the central desert area regularly go without rain for years at a time.

46. Approximately half of South America's population lives in Brazil.

47. The ancient kingdom of Tiwanaku was a major Indian civilization in the Andes Mountains of South America.

48. Aztec Emperor Montezuma is said to have drunk 50 goblets of chocolate, flavored with chili peppers, every day.

49. The world's widest road is Brazil's Monumental Axis.

50. It could hold 160 cars side-by-side.

25 Earth-quaking Facts

1. Earthquakes occur when a fault (where Earth's tectonic plates meet) slips, releasing energy in waves that move through the ground.

2. Earth has 16 tectonic plates.

3. Earthquakes occur almost continuously. Fortunately, most are undetectable by humans.

4. An estimated 500,000 earthquakes are detected in the world each year.

5. Scientists measure earthquake magnitude based upon the amount of energy released by the rock movements.

6. An earthquake with a Richter magnitude of 2 is about the smallest earthquake that can be felt by humans without the aid of instruments.

7. The highest magnitude earthquake ever recorded, a 9.5, was centered off the coast in Chile in 1960.

8. Seismographs recorded seismic waves that traveled all around Earth for days after the Chilean earthquake in 1960.

9. More earthquakes occur in Alaska than in any other U.S. state.

10. The San Andreas Fault, an 800-mile fracture in Earth's crust that stretches along the California coast, moves at the same rate your fingernails grow—about 2 inches per year.

11. If that rate keeps up, scientists project that Los Angeles and

San Francisco will be adjacent to one another in approximately 15 million years.

12. Shifting tectonic plates will someday cause the Horn of Africa to break apart from the rest of the continent.

13. Moonquakes are earthquakes on the moon.

14. Antarctica has icequakes, which are little earthquakes within the ice sheet.

15. Indonesia has more earthquakes than any other country.

16. The Loma Prieta earthquake, which struck the San Francisco area in 1989, delayed the World Series between the San Francisco Giants and the Oakland Athletics by 10 days.

17. The most powerful tremor in U.S. history—lasting three minutes and measuring 9.2 on the Richter scale—struck Prince William Sound in Alaska on March 28, 1964.

18. The December 2004 9.1-magnitude earthquake that struck off the coast of the Indonesian island of Sumatra, and the tsunami that followed, killed at least 230,000 people.

19. Scientists say the tremor was so strong that the December 26, 2004, Indonesian quake wobbled Earth's rotation on its axis by almost an inch.

20. A seiche is the sloshing of water that happens during and after an earthquake.

21. The swimming pool at the University of Arizona–Tucson lost water from sloshing caused by the 1985 Michoacán, Mexico, earthquake 1,240 miles away.

22. Seismology is the study of earthquakes.

23. Alaska experiences about 60 earthquakes per day.

24. On February 6, 2023, a magnitude 7.8 earthquake struck parts of Turkey and Syria.

25. The deadly quake killed at least 57,000 people and caused widespread damage.

52 Amazing Architecture Facts

1. Abrāj al-Bayt, also called Makkah Royal Clock Tower, is the world's tallest building with a clock face.

2. The multitowered skyscraper complex is adjacent to the Great Mosque in Mecca, Saudi Arabia.

3. The central clock tower, including its spire, rises to a height of 1,972 feet.

4. Taliesin is regarded as a prime example of Frank Lloyd Wright's organic architecture.

5. The house and 600-acre estate is in Spring Green, Wisconsin.

6. The Sydney Opera House in Australia is a complex of theaters and halls linked beneath its famous white roof, the design of which suggests billowing sails.

7. The largest "sail" is as tall as a 20-story building.

8. Originally built for the 1962 World's Fair, Seattle's 605-foot Space Needle was the tallest building west of the Mississippi when it was completed.

9. The futuristic blueprints for the Space Needle evolved from artist Edward E. Carlson's visionary doodle on a placemat.

10. The resulting design looks like a flying saucer balanced on three giant supports.

11. The Space Needle was built to withstand winds of up to 200 miles per hour.

12. The Krzywy Domek (Crooked House) in Sopot, Poland, isn't a home, but part of the Rezydent shopping center.

13. Its wavy architecture looks like a fun house mirror reflection.

14. The Crooked House's bent lines and distorted walls and doors were inspired by the fairy tale illustrations of Jan Marcin Szancer and Per Dahlberg.

15. The Empire State Building in New York City was the world's

tallest building when it opened in 1931, soaring 1,454 feet to the top of its lightning rod.

16. More than 3,000 workers took less than 14 months to build the Empire State Building, the framework erected at a pace of 4.5 stories per week.

17. The pyramids of Giza, Egypt, are the only wonders of the ancient world that still stand.

18. They were built between about 2650 and 2500 B.C.

19. The Great Pyramid is the largest pyramid ever built.

20. It is 482 feet tall and contains some 2.3 million stone blocks, each weighing an average 2.7 tons.

21. At 1,483 feet, the Petronas Twin Towers in Kuala Lumpur, Malaysia, became the first skyscrapers to top any U.S. skyscrapers when they were completed in 1998.

22. Built in the 1740s, Drayton Hall in South Carolina is America's oldest unrestored plantation house that is open to the public.

23. Drayton Hall is considered one of the finest examples of Georgian Palladian architecture in the nation.

24. There are 10,000 panes of glass in California's Crystal Cathedral.

25. At 2,080 feet, Japan's Tokyo Skytree is the world's tallest free-standing communications tower.

26. The Colosseum in Rome, Italy, once seated 50,000 people.

27. Mazes beneath the main floor kept wild animals contained in preparation for events.

28. Trapdoors allowed gladiators to make surprise appearances and the dead bodies to be hauled away.

29. At 630 feet tall, the Gateway Arch in St. Louis, Missouri, is the nation's tallest monument.

30. The distance between its two legs is equal to its height.

31. The towering steel structure was designed by Finnish-American architect Eero Saarinen in 1947 and completed in 1965.

32. The top of the Gateway Arch is accessible by two trams—one in each leg—that are made up of eight cylindrical, five-seat compartments.

33. The 110-story Willis Tower (formerly Sears Tower) in Chicago was the tallest building in the world from 1974 until 1998.

34. Willis Tower visitors can take a 45-second elevator ride to the 103rd floor, 1,353 feet up, and stand on the Ledge, a series of windows that extend from the building.

35. The Ledge is made of three layers of half-inch thick glass laminated into one seamless unit. It is built to withstand four tons of pressure and can hold 10,000 pounds.

36. In the ancient Mayan city of Chichén Itzá, Mexico, is the Temple of Kukulcán ("El Castillo").

37. Each side of the pyramid has a staircase made of 91 steps that leads to the top.

38. The 1,776-foot One World Trade Center in New York City is the tallest building in the Western Hemisphere.

39. The Shanghai Tower is the world's second-tallest building at 2,073 feet, after the 2,717-foot Burj Khalifa.

40. The Shanghai Tower's elevator travels 3,543 feet per minute.

41. Completed in 1439, the Strasbourg Cathedral in France was the world's tallest building until 1874.

42. The first skyscraper was pioneered in Chicago with the 138-foot-tall Home Insurance Building in 1885.

43. China has more skyscrapers than any other country.

44. Barcelona, Spain's Casa Milà, known as La Pedrera (or, "The Quarry"), expresses Antoni Gaudí's personal architectural style.

45. The façade of Casa Milà was built without flat surfaces, straight lines, or symmetry.

46. The roofline undulates, punctuated by chimneys that look squeezed from a pastry tube.

47. India's Taj Mahal was commissioned by Shah Jahan to honor his wife Mumtaz, who died in childbirth.

48. It took 20,000 workers and 1,000 elephants 22 years to build the Taj Mahal.

49. The structure and its grounds cover 42 acres.

50. The terms "log house" and "log cabin" actually denoted two distinct types of dwellings.

51. Log cabin timbers were left round while log houses were made from notched, square-hewn logs.

52. Cupboards originated with pioneers and were truly "cup boards"—a shelf built from a single board to hold cups and dishes. It was only later that they acquired sides and fronts.

18 Mariana Trench Facts

1. The Mariana Trench in the Pacific Ocean is the deepest location on planet Earth.

2. It is named after the nearby Mariana Islands, which themselves were named after the Spanish Queen Mariana of Austria.

3. Guam, a territory of the U.S., is less than 200 nautical miles away from the trench, giving the U.S. jurisdiction over it.

4. The trench was designated a U.S. national monument in 2009.

5. The Mariana Trench marks the tectonic boundary between two plates.

6. Because the Pacific plate is composed of older crustal material, it is colder and denser than the younger Mariana plate. As a result, it subducts underneath.

7. Mariana Trench's deepest point, called Challenger Deep, measures 36,201 feet deep. This is nearly 7 miles below sea level.

8. If you placed Mount Everest at the bottom of Challenger Deep, its peak would still be 7,000 feet underwater.

9. Crescent shaped, the trench is over 1,550 miles long and 43 miles wide.

10. Pressure increases with depth. At the bottom of the trench, the water column exerts a pressure of more than a thousand times the standard atmospheric pressure at sea level.

11. The depth and perpetual darkness of the Mariana Trench gives it a temperature just a few degrees above freezing: about 34 to 39°F.

12. The first global oceanographic cruise, the 1875 H.M.S. *Challenger* expedition, used weighted sounding rope to measure the trench.

13. Later in 1951, H.M.S. *Challenger II* surveyed the trench again using echo sounding. This was an easier and more accurate way to measure depth.

14. 1960 was the first and only time humans have descended into the Challenger Deep. Jacques Piccard and Navy Lt. Don Walsh boarded the U.S. Navy bathyscaphe called the *Trieste*.

15. While their descent into the depths took five hours, only 20 minutes were spent exploring the bottom.

16. Due to the silt stirred up by their submersible, photography was impossible.

17. In 2012, Canadian filmmaker James Cameron piloted the *Deepsea Challenger* to a depth of 35,756 feet. This set a world record for a solo descent.

18. Data suggests that microbial life-forms occupy the bottom of the Mariana Trench, and scientists believe there are many new species awaiting discovery.

54 Plant Facts

1. There are about 400,000 known plant species.

2. The giant corpse flower is named for its smell.

3. It secretes cadaverene and putrescine, odor compounds responsible for its rotting flesh smell.

4. The first type of aspirin came from the tree bark of a willow tree.

5. Bamboo can grow 35 inches in a single day.

6. Corn is the most planted crop in the United States.

7. Wheat is the third-most-produced grain in the world after corn and rice.

8. There are about 40,000 types of rice.

9. Cacti spines can be up to six inches long.

10. The baobab tree is the world's largest succulent, reaching heights of 75 feet.

11. It can live for several thousand years.

12. Dendrochronology is the science of calculating a tree's age by its rings.

13. The first potatoes were cultivated in Peru about 7,000 years ago.

14. Strawberries are the only fruit that bears seeds on the outside.

15. The average strawberry has 200 seeds.

16. Poison ivy produces a skin irritant called urushiol.

17. Moon flowers open in the evening and close before midday.

18. Moon flower vines always coil clockwise.

19. Morning glory blooms open from dawn to midmorning.

20. The California redwood (coast redwood and giant sequoia) are the tallest and largest living organism in the world.

21. A sunflower looks like one large flower, but each head is composed of hundreds of tiny flowers called florets, which ripen to become the seeds.

22. Sunflowers range in height from 1-foot-tall dwarfs to 15-foot-high giants.

23. The lemon tree is the smallest evergreen tree that is native to Asia.

24. Most coal is formed from ancient plant remains.

25. The *Rafflesia arnoldii* plant has the largest single bloom of any plant, measuring three feet across.

26. Also known as "meat flower," the parasitic *Rafflesia arnoldii* can hold several gallons of nectar, and its smell has been compared to "buffalo carcass in an advanced stage of decomposition."

27. Aquatic duckweed is the smallest flowering plant on Earth.

28. The plant is only 0.61 millimeter long, and the edible fruit is about the size of a grain of salt.

29. Stinging nettle leaves contain a mixture of chemicals that it injects into the skin of animals.

30. Dutchman's breeches are named for their flowers, which look like upside-down pants.

31. Thistles have spines on their stems and leaves that keep most animals away from the plant.

32. Bamboo is the world's tallest grass, growing as much as 35.4 inches in a single day.

33. Vanilla comes from a type of orchid.

34. Although tulips are associated with Holland, they are actually not native there.

35. Tulips descend mostly from species originating in the Middle East.

36. During the 1600s, tulip bulbs were worth more than gold in Holland.

37. The craze was called tulip mania and crashed the Dutch economy.

38. The Jerusalem artichoke is not really an artichoke.

39. The passionflower gets its name from the complex structure of its flowers, which can be seen as symbols of the crucifixion of Christ.

40. Turkey and Bulgaria are the largest producers of the Damask rose (*Rosa damascena*).

41. Although other plants die if they lose 8 to 12 percent of their water content, the resurrection fern can survive despite losing 97 percent of its water content.

42. The baobab tree can store 1,000 to 120,000 liters of water in its swollen trunk.

43. Spanish moss is neither moss nor Spanish.

44. It's related to the pineapple and has been used to stuff furniture, car seats, and mattresses.

45. The technical term for air plant is epiphyte.

46. These plants get moisture and nutrients from the air.

47. The top of the *Hydnora africana* flower looks like a gaping, fang-filled mouth.

48. It emits a putrid scent to attract dung or carrion beetles.

49. The carnivorous King Monkey Cup traps prey such as scorpions, mice, rats, and birds in pitchers up to 14 inches long and 6 inches wide.

50. It then digests them in a half gallon of enzymatic fluid.

51. Ancient Greeks gave hemlock to prisoners who were condemned to die, including Socrates.

52. Ingesting hemlock will eventually paralyze the nervous system, causing death from lack of oxygen to the brain and heart.

53. California boasts the oldest known living tree—a Bristlecone Pine named "Methuselah."

54. According to tree-ring data, it is 4,853 years old.

62 Silly City Names

1. Accident, Maryland

2. Angel Fire, New Mexico

3. Bear, Delaware

4. Benevolence, Georgia

5. Blue Earth, Minnesota

6. Boogertown, North Carolina

7. Boring, Maryland

8. Bread Loaf, Vermont

9. Bumble Bee, Arizona

10. Busti, New York

11. Buttermilk, Kansas

12. Buttzville, New Jersey

13. Cheddar, South Carolina

14. Chipmunk, New York

15. Church, Iowa

16. Climax, Georgia

17. Conception, Missouri

18. Convent, Louisiana

19. Deadman Crossing, Ohio

20. Devil Town, Ohio

21. Devils Den, California

22. Devil's Slide, Utah

23. Ding Dong, Texas

24. Embarrass, Wisconsin *and* Minnesota

25. Energy, Illinois

26. Fanny, West Virginia

27. Frogtown, Virginia

22. Gray Mule, Texas

23. Half Hell, North Carolina

24. Hell, Michigan

25. Hooker, Oklahoma

26. Horneytown, North Carolina

27. Hot Coffee, Mississippi

28. Hygiene, Colorado

29. Intercourse, Pennsylvania

30. Jackass Flats, Nevada

31. Lizard Lick, North Carolina

32. Monkeys Eyebrow, Kentucky

33. Mosquitoville, Vermont

34. Nimrod, Minnesota

35. Nirvana, Michigan

36. Normal, Illinois

37. Oatmeal, Texas

38. Oblong, Illinois

39. Ordinary, Virginia

40. Peculiar, Missouri

41. Pie, West Virginia

42. Pray, Montana

43. Rainbow City, Alabama

44. Red Devil, Alaska

45. Sandwich, Massachusetts

46. Santa Claus, Indiana

47. Spider, Kentucky

48. Success, Missouri

49. Suck-Egg Hollow, Tennessee

50. Tea, South Dakota

51. Ticktown, Virginia

52. Tightwad, Missouri

53. Toad Suck, Arkansas

54. Toast, North Carolina

55. Truth or Consequences, New Mexico

56. Turkey Foot, Florida

57. Vixen, Louisiana

58. Waterproof, Louisiana

59. What Cheer, Iowa

60. Why, Arizona

61. Wide Awake, Colorado

62. Yum Yum, Tennessee

43 Amazing Facts About Asia

1. Asia is the most populous continent on Earth, with an estimated 4.7 billion people.

2. Roughly 60 percent of the world's population lives in Asia.

3. Asia is the largest continent with a land area of 17,207,994 square miles.

4. Turkey's capital city of Istanbul is located in both Asia and Europe.

5. Indonesia has the longest coastline of any Asian country at over 33,999 miles.

6. Nepal has the only national flag that is not a rectangle or a square. Its shape evokes the Himalayan peaks.

7. Kyoto, Japan, is home to about 1,660 Buddhist temples and more than 400 Shinto shrines.

8. The Maldives is Asia's smallest country by landmass (115 square miles) and population (425,000).

9. China is the most populous country in Asia and the world with around 1.4 billion people.

10. The overwhelming majority of the Chinese population is found in the eastern half of the country; the west remains sparsely populated.

11. Though ranked first in the world in total population, China's

overall density is less than that of many other countries in Asia and Europe.

12. India is the most populous democracy in the world with some 1.3 billion people.

13. Cows are sacred in India.

14. Mount Everest on the China–Nepal border is the tallest peak on Earth at 29,035 feet measured from sea level.

15. The Himalayan jumping spider, which lives on Mount Everest, is the world's highest-dwelling creature.

16. The Dead Sea, between Israel and Jordan, is the world's lowest point at 1,414 feet below sea level.

17. There are more horses than people in the country of Mongolia.

18. At 3,977 miles, the Great Wall of China is the world's longest artificial structure.

19. Borneo is Asia's largest island at 290,321 square miles.

20. The island is shared by three countries: Brunei, Indonesia, and Malaysia.

21. The Forbidden City in Beijing, China, was built with 3.1 million bricks.

22. Bhutan is the world's only Buddhist kingdom.

23. Many of the world's tallest skyscrapers are in Asia, including the 2,717-foot-tall Burj Khalifa building in the United Arab Emirates.

24. The United Arab Emirates is composed of seven sheikhdoms, or emirates: Abu Dhabi, Ajman, Dubai, Fujairah, Ras al Khaimah, Sharjah, and Umm al Quwain.

25. The mango is the national fruit of India.

26. The Pamir and the Tian Shan are among the mountain ranges that cover more than 90 percent of Tajikistan.

27. At 28,251 feet, K2 is Earth's second highest mountain.

28. It was named K2 because it was the second peak measured in the Karakoram Range.

29. Macau, a special administrative region of China, has the world's highest population density with 56,059 people per square mile.

30. Portions of Azerbaijan, Georgia, Kazakhstan, Russia, and Turkey fall within both Europe and Asia, but the larger section of each is in Asia.

31. Although 77 percent of Russia's landmass is in Asia, most Russians live on the European side.

32. Lake Baikal in southern Siberia is Earth's deepest lake at 5,315 feet.

33. It holds approximately one-fifth of the world's fresh surface water.

34. Japan has an average life expectancy of 84.62 years, one of the world's highest.

35. The Son Doong cave in Vietnam's Phong Nha-Ke Bang National Park is the world's largest cave.

36. One of the world's tallest flagpoles is found in Jeddah, Saudi Arabia. It stands 561 feet tall.

37. The national drink of Kyrgyzstan is kumis, made from fermented horse milk.

38. The Philippine archipelago includes over 7,000 islands.

39. There are 12 landlocked countries in Asia: Uzbekistan, Armenia, Turkmenistan, Kyrgyzstan, Afghanistan, Tajikistan, Laos, Mongolia, Kazakhstan, Nepal, Bhutan, and Azerbaijan.

40. Asia is home to more roller coasters than any other continent.

41. Devout Muslims throughout the world turn toward Mecca, Saudi Arabia, in prayer five times each day.

42. The Dome of the Rock, a shrine in Jerusalem that dates to the late 7th century A.D., is the oldest Islamic monument still in existence.

43. The rock over which the shrine was built is sacred to both Muslims and Jews.

43 Cave Facts

1. According to the National Cave and Karst Research Institute, caves are naturally occurring hollow spaces in the ground.

2. They are large enough for a person to enter, and have multiple rooms or passageways to explore.

3. *Speleology* refers to the scientific study and exploration of caves.

4. *Spelunking,* however, refers to the recreational pastime of exploring cave systems.

5. This is also called caving, or potholing, in the United Kingdom.

6. There are several different types of cave, each formed by different natural processes.

7. Lava tubes are formed through volcanic activity.

8. As lava flows down the flanks of volcanoes, the surface of the lava flow, where the molten rock meets the air, cools and hardens first. These sections insulate the lava flowing inside them.

9. When the eruption stops, a hollow tube is left behind.

10. Sometimes, the last remaining bits of lava create stalactites and stalagmites.

11. Just off the southern coast of South Korea, the multicolored carbonate speleothems of the lava tube cave system in Jeju Volcanic Island attracts many visitors.

12. The air route from Seoul to Jeju Island served over 11 million passengers in 2015.

13. Sea caves occur on nearly every cliffed coast, or coasts where wave action breaks onto cliff faces.

14. Waves are a type of mechanical weathering.

15. They crash into weak zones in sea cliffs and slowly break apart the rock, creating a larger and larger cavity over time.

16. When the same process occurs along lakes, it creates what is known as *littoral caves.*

17. Some sea caves form below sea level, but later emerge above the water due to uplift.

18. Similar processes form eolian caves, which are carved by wind.

19. Small particles of sediments like sand or silt are carried by winds and blasted against cliff faces.

20. A well-known example of an eolian cave is Mesa Verde National Park in Colorado.

21. The homes inside the cave were built by the Ancestral Pueblo people.

22. Glacier caves are created in glaciers because of melting ice and flowing water.

23. The caves tend to be unstable as their shapes and forms change from year to year.

24. Ice caves are different from glacial caves. Ice caves are rock caves that have permanent ice deposits.

25. Solution caves are the most common type of cave.

26. They form in karst, a type of landscape made from soluble rocks like carbonate limestone, dolomite, or marble, or evaporite gypsum, anhydrite, or halite (salt).

27. Rainwater collects carbon dioxide from the atmosphere and soil and forms a weak carbonic acid. The acid seeps through fractures in the ground and dissolves the surrounding rock.

28. Over time, the cracks become large enough to form caves.

29. Cave formations are called *speleogens* and *speleothems.*

30. Speleogens are created during the formation of the cave.

31. Speleothems are formed later by mineral deposits.

32. Stalactites are icicle-shaped deposits that grow downward from a cave's ceiling.

33. Stalagmites are similar, but grow upward from the floor.

34. Stalactites and stalagmites will occasionally connect, forming columns.

35. Flowstones are sheetlike deposits from which water flows down the walls or along the floors of a cave.

36. Mammoth Cave in Kentucky is the world's largest network of caverns.

37. There are more than 350 miles of underground passages on five different levels. New caves and passageways are still being discovered.

38. Cave of the Crystals is a cave of large selenite crystals underneath the Sierra de Naica Mountain in Chihuahua, Mexico.

39. It was discovered by brothers Juan and Pedro Sanchez.

40. The largest of the cave's crystals is over 37 feet long.

41. In 2018, 12 members of the Wild Boars soccer team and their 25-year-old assistant coach became trapped in Thailand's Tham Luang Nang Non cave.

42. Soon after they entered, heavy rainfall began to flood the cave system, blocking the exit.

43. Monsoons were not expected to start for another month. The boys and their coach were eventually rescued by cave divers.

47 Facts About Africa

1. Africa is Earth's second largest continent with 11,608,161 square miles.

2. With an area of 226,917 square miles, Madagascar is Africa's largest island.

3. Many kinds of wild guinea fowl are found in Africa.

4. The birds derive their name from a section of Africa's west coast.

5. Two-thirds of Africa lies in the Northern Hemisphere.

6. The world's deepest mine is Harmony Gold's Mponeng gold mine, near Johannesburg, South Africa.

7. It goes nearly two and a half miles deep.

8. Africa has over 25 percent of the world's bird species.

9. The Nile is the world's longest river.

10. It flows for 4,132 miles.

11. The first great civilization in Africa arose 6,000 years ago on the banks of the lower Nile River.

12. Timbuktu, Mali, is home of one of the oldest universities in the world, established in 982 B.C.

13. African elephants, the world's largest living land animal, have ears shaped like the continent of Africa.

14. More than one million western white-bearded wildebeest inhabit the Serengeti Plains and acacia savanna of northwestern Tanzania and adjacent Kenya.

15. Some 1,600 languages are spoken in Africa—more than any other continent.

16. Ethiopia is the only African country with its own indigenous alphabet.

17. The world's largest frog, the Goliath frog, is found in Equatorial Guinea and Cameroon.

18. Libya has no natural rivers.

19. Lake Tanganyika is the world's second deepest lake at 4,820 feet.

20. In 1871, the missing explorer David Livingstone was found by Henry Morton Stanley on the shore of Lake Tanganyika.

21. South Africa is home to the world's oldest mountains.

22. The Barberton Greenstone Belt, also known as the Makhonjwa Mountains, is formed of rocks dating back 3.6 billion years.

23. Zanzibar, an island off the east coast of Africa, is made up entirely of coral.

24. The highest point in Africa is Mount Kilimanjaro in Tanzania at 19,340 feet high.

25. Africa's lowest point is Lake Assal in Djibouti at 512 feet below sea level.

26. The most populous African country is Nigeria with around 206 million people.

27. Africa's longest mountain range is the Atlas Mountains, which stretch 1,553 miles across Morocco, Algeria, and Tunisia.

28. The kingdom of Lesotho is completely encircled by South Africa.

29. Gorée Island, Senegal, was the site of one of the earliest European settlements in Western Africa and long served as an outpost for slave trading.

30. The Valley of the Kings in Egypt was the burial site of almost all the pharaohs of the 18th, 19th, and 20th dynasties.

31. While Egypt is famous for its pyramids, Sudan has double the number of pyramids in Egypt.

32. Lake Victoria is the largest lake in Africa and the second-largest freshwater lake in the world.

33. Victoria Falls, on the border between Zambia and Zimbabwe, is 355 feet high.

34. All of Africa was colonized by foreign powers during the "scramble for Africa" except Ethiopia and Liberia.

35. The World Cup was played for the first time on the African continent in 2010 when South Africa hosted the event.

36. Africa's largest country is Algeria, at 919,595 square miles.

37. The Seychelles is the smallest country in Africa, with an area of just 176 square miles.

38. Deforestation rates in Africa are twice the average for the rest of the world.

39. The vast Etosha Pan, the extremely flat salt pan covering an area of approximately 1,900 square miles, is the largest of its kind in Africa.

40. The largest impact crater on Earth is the Vredefort Crater, near Johannesburg, South Africa, with an estimated diameter of 155 to 186 miles.

41. South Sudan, the newest country in Africa, gained independence from the Republic of Sudan in 2011.

42. Africa has more countries than any other continent with 54.

43. In 1974, scientists discovered the oldest known human ancestor in Ethiopia.

44. The 3.2-million-year-old skeleton was named "Lucy" after the Beatles' song "Lucy in the Sky with Diamonds."

45. The Strait of Gibraltar, which separates Africa (Morocco) from Europe (Spain), is just under nine miles wide at its narrowest point.

46. Monrovia, the capital of Liberia, is the only foreign capital city named after a U.S. president.

47. James Monroe was president when the city was founded in 1822 as a refuge for freed slaves.

24 Volcanic Facts

1. There are some 1,900 active volcanoes on Earth.

2. The Ring of Fire is a seismically active belt of volcanoes and tectonic plate boundaries that stretches 24,900 miles around the Pacific Ocean.

3. About 75 percent of the world's volcanoes occur within the Ring of Fire.

4. The word "volcano" comes from the name Vulcan, the Roman god of fire.

5. Mauna Loa in Hawaii is the world's largest active volcano, rising approximately 30,000 feet from the base at the bottom of the ocean to its peak.

6. Molten rock is called magma when it's still under the earth, and lava when it comes spewing out of a volcano.

7. Fresh lava can reach temperatures as high as 2,200°F.

8. Nevado Ojos del Salado on the Argentina–Chile border is the highest active volcano in the world at 22,615 feet above sea level.

9. The most famous eruption of Mount Vesuvius, in southern Italy, occurred in A.D. 79, when lava and ashes buried the towns of Pompeii and Herculaneum.

10. The Hawaiian Islands are composed of shield volcanoes that have built up from the seafloor to the surface.

11. The deadliest known volcano eruption occurred in Indonesia when the 13,000-foot Mount Tambora blew two million tons of debris into the air in April 1815.

12. The debris in the atmosphere released by Mount Tambora's eruption darkened skies all around the world and continued to block sunlight for years afterward.

13. In 1816, parts of the United States saw snow in June and July, thanks to the persistent cold caused by the eruption on the other side of the world.

14. All told, the Mount Tambora eruption claimed more than 70,000 lives.

15. Plinian are the most explosive type of eruption.

16. Icelandic eruptions are the mildest type and consist of a flow of lava from long fissures, or cracks, in the ground.

17. Some lavas are liquid enough to flow downhill at 35 miles per hour; others move at the rate of only inches per day.

18. The speed depends on the temperature and composition.

19. Since 1983, the Hawaiian volcano Mount Kilauea has been in almost continual eruption, producing rivers of lava that flow to the sea 30 miles away.

20. The great clouds of ash from the 1980 eruption of Mount St. Helens drifted into Idaho and increased crop yields by an average of 30 percent throughout the 1980s.

21. When the top of a volcano collapses, a pit called a caldera is formed.

22. When the volcano Krakatau (in Indonesia) erupted in 1883, people in China—3,000 miles away—heard the sound.

23. Mars is home to the largest volcano in the solar system, Olympus Mons.

24. It's roughly three times the height of Mount Everest and covers an area nearly the size of Arizona.

42 Facts About Cities

1. More than half of the world's population currently lives in cities.

2. Megacities are defined as urban areas with a population of more than 10 million people.

3. Tokyo, Japan, is the largest city in the world, with a population of more than 37 million people.

4. Large men known as *oshiya* ("pushers") are employed to cram commuters onto Tokyo's overcrowded subways and trains.

5. Chicago has more movable bridges than any other city in the world.

6. Delhi, India, is the second-largest city in the world with a population over 32 million.

7. Delhi and New Delhi are the same city.

8. New Delhi is actually a district within India's capital city of Delhi which has government buildings.

9. While Tokyo only gained 559,000 people between 2010 and 2020, Delhi gained over 8 million people in the same time frame.

10. Tokyo boasts the world's first public monorail line, built to coincide with the 1964 Olympics. It runs between downtown and Haneda International Airport.

11. Shanghai is the world's third-largest city with a population of 28 million.

12. The complete name of Los Angeles is technically El Pueblo de Nuestra Señora la Reina de los Ángeles de Porciúncula.

13. Dhaka is the capital of Bangladesh and the fourth most populous city with nearly 22.5 million residents.

14. The people of Dhaka speak English and Bengali.

15. St. Augustine, Florida, is the oldest permanently inhabited city in the United States.

16. Founded in 1565, St. Augustine is filled with reminders of its early Spanish history.

17. The fifth most populous city in the world is São Paulo, Brazil, with some 22,430,000 people.

18. Mexico City ranks as the sixth-largest city in the world. More than 22,085,000 souls call the Mexico City metro area home.

19. Because it was built on a soft lake bed, Mexico City is sinking at a rate of nearly four inches per year.

20. Pittsburgh has more city-maintained steps than any other city in the world. If stacked up, they would reach a height of 26,000 feet.

21. Cairo, Egypt, is Africa's most populous city with some 21,750,000 residents. Lagos, Nigeria, and Kinshasa, Democratic Republic of Congo, follow.

22. Rio de Janeiro's name means "River of January" in Portuguese.

23. The trains in Mumbai, India, carry around six million people per day—that's more than the entire population of Finland.

24. Osaka, Japan, has about 19 million residents.

25. New York City is the largest city in the U.S. with a population over 8 million in the city proper.

26. A visitor walking the Las Vegas Strip could see everything from replicas of the Eiffel Tower and the New York skyline to dazzling water fountains that dance to music.

27. The heaviest building in the world is in Bucharest, Romania. The Palace of the Parliament weighs over 700,000 tons.

28. The fastest growing cities in the world are in Asia and Africa.

29. Savannah, Georgia, was America's first planned city.

30. James Edward Oglethorpe founded Savannah in 1733, centering the city around 24 squares. His layout remains intact today.

31. Athens, Greece, is Europe's oldest capital city.

32. The city of Amsterdam in the Netherlands has more bicycles than people.

33. Kolkata (formerly Calcutta) ranks as India's third most populous city, after Delhi and Mumbai.

34. Helsinki's sidewalks get heated from underground.

35. Kuala Lumpur means "river junction" in Malay. This Malaysian city sits where the Klang and Gombak rivers flow together.

36. London's famous indoor marketplace Harrod's sold cocaine and cocaine-related products to shoppers until 1916.

37. Manila, Philippines, has some 14.4 million inhabitants. Its population density is one of the greatest of any city in the world.

38. Buenos Aires, Argentina, is South America's second most populous city after São Paulo, Brazil.

39. Rome, Italy, has a museum dedicated to pasta.

40. An abandoned network of underground stone quarries beneath Paris hold the remains of six million people.

41. Istanbul, Turkey, has more than 15 million inhabitants and straddles two continents.

42. Vienna, Austria, boasts the world's oldest zoo, dating to 1752.

19 Eiffel Tower Facts

1. Gustave Eiffel designed his monument to the French Revolution in 1887 as a grand entranceway to the 1889 International Exposition in Paris.

2. A crew of 300 workers assembled the tower in two years, two months, and five days.

3. They came in under budget and on time for the start of the fair.

4. The Eiffel Tower was assembled from 18,038 iron parts.

5. Every seven years, at least 25 workers use approximately 60 tons of paint to rustproof the tower.

6. The Eiffel Tower stands 1,063 feet tall.

7. It was the world's tallest building until 1931 when it was topped by the Empire State Building in New York City.

8. On a clear Parisian day, a person at the top of the Eiffel Tower can see about 42 miles in every direction.

9. In just one year, the tower recouped nearly the entire cost of its construction—thanks to elevator ticket sales.

10. The tower was one of the first tall structures in the world to use passenger elevators.

11. On the four sides of the tower, the names of 72 famous French scientists and engineers are engraved to honor their national contributions.

12. There are 1,665 steps to the top of the tower, though it's widely thought there are 1,792, representing the year of the First French Republic.

13. There are 2.5 million rivets (short metal pins) in the Eiffel Tower.

14. Heat from the sun can cause the tower to expand up to three-fourths of an inch.

15. During the cold winter months, the tower shrinks approximately six inches.

16. The Eiffel Tower weighs more than 10,000 tons.

17. Today, the tower attracts about 7 million visitors per year.

18. The French landmark was nearly demolished in 1909, when its original 20-year permit expired.

19. But because its antennae were used for telecommunications, the Eiffel Tower was spared.

46 Facts About Europe

1. Europe is Earth's second smallest continent.

2. Germany is the most populous country entirely in Europe with an estimated 83.2 million people.

3. The largest country entirely in Europe is Ukraine with a total area of 233,032 square miles.

4. The "Running of the Bulls" in Pamplona, Spain, can be traced back to the 1300s.

5. Mount Elbrus in Russia is the highest peak in Europe at 18,510 feet above sea level.

6. Amsterdam (population 1,166,000) has more bikes than people.

7. Europe is likely named after the Phoenician princess Europa who was seduced by the Greek god Zeus and taken to the island of Crete.

8. Hamburgers got their name from Hamburg, Germany.

9. The Pyrenees Mountains form a natural barrier between France, Spain, and Portugal.

10. The longest town name in Europe is Lanfairpwllgwyngyllgogerychwyrndrobwllllantysiliogo.

11. It's located in Wales.

12. The national anthem of Spain has no words.

13. The longest reign of any European monarch was that of Afonso I Henrique of Portugal who ruled for 73 years and 220 days.

14. Hum, Croatia, is Europe's smallest town with just 30 inhabitants.

15. The highest toilet in Europe is located on Mont Blanc ("White Mountain") at an elevation of over 13,780 feet.

16. Portugal is the world's leading producer of cork.

17. The Volga is Europe's longest river and the principal waterway of Russia.

18. The river flows for 2,325 miles.

19. Liechtenstein is double landlocked: it's landlocked by countries which are also landlocked.

20. The European continent is located completely in the Northern Hemisphere and mostly in the Eastern Hemisphere.

21. Vatican City (Holy See) is the least populated country in Europe with an estimated 800 people.

22. About 10 percent of the planet's population lives in Europe.

23. The oldest known European cave drawings are found in Chauvet Cave in southern France.

24. The drawings are estimated to be 32,000 to 35,000 years old.

25. Greek is one of the oldest languages in Europe.

26. You can get rained on from above and below at the Cliffs of Moher in Ireland.

27. The Rhine River is the busiest waterway in Europe.

28. The Rhine runs through the most populated part of Europe.

29. Europe's coastline measures more than one and a half times the length of the equator.

30. Helsinki, Finland, is the northernmost capital in Europe.

31. Mount Etna on the Italian island of Sicily is the tallest and most active volcano in Europe.

32. Bakken is the oldest operating amusement park in the world.

33. Located north of Copenhagen, Denmark, the park opened in 1583.

34. Paris, France, is home to the largest triumphal arch in the world, the Arc de Triomph.

35. Vatican City is the smallest country in the world with an area under 0.2 square miles.

36. Norway has the longest coastline in Europe with over 62,000 miles.

37. The port of Rotterdam, Netherlands, is the largest and busiest container port in Europe.

38. Budapest, Hungary, consists of two parts: Buda, located on the west bank of the Danube River, and Pest, located on the east bank.

39. The three Baltic States of Estonia, Latvia, and Lithuania were once Soviet Republics.

40. The oldest human DNA found and analyzed was taken from the thighbone of a 400,000-year-old human-like species whose remains were found in northern Spain.

41. Monaco is the most densely populated country in Europe with 49,000 people per square mile.

42. Scandinavia consists of the countries of Norway, Sweden, and Denmark.

43. Some people also consider Iceland, the Faeroe Islands, and Finland to be part of Scandinavia.

44. The Italian city of Pisa is home to the famous bell tower called the Leaning Tower of Pisa.

45. Europe is home to five of the ten smallest countries in the world: Vatican City, Monaco, San Marino, Liechtenstein, and Malta.

46. The Wieliczka salt mine in southern Poland has been in operation since the 13th century.

41 Facts to Rock Your World

1. Rocks are made up of a mix of minerals.

2. There are more than 4,000 known minerals.

3. Rubies and sapphires are made of the same mineral—corundum.

4. Humans have used the metals and minerals in rock since the beginning of civilization.

5. The mineral salt was so valuable it was once traded ounce for ounce for gold.

6. There are three main types of rocks: igneous, sedimentary, and metamorphic.

7. They are classified depending upon how they were formed in the different layers of Earth.

8. The word "igneous" comes from the Latin word "*ignis*" which means "of fire."

9. Sedimentary rocks form layers at the bottoms of oceans and lakes.

10. Layers of sedimentary rocks are called strata.

11. Marble is a metamorphic rock formed when limestone is exposed to high heat and pressure within Earth.

12. Geologists use the Mohs scale to measure how hard a rock is. The higher the number, the harder the rock.

13. Diamonds, the most famous gem and the hardest known material on Earth, are a 10 on the Mohs scale.

14. Diamonds make great scalpels because their sharp edges never dull, and their hydrophobic surfaces ensure that wet tissue never sticks to them during surgery.

15. Diamonds come in a variety of colors, known in the biz as "fancies."

16. Most diamonds are more than 3 billion years old.

17. Rocks that are rich in metals are called ores.

18. Opal is Australia's national gemstone.

19. Amber is an orange gem that comes from the dried resin of prehistoric trees.

20. Of all gems, amber is the lightest.

21. Genuine amber floats in saltwater.

22. When rubbed, amber can acquire an electric charge.

23. Thus, the Greek word for amber is electrum, which is the origin of the word electricity.

24. The marble quarry in Carrara, Italy, is prized for its brilliant white marble.

25. The artist Michelangelo carved his famous statue of David from a single block of Carrara marble.

26. *Fulgurite* (from the Latin word for "thunderbolt") forms when lightning strikes sand or rock.

27. South Dakota's Mount Rushmore was carved from igneous rock.

28. It took 14 years to carve the faces of George Washington, Thomas Jefferson, Theodore Roosevelt, and Abraham Lincoln out of the granite cliff.

29. A type of volcanic rock called pumice can float on water.

30. Fluorite can glow under ultraviolet light.

31. A rock expert is called a petrologist.

32. Meteorites are rocks from space.

33. Most sandy beaches are actually billions of broken quartz crystals.

34. Located close to the village of Ellora in western India is a series of 34 magnificent temples cut from basaltic cliffs.

35. Off the southwest coast of the Australian state of Victoria, the

Twelve Apostles, a spectacular formation of limestone sea stacks, are part of Port Campbell National Park.

36. Mammoth Cave in Kentucky is made up of limestone.

37. The dazzling array of stalagmites, stalactites, and columns are the product of water seeping downward for millennia.

38. Chimney Rock rises almost 325 feet above the North Platte River Valley in Nebraska.

39. It was a landmark for pioneers traveling the Mormon, California, and Oregon trails.

40. Though the Hope Diamond is more famous, the Cullinan is the largest diamond ever found.

41. Unearthed in South Africa in 1905, this 3,100-carat monster was cut into several stones that are still part of the British Crown Jewels.

41 National Parks Facts

1. The U.S. National Park Service administers more than 400 sites of historic, scientific, cultural, scenic, or other interest.

2. Established in 1872, Yellowstone National Park (in Wyoming, Montana, and Idaho) is the oldest national park in the world.

3. Yellowstone is home to more than half of the world's thermal features.

4. The famous Old Faithful geyser at Yellowstone erupts every 35 to 120 minutes.

5. Although Yellowstone was the first national park, it wasn't the first area set aside as a park.

6. That honor is shared by the National Capital Parks, the White House, and the National Mall, all designated on July 16, 1790.

7. National park status doesn't last forever.

8. A couple dozen sites have been turned over to the states or to other federal departments.

9. Denali National Park and Preserve in Alaska is home to Denali (Mount McKinley), the highest point on the North American landmass.

10. The lowest point in a national park is Badwater Basin in California's Death Valley National Park, which is 282 feet below sea level.

11. From 1931 to 1934, just over half an inch of rain fell there, and summer temperatures can exceed 130°F.

12. Death Valley National Park also has a mountain, Telescope Peak, which tops out at 11,049 feet above sea level.

13. From the top of this mountain to Badwater Basin, it's twice the vertical drop of the Grand Canyon.

14. Arizona's Grand Canyon is 227 miles long, 18 miles wide, and a mile deep.

15. The Grand Canyon has a lesser-known peer: Colorado's Black Canyon of the Gunnison National Park.

16. The dark gneiss walls of the canyon (more than a half mile deep) narrow to a quarter mile wide, presenting a majestic, shadowy abyss.

17. In Alaska's Kobuk Valley National Park, there are approximately 25 square miles of rolling sand dunes, and summer temperatures can hit 100°F.

18. Dinosaurs once roamed what is today Badlands National Park in South Dakota.

19. The term badland was first coined by French-Canadian trappers from the French phrase for "bad lands to cross."

20. Oregon's Crater Lake National Park is a corpse—a volcanic corpse, that is.

21. After the volcano's walls collapsed, precipitation fell and melted to form a shimmering six-mile-wide lake.

22. Joshua Tree National Park in California is home to huge tarantulas.

23. The northernmost point in a national park is Inupiat Heritage Center at Barrow, Alaska.

24. The southernmost point is the National Park of American Samoa, below the equator.

25. Found at Great Sand Dunes National Park in Colorado, the tallest dunes in North America are 75 stories tall.

26. The longest arch at Arches National Park in Utah measures 306 feet from base to base. A football field is 300 feet long.

27. Mammoth Cave National Park in Kentucky is home to the world's largest cave system. There are more than 360 miles of its underground passages on five different levels.

28. Devils Tower in Wyoming began its history buried—a tall pillar of magma that leaked or burned through the other rock in the area, then cooled underground.

29. Millions of years of erosion laid bare the 867-foot tower.

30. There are more than 500 islands in Minnesota's Voyageurs National Park.

31. Florida's Everglades National Park is the only place where the American crocodile and the American alligator coexist.

32. Sleeping Bear Dunes National Lakeshore in western Michigan is one of the world's largest complexes of shifting dunes. The dunes may rise to almost 500 feet above the lake.

33. Dry Tortugas National Park in Florida is known for its world-renowned concentration of turtles.

34. The farthest national park from Washington, D.C., is War in the Pacific National Historical Park in Guam. It's far closer to the Philippines than to Hawaii.

35. While a portion of California's Yosemite National Park has all the conveniences of a small city, 95 percent of the park is designated as wilderness.

36. The glaciers that carved out Yosemite Valley left behind huge granite domes and peaks.

37. The two largest are El Capitan, which towers 3,600 feet from the valley floor, and Half Dome, which rises some 4,800 feet.

38. Everglades National Park constitutes the largest subtropical wilderness left in the United States.

39. Mesa Verde National Park in Colorado preserves a large complex of cliff dwellings that were built hundreds of years ago by the Anasazi people.

40. Great Smoky Mountains National Park, located along the border of Tennessee and North Carolina, is consistently the most visited park in the system.

41. Hot Springs National Park protects 47 different hot springs, as well as the eight historic bathhouses in the town of Hot Springs, Arkansas.

31 Watery Facts

1. About 97 percent of the water on Earth is salt water.

2. The amount of water on Earth hasn't changed since the planet was formed.

3. Fresh water makes up only 3 percent of Earth's water.

4. Earth has five oceans: the Atlantic, Pacific, Indian, Southern (or Antarctic), and Arctic Oceans.

5. The Pacific Ocean is the world's largest ocean, covering approximately 30 percent of Earth's surface.

6. Most of Earth's fresh water is frozen in polar ice caps.

7. Salt in the ocean comes from two sources: runoff from the land and openings in the seafloor.

8. The Dead Sea is reported to be six times saltier than an ocean, and salt provides buoyancy for swimmers.

9. The Red Sea is actually blue.

10. Some believe whoever named it saw the Red Sea while the red- and pink-hued *Trichodesmium erythraeum*, a type of marine algae, was in full bloom.

11. The oceans contain enough gold that if it were mined, every person on Earth would receive nine pounds.

12. The Great Lakes contain 20 percent of the world's fresh water.

13. At some time, most of it flows over Niagara Falls.

14. Russia's Lake Baikal is Earth's oldest lake.

15. It holds more water than all of the Great Lakes combined.

16. The Pacific Ocean contains some 25,000 islands, far more than other oceans.

17. The Atlantic is Earth's second-largest ocean.

18. Frozen water is lighter than liquid water, which is why ice floats.

19. Glacial ice is blue rather than white.

20. The Indian Ocean, the third-largest ocean in the world, covers about 14 percent of Earth's surface.

21. Oceans are home to over 90 percent of all living species on Earth.

22. Humans have explored less than 10 percent of Earth's oceans.

23. The Pacific Ocean is wider than the moon.

24. Cold water weighs more than hot water.

25. The Atlantic Ocean's name comes from Atlas, the Greek god of navigation and astronomy.

26. Most volcanic activity occurs beneath the water in the ocean.

27. The Pacific is the oldest of Earth's current oceans.

28. About 50 percent of the United States lies beneath the ocean.

29. In the deep ocean, the pressure is as much as 9,000 pounds per square inch.

30. Earth's longest chain of mountains, the Mid-Ocean Ridge, is almost entirely beneath the ocean.

31. The oceans are divided into different zones which are distinguished based on the amount of sunlight they receive: sunlight, twilight, midnight, the abyss, and the trenches.

39 Facts About Oceania

1. Oceania includes 14 countries: Australia, Micronesia, Fiji, Kiribati, Marshall Islands, Nauru, New Zealand, Palau, Papua New Guinea, Samoa, Solomon Islands, Tonga, Tuvalu, and Vanuatu.

2. Nauru is the most densely populated country in Oceania, with 1,402 people per square mile.

3. Sheep far outnumber humans in both Australia and New Zealand.

4. The Oceania–Asia border cuts the island of New Guinea in half.

5. Papua New Guinea (in Oceania) makes up half the island and Indonesia (in Asia) claims the other half.

6. Australia is the sixth-largest country in the world with an area of 2,988,901 square miles.

7. Oceania is the world's smallest continent in terms of land area.

8. About one third of New Zealand's land area is devoted to national parks, wilderness areas, and other conservation efforts.

9. Only 45 of Tonga's 171 islands are inhabited.

10. Australia is wider than the moon.

11. Dutch explorer Abel Tasman was the first European to reach the Australian island that was later named Tasmania in his honor.

12. Thousands of years ago, a land bridge connected the island of Tasmania to Australia.

13. New Zealand's South Island has 18 peaks that top 9,800 feet.

14. Kiribati is the only country in the world situated in all four hemispheres.

15. Despite having a land area of just 313 square miles, Kiribati's coral atolls and islands are dispersed over some 1.35 million square miles.

16. Australia is Oceania's most populous country with a population of 25,978,935 people.

17. Lake Eyre in South Australia is the continent's largest lake.

18. It's also Oceania's lowest point at 50 feet below sea level.

19. Aboriginal Australians have lived on the continent for more than 60,000 years.

20. British explorer James Cook was the first European to reach Australia in 1770.

21. Starting in 1788, Britain established penal colonies in New South Wales, Tasmania, and Western Australia.

22. Rugby is popular in Oceania. It's the national sport in New Zealand, Samoa, Tonga, and Fiji.

23. Cricket is the national sport of Australia.

24. New Zealand's North Island is home to a collection of geothermal attractions.

25. The Champagne Pool (a hot, steamy, bubbly lake) spills over to create a bright, sulfury, yellow-green pond called the Devil's Bath.

26. There are more than 140 species of marsupials in Australia.

27. Australia is the largest country entirely in the Southern Hemisphere. (Brazil is larger, but some spills into the Northern Hemisphere.)

28. Fiji and Vanuatu are popular with tourists because of their coral reefs and unspoiled beaches.

29. Aoraki-Mount Cook on the South Island of New Zealand is one of the highest island peaks in the world at 12,218 feet.

30. Australia is the largest country in the world without any land borders.

31. Tarawa—the capital of Kiribati—is about halfway between Hawaii and Australia.

32. Uluru in Northern Territory, Australia, is the largest rock monolith in the world.

33. It rises more than 1,100 feet from the desert floor and is about two miles wide.

34. Spanish explorer Alonso de Salazar was the first European to spot the Marshall Islands in 1529.

35. The sea breeze known as the "Fremantle Doctor" affects the city of Perth on the west coast of Australia and is one of the most consistent winds in the world.

36. Aside from Australia and New Zealand, the roughly 25,000 islands of Oceania make up three large cultural regions: Melanesia, Micronesia, and Polynesia.

37. There are more kangaroos than people in Australia.

38. The Outback is the vast, dry, and remote inland parts of Australia.

39. Australian toilets are designed to flush counterclockwise.

18 Lightning Facts

1. Lightning isn't only associated with thunderstorms.

2. Volcanic eruptions, snowstorms, and even large forest fires have been associated with lightning discharges.

3. How hot is a lightning bolt? 54,000°F!

4. A bolt of lightning is about as thick as your thumb.

5. You can see it from so far away because of its luminescence.

6. Ball lightning, also called globe lightning, is a rare aerial phenomenon in the form of a brilliant sphere several inches in diameter.

7. There are about 3,000 lightning flashes on Earth every minute.

8. Florida is considered the "lightning capital" of the U.S., with more lightning injuries and deaths than any other state.

9. Florida has 3,500 cloud-to-ground lightning flashes per day and 1.2 million flashes per year.

10. The odds of being struck by lightning in the United States in any given year: about 1 in 700,000.

11. The odds of being struck by lightning in your lifetime: about 1 in 3,000.

12. About 90 percent of all lightning strike victims survive with varying degrees of disability.

13. Lightning can strike up to 20 miles away from the originating storm.

14. These bolts, called "Positive Giants," seem to randomly come from a clear sky.

15. They are usually much more destructive than ordinary lightning.

16. Cloud-to-cloud lightning can stretch over amazing distances. Radar has recorded at least one of these "crawlers" that was more than 75 miles long.

17. As ice particles swirl around in a storm cloud, they collide and cause separation of electrical charges, which can cause lightning.

18. If your hair stands on end, you hear high-pitched or crackling noises, or see a blue halo (referred to as St. Elmo's fire) around objects, there is electrical activity near you.

45 City Nicknames

1. Anchortown (Anchorage)
2. ATL (Atlanta)
3. Beantown (Boston)
4. The Big Apple (New York City)
5. Big D (Dallas)
6. The Big Easy (New Orleans)
7. Bride of the Sea (Venice)
8. Charm City (Baltimore)
9. Chi-Town (Chicago)
10. Cincy (Cincinnati)

11. City of Angels (Los Angeles)

12. City of Brotherly Love (Philadelphia)

13. City of Kings (Lima)

14. City of Light (Paris)

15. City of Roses (Pasadena)

16. City of 100 Spires (Prague)

17. City of Trees (Sacramento)

18. Dogtown (Santa Monica)

19. The Emerald City (Seattle)

20. The Empire City (New York City)

21. The Eternal City (Rome)

22. The Fair City (Dublin)

23. Fog City (San Francisco)

24. Hotlanta (Atlanta)

25. Iron City (Pittsburgh)

26. Jet City (Seattle)

27. La-La-Land (Los Angeles)

28. Mile-High City (Denver)

29. Motor City (Detroit)

30. Music City (Nashville)

31. NOLA (New Orleans)

32. The Old Pueblo (Tucson)

33. Philly (Philadelphia)

34. Ra-Cha-Cha (Rochester)

35. Rocket City (Huntsville)

36. Rose City (Portland)

37. Rubber City (Akron)

38. Sin City (Las Vegas)

39. Space City (Houston)

40. The Steel City (Gary)

41. Surf City, USA (Huntington Beach)

42. Valley of the Sun (Phoenix)

43. Venice of the North (Amsterdam)

44. Vice City (Miami)

45. Windy City (Chicago)

13 Facts About Big Ben

1. Big Ben is the largest clock in Britain.

2. The name Big Ben originally referred only to the clock's bell, but has come to represent the entire clock.

3. Historians believe that the clock's nickname comes from Sir Benjamin Hall, who was the commissioner of works.

4. The world-famous clock, and the tower itself, are both known by the same name, although the tower is official named Elizabeth Tower.

5. Big Ben's clock tower is 320 feet high and has 393 steps.

6. The four faces of the clock are lit up at night so passersby can easily read the time.

7. The clock has been keeping exact time since it was first set up in 1859.

8. The huge bell weighs 13.5 tons.

9. The first bell made for the clock was damaged during a test ringing.

10. The bell does not swing or have a clapper; instead it is hit by a hammer on the outside to chime.

11. The clock's hour hands are 9 feet long and the minute hands are 14 feet long.

12. Big Ben's bells were silenced for a time during World War I.

13. Big Ben survived air raids during World War II, which damaged many of London's other buildings.

38 Castle Facts

1. The word castle comes from a Latin word meaning "fortress."

2. Castles were dark and cold, but they offered protection from enemies.

3. Motte and bailey castles were an early type of castle made of wood.

4. Later stone castles were stronger but more expensive and took an average of ten years to build.

5. Stone castle walls could be up to 33 feet thick. Once the cannon was invented, no castles could withstand that kind of attack.

6. Many medieval castles, such as the 14th-century Bodiam Castle in England, had a wide moat filled with water.

7. The drawbridge was invented late in the medieval-castle era.

8. Spis Castle in eastern Slovakia is one of the largest medieval castles in Europe.

9. Most medieval castles had tight clockwise spiral stairs (from the ascender's perspective), created to disadvantage right-handed invaders.

10. Bathrooms in medieval castles were called "garderobes."

11. They had holes that led out of the castle into moats.

12. Europe's oldest standing castle is Château de Doué-la-Fontaine, built around A.D. 950.

13. Medieval castles were also built in India, Korea, and Japan.

14. Sitting on a hilltop in Austria, Hohensalzburg Castle was originally built in 1077, but was greatly expanded in the late 15th century.

15. Caerphilly Castle in Wales has a tower that leans more than the Leaning Tower of Pisa.

16. The castle on Mont St. Michel, a French island, can be reached by land only during low tide.

17. The Tower of London has served as a castle, a prison, a treasury, and even a zoo.

18. Built by William the Conqueror after he became ruler of England, Windsor Castle is still the primary residence of English royalty.

19. Arrow slits cut into castle walls allowed archers to shoot arrows at attackers without getting hit themselves.

20. Malbork Castle in Poland was built by Teutonic Knights in 1274.

21. The keep was a large tower and the last place of defense in a castle.

22. Ale, not butter, was kept in a room called the buttery.

23. Situated on an extinct volcanic crag, Edinburgh Castle dominates the skyline of the Scottish city.

24. Iolani Palace in Honolulu served as the home to Hawaii's last two monarchs—King Kalakaua and his sister, Queen Lili'uokalani, who succeeded him.

25. Built between 1879 and 1882, Iolani Palace featured modern technology, including indoor plumbing, telephones, and gas lighting.

26. Prague Castle in the Czech Republic covers an area of nearly 750,000 square feet.

27. Bran Castle in Romania is known as Dracula's castle.

28. While Vlad III Dracula, better known as Vlad the Impaler, is thought to be the inspiration for Bram Stoker's title character, most historians agree Vlad III Dracula never stepped foot in Castle Bran.

29. The lord and lady of a medieval castle slept and lived in rooms called "solar chambers," usually on an upper story.

30. Castles were home to lords and ladies, but also servants, spinners, craftspeople, and other staff.

31. Farming peasants or serfs who grew food for the castle's inhabitants usually lived in cottages on the lord's estate.

32. Neuschwanstein Castle in Germany served as the main inspiration for the castle in Disney's *Sleeping Beauty*.

33. Richard the Lionheart built the Chateau Gaillard in France.

34. Concentric castles, such as Krak des Chevaliers (Syria) and Beaumaris Castle (Wales), have a series of defensive walls surrounding them.

35. In 1900, George Boldt, owner of the Waldorf–Astoria Hotel, bought one of the Thousand Islands in the St. Lawrence River and began construction on a six-story, 120-room castle in honor of his wife, Louise.

36. When Louise died in 1904, Boldt abandoned the castle.

37. The Thousand Islands Bridge Authority purchased Boldt castle in 1977, restored it, and opened it to the public.

38. Built in the 1920s by artist Waldo Ballard, the Herreshoff Castle in Marblehead, Massachusetts, was patterned after Erik the Red's castle in Qagssiarssuk, Greenland.

36 Facts About Lady Liberty

1. The statue's real name is "Liberty Enlightening the World."

2. Construction of the statue began in France in 1875.

3. Lady Liberty was sculpted by Frédéric Auguste Bartholdi.

4. Alexandre Gustave Eiffel was the structural engineer.

5. The statue was completed in Paris in June 1884, and given to the American people on July 4, 1884.

6. Lady Liberty was reassembled and dedicated in the United States on October 28, 1886.

7. The model for the face of the statue is reputed to be the sculptor's mother, Charlotte Bartholdi.

8. A quarter-scale bronze replica of Lady Liberty was erected in Paris in 1889 as a gift from Americans living in the city.

9. The replica statue stands about 35 feet high and is located on a small island in the River Seine, about a mile south of the Eiffel Tower.

10. There are 25 windows and 7 spikes in Lady Liberty's crown.

11. The spikes are said to symbolize the seven seas.

12. The inscription on the statue's tablet reads the adoption date of the Declaration of Independence: July 4, 1776 (in Roman numerals).

13. More than four million people visit the Statue of Liberty in an ordinary year.

14. Symbolizing freedom and the opportunity for a better life, the Statue of Liberty greeted millions of immigrants as they sailed through New York Harbor on their way to nearby Ellis Island.

15. The Statue of Liberty is technically located in New Jersey, not New York.

16. Lady Liberty is 152 feet 2 inches tall from base to torch and 305 feet 1 inch tall from the ground to the tip of her torch.

17. The statue's hand is 16 feet 5 inches long and her index finger is 8 feet long.

18. Her fingernails are 13 inches long by 10 inches wide and weigh approximately 3.5 pounds each.

19. Lady Liberty's eyes are each 2 feet 6 inches across.

20. She weighs about 450,000 pounds (225 tons).

21. Lady Liberty's sandals are 25 feet long, making her shoe size 879.

22. There are 192 steps from the ground to the top of the pedestal and 354 steps from the pedestal to the crown.

23. The statue functioned as an actual lighthouse from 1886 to 1902.

24. There was an electric plant on the island to generate power for the light, which could be seen 24 miles away.

25. The Statue of Liberty underwent a multimillion dollar renovation in the mid-1980s before being rededicated on July 4, 1986.

26. During the renovation, Lady Liberty received a new torch because the old one was corroded beyond repair.

27. The Statue of Liberty's original torch is now on display at the monument's museum.

28. A plaque at the pedestal's entrance is inscribed with a sonnet, "The New Colossus" (1883) by Emma Lazarus.

29. The sonnet was written to help raise money for the pedestal. It includes the famous lines, "Give me your tired, your poor, Your huddled masses yearning to breathe free."

30. The site was added to UNESCO's World Heritage List in 1984.

31. Lady Liberty has a 35-foot waistline.

32. Her most famous cinematic appearance was in the 1968 film *Planet of the Apes*, where she is seen half buried in sand.

33. The statue is also destroyed in the films *Independence Day* and *The Day After Tomorrow*.

34. The statue has an iron infrastructure and copper exterior which has turned green due to oxidation.

35. Although it's a sign of damage, the patina (green coating) also acts as a form of protection from further deterioration.

36. The Statue of Liberty's foundation weighs 54 million pounds.

12 Brinicle Facts

1. Sometimes called a sea stalactite, brinicles are hollow tubes of ice that form downward from floating ocean ice blocks.

2. While icicles form when fresh water freezes as it drips off an object, brinicles develop entirely underwater.

3. Unlike freshwater ice, ice in the ocean is composed of two main elements: water and salt. As seawater freezes, its salt brine concentrates are expelled out.

4. The process of forcing salt impurities out during freezing causes sea ice to be porous and sponge-like, unlike its freshwater counterpart.

5. The expelled saline water is able to stay liquid due to its lowered freezing point.

6. Air temperature affects the salinity of brinicle water; the lower the temperature, the greater its concentration of salt.

7. When the salty brine comes into contact with neighboring ocean water, its extremely low temperature causes ice to instantly form around it.

8. Because brine is heavier than the water around it, it flows downward toward the ocean floor. This creates a hollow ice stalactite.

9. Occasionally, brinicle tubes reach the seafloor and freeze bottom-dwelling creatures like sea urchins and starfish.

10. For this reason, brinicles are sometimes referred to as "ice fingers of death."

11. Rarely observed, a brinicle formation was filmed for the first time in 2011.

12. Producer Kathryn Jeffs and cameramen Hugh Miller and Doug Anderson managed to capture the phenomenon for the BBC series *Frozen Planet*.

37 Facts About Antarctica

1. Antarctica is the fifth largest continent with an area of about 5,482,650 square miles.

2. The Southern Ocean surrounds Antarctica.

3. Around 90 percent of the ice on Earth is found in Antarctica.

4. Antarctica is the windiest continent on average with winds in some places reaching 200 miles per hour.

5. The name Antarctica comes from a Greek word meaning "opposite to the north."

6. The average thickness of Antarctic ice is about 1 mile.

7. The highest point on Antarctica is Vinson Massif at 16,066 feet.

8. Antarctica is almost one and half times the size of the U.S.

9. The McMurdo Dry Valleys of Antarctica are Earth's driest place.

10. NASA has tested equipment for Mars missions in the Dry Valleys region because the conditions are so similar to Mars.

11. More than 99 percent of Antarctica is covered by ice.

12. Antarctica is the only continent with no countries, economy, or permanent population.

13. Antarctica lies almost entirely within the Antarctic Circle.

14. There is just one insect native to Antarctica, the *Belgica Antarctica*.

15. Antarctica is the only continent without a time zone.

16. Most research stations keep the time of their own home country.

17. At 2,175 miles long, the Transantarctic Mountain range is the longest in Antarctica.

18. There are no polar bears in Antarctica, only in the Arctic.

19. Antarctica was discovered in the 1820s, but no one would set foot on it for another 75+ years.

20. Only 1 percent of Antarctica is permanently ice-free.

21. Antarctica is home to Mount Erebus, the planet's southernmost active volcano.

22. The highest temperature ever recorded in Antarctica was 69.3°F in February 2020.

23. In 1911, Norwegian explorer Roald Amundsen was the first human to reach the South Pole.

24. Antarctica is the best place in the world to find meteorites.

25. Antarctica is the only continent with no native species of reptiles.

26. Denman Glacier is Earth's lowest point on land at more than 11,482 feet below sea level.

27. The most abundant land animal on Antarctica is not the penguin, but the tiny nematode worm.

28. The male Emperor penguin is the only warm-blooded animal (besides human researchers) that remains on the Antarctic continent through the winter.

29. In 1977, Argentina sent a pregnant mother to Antarctica in an effort to claim a portion of the continent.

30. The boy became the first person born in Antarctica.

31. Seven nations have territorial claims on Antarctica, including the United Kingdom, France, Norway, Australia, New Zealand, Argentina, and Chile.

32. The Ross Ice Shelf is Antarctica's largest ice shelf at some 197,000 square miles.

33. The hole in the ozone layer over Antarctica is twice the size of Europe.

34. If Antarctica's ice sheets melted, all of the world's oceans would rise by 200 feet.

35. Lake Vostok is a pristine freshwater lake buried beneath 2.5 miles of solid ice.

36. There are more than 200 liquid lakes beneath Antarctica that are prevented from freezing by the warmth of Earth's core.

37. There are more than 40 airports in Antarctica.

40 Unusual U.S. Facts

1. The city of Beaver, Oklahoma, is the cow-chip-throwing capital of the world. Its annual competition is held in April.

2. When the London Bridge (the version built in 1831) started to disintegrate, it was taken apart stone by stone, shipped to the U.S., and lovingly reconstructed in Lake Havasu City, Arizona.

3. Cows outnumber people in Montana.

4. In the United States, there are more statues of Lewis and Clark's Native American guide Sacagawea than of any other woman.

5. Honolulu is the only place in the U.S. that has a royal palace.

6. Middlesboro, Kentucky, is built entirely within a meteor crater.

7. The only active diamond mine in the U.S. is in Arkansas.

8. Delaware has only three counties—the fewest of any state. (Even Rhode Island, the smallest state, has five.)

9. Louisiana doesn't have counties; it's subdivided into "parishes."

10. Alaska is divided into 19 organized boroughs and one unorganized borough.

11. Hawaii is the most isolated population center on Earth.

12. It is 2,390 miles from California, 3,850 miles from Japan, and 4,900 miles from China.

13. Utah is the only state with a three-word capital: Salt Lake City.

14. If you stand at the point known as Four Corners in the southwestern U.S., you can reach into four states—Utah, Colorado, New Mexico, and Arizona—without moving your feet.

15. Maine is the only one-syllable state.

16. Texas is the only state in the country that has had the flags of six nations flying over it: Mexico, Spain, France, the Confederate States, the Republic of Texas, and the United States.

17. Oregon is the only state with a reversible flag: each side has a different design.

18. Schenectady, New York, has the zip code 12345.

19. Virginia is the birthplace of more U.S. presidents than any other state (eight), followed by Ohio (seven).

20. Oklahoma's nickname as the "Sooner State" refers to the first day that homesteading was allowed.

21. On April 22, 1889, more than 50,000 people showed up to stake their claim, but some folks jumped the fence before the noon starting gun. They were called "sooners."

22. Elfreth's Alley in Philadelphia is the oldest residential street in America, where people have lived since 1702.

23. Maine is the only state that shares a border with only one other state—New Hampshire.

24. The only letter in the alphabet that does not appear in the name of a U.S. state is Q.

25. The 56-foot statue of Vulcan in Birmingham, Alabama, is the largest cast-iron statue in the world.

26. Birmingham began as a mining town for coal, limestone, and iron ore, which were forged to make steel.

27. The 24-foot replica of Leonardo da Vinci's *Il Gavallo* in Grand Rapids, Michigan, is the largest equestrian bronze statue in the Western Hemisphere.

28. The state of Florida is larger than England.

29. Michigan can claim the only floating post office in the world.

30. The *J. W. Westcott II* mail ship delivers freight and correspondence to boats ferrying on the Detroit River.

31. The Haskell Free Library and Opera House is an official heritage site in both the state of Vermont and province of Quebec—the international border runs right through the building.

32. North Dakota has more golf courses per capita than any other state.

33. Ripon, Wisconsin, is known as the birthplace of the Republican Party.

34. When the Crazy Horse Memorial near Rapid City, South Dakota, is completed, it will be taller than the Washington Monument and larger than the Sphinx in Egypt.

35. The Atlantic City Boardwalk is the oldest in the United States and the longest in the world, now spanning 5.5 miles.

36. Niagara Falls State Park in New York is the oldest state park in the United States.

37. Approximately 140 of the park's 400-plus acres are underwater.

38. The United States paid Russia $7.2 million for Alaska, less than 2 cents per acre.

39. Carhenge, the funky automotive version of England's mystical stone circle, is found in Alliance, Nebraska.

40. In 1791, Vermont became the first state to be added to the United States since the original 13.

Chapter 3

History

43 Facts About Ancient Civilizations

1. Mesopotamia, located between the Tigris and Euphrates rivers in modern-day Iraq, has been home to many great civilizations: Sumer, Babylonia, and Assyria.

2. Mesopotamia means "between two rivers" in Greek.

3. Most Western scholars agree that the Sumerian civilization in Mesopotamia was the world's first civilization.

4. Ancient Sumer is believed to have begun around 4000 B.C.

5. Archaeological evidence suggests that "pre-civilized" cultures lived in the Tigris and Euphrates river valleys long before the emergence of Sumer.

6. The great city of Ur, associated with Sumer, is possibly the world's first city.

7. The Harappan civilization developed in the Indus River Valley in modern-day Pakistan and India, beginning around 3500 B.C.

8. However, it is clear that agricultural communities had inhabited the Indus Valley area since at least 9000 B.C.

9. The Harappan civilization (circa 3300–1600 B.C.) boasted a network of earthenware pipes that carried water from people's homes into municipal drains and cesspools.

10. Nearly every home in the cities of Harappa and Mohenjo-Daro had a toilet connected to a sophisticated sewage system.

11. Located in Africa's Nile Valley, ancient Egypt is generally cited as beginning in 3200 B.C., though agricultural societies had settled in the Nile River Valley since the tenth millennium B.C.

12. Ancient Egyptian doctors had specialties.

13. Dentists were known as "doctors of the tooth," while the term for proctologists literally translates to "shepherd of the anus."

14. The Elamite kingdom in modern-day Iran began around 2700 B.C., though recent evidence suggests that a city existed in this area at a far earlier date—perhaps early enough to rival Sumer.

15. Meanwhile, the ancient Chinese civilizations, located in the Yangtze and Yellow river valleys, are said to have begun around 2200 B.C.

16. The earliest paved roads go back all the way to 4000 B.C., in the Mesopotamian cities of Ur and Babylon.

17. In the southern region of Sumer, brick makers would set their bricks in place with a substance called bitumen, a.k.a. asphalt.

18. The Akkadians formed the first united empire where the city-states of Sumer were united under one ruler.

19. The earliest writing, cuneiform, was inscribed on clay tablets and dates to around 3200 B.C.

20. Over a period of 3,000 years, it was used to record about 15 different languages, including Sumerian, Akkadian, Babylonian, and Assyrian.

21. Ancient Mesopotamians built terraced pyramids called ziggurats to honor the gods.

22. The ancient city of Babylon, located on the modern-day site of Al Hillah, Iraq, became the most powerful city in Mesopotamia.

23. The ancient Babylonians worshipped a half-human/half-fish creature named Oannes who gave them the gift of civilization.

24. The saying "an eye for an eye" comes from the Code of Hammurabi—a system of 282 laws created by Babylonian King Hammurabi in 1750 B.C.

25. Babylon was said to have been surrounded by walls 100 feet wide, 300 feet high, and 56 miles in length.

26. The walls were built with some 15 million bricks.

27. According to legend, King Nebuchadnezzar II built the Hanging Gardens of Babylon around 600 B.C. as a present for his wife.

28. There is some speculation over whether or not the Hanging Gardens of Babylon, one of the seven wonders of the ancient world, ever existed.

29. The oldest known furnace for copper smelting was discovered at Timna in southern Israel, and dated to approximately 4200 B.C.

30. It was simply a small hole in the ground that could be covered with a flat rock.

31. The Sumerians are often credited with inventing the wheel.

32. Ancient Sumerian kings and queens played board games.

33. The ancient Maya developed a picture-based written language around 300 B.C.

34. The Chavín, the earliest ancestors of the Inca, arrived in modern-day Peru around 1200 B.C.

35. Australian Aborigines, the world's oldest living culture, have existed for at least 50,000 years.

36. The earliest writing in ancient China dates to around 1200 B.C. It was carved into animal bones.

37. More than 10,000 individual pieces of graffiti have been found in the ancient Roman city of Pompeii, on surfaces ranging from bathroom walls to the outer walls of private mansions.

38. Ancient Romans ran cold aqueduct water in pipes through their houses in an early form of air conditioning.

39. Between 750 and 550 B.C., the ancient Greeks established 200 colonies throughout the Mediterranean.

40. Ireland's Newgrange, an ancient Celtic site, is 60 years older

than Egypt's Giza Pyramids and 1,000 years older than England's Stonehenge.

41. Ancient Persian engineers built a type of evaporative cooler called "Yakhchāl" that could store ice even in the middle of summer.

42. According to the ancient Greek historian Herodotus, ancient Persians would get drunk after making an important decision to see if they felt the same way about it when intoxicated.

43. The earliest coins were used in the ancient kingdom of Lydia (modern-day Turkey) in the seventh century B.C.

11 of History's Shortest Wars

1. **Anglo-Zanzibar War (9:02–9:40 A.M., Aug. 27, 1896, Great Britain vs. Zanzibar):** After Khalid bin Barghash acceded to the sultanate of Zanzibar (an island off modern Tanzania) without prior British permission, the Royal Navy gave Zanzibar a taste of British anger. Bin Barghash tapped out after just 38 minutes of shelling in what is history's shortest recorded war.

2. **Spanish-American War (Apr. 25–Aug. 12, 1898, U.S. vs. Spain):** Spain once had an empire, some of which was near Florida. After months of tension, the USS *Maine* mysteriously blew up in Havana Harbor. Though no one knew why it exploded, the U.S. declared war anyway. A few months later, Spain had lost Cuba, Guam, the Philippines, and Puerto Rico.

3. **Nazi-Polish War (Sep. 1–Oct. 6, 1939, Nazi Germany and Soviet Union vs. Poland):** After Russian and German negotiators signed a secret agreement in August dividing Poland, the Nazis invaded in vicious armored thrusts with heavy air attacks. Polish forces fought with valor, but their strategic position was impossible. Russian troops entered from the east on September 17, and Poland became the first European nation conquered in World War II.

4. **Nazi-Danish War (4:15–9:20 A.M., Apr. 9, 1940, Nazi Germany vs. Denmark):** Arguably the biggest mismatch of World War II (unless one counts Germany's invasion of

Luxembourg). Sixteen Danish soldiers died before the Danish government ordered the resistance to cease.

5. **Suez/Sinai War (Oct. 29–Nov. 6, 1956, Israel, Britain, and France vs. Egypt):** The Egyptians decided to nationalize the Suez Canal, which seems logical today given that the Suez is entirely in Egypt. British and French companies operating the canal didn't agree. The Israelis invaded by land, the British and French by air and sea. The invaders won a complete military victory, but withdrew under international pressure.

6. **Six-Day War (June 5–10, 1967, Israel vs. Egypt, Syria, and Jordan):** Israelis launched a sneak attack on Egyptians, destroying the Egyptian air force on its airfields and sending the Egyptians reeling back toward the Suez Canal. Jordanians attacked the Israelis and immediately regretted it. The Israelis attacked Syria and seized the Golan Heights.

7. **Yom Kippur War (Oct. 6–25, 1973, Egypt and Syria vs. Israel):** Egyptians and Syrians, still annoyed and embarrassed over the Six-Day War, attacked Israelis on a national religious holiday. Israeli forces were caught napping at first but soon regained the upper hand—they struck within artillery range of Damascus and crossed the Suez Canal. The United Nations' ceasefire came as a major relief to all involved.

8. **Soccer War (July 15–19, 1969, El Salvador vs. Honduras):** Immigration was the core issue, specifically the forced expulsion of some 60,000 Salvadoran illegal immigrants from Honduras. When a soccer series between the two nations fueled tensions, each managed to insult the other enough to start a bloody yet inconclusive war.

9. **Falklands War (Mar. 19–June 14, 1982, Argentina vs. UK):** Argentina has long claimed the Falkland Islands as Las Islas Malvinas. In 1982, Argentina decided to enforce this claim by invading the Falklands and South Georgia. Although the Argentines had a surprise for the Royal Navy in the form of air-launched antiship missiles, the battle for the islands went heavily against Argentina. Its survivors, including most of its marines, were shipped home minus their weaponry.

10. Invasion of Grenada (Oct. 25 to mid-Dec., 1983, U.S. vs. Grenada and Cuba): Concerned about a recent Marxist takeover on the Caribbean island of Grenada, elite U.S. forces invaded by air and sea. Some 1,000 Americans (including 600 medical students) were in danger in Grenada; after the relatively recent Iran hostage crisis, the lives of U.S. citizens overseas were a powerful talking point in domestic politics.

11. First Gulf War (Jan. 16–Mar. 3, 1991, Allies vs. Iraq): Saddam Hussein misjudged the world's tolerance for military adventurism (at least involving oil-rich Western-friendly Arab emirates) by sending his oversized military into Kuwait. President George H.W. Bush led a diverse world coalition that deployed into Saudi Arabia and started bombing the Iraqi military. Most remarkable: The primary land conflict of the war lasted just 100 hours, February 23–27.

52 Facts About Ancient Egypt

1. Ancient Egypt was one of the most advanced and stable civilizations in human history, lasting for more than 3,000 years.

2. Ancient Egypt lasted from around 3100 B.C. to 30 B.C., when it was conquered by the Romans.

3. Ancient Egyptians called their picture-based written language *medu netjer*, which means "words of god."

4. The ancient Greeks renamed the writing system hieroglyphs, or "sacred carvings."

5. The names of important people were written inside an oval called a cartouche.

6. As the breadbasket of the eastern Mediterranean and a gateway to African trade, geography guaranteed that major powers would covet Egypt.

7. Ramesses II, also known as Ramesses the Great, made the world's oldest known peace treaty after a long war against the Hittites ended in a draw in 1274 B.C.

8. Some of Egypt's greatest rulers and heroes were women.

9. The best example is Hatshepsut, who reigned as Pharaoh of Egypt rather than as queen.

10. Ancient reports during a military campaign in Nubia tell of the homage Hatshepsut received from rebels defeated on the battlefield.

11. Pharaohs wore a cobra symbol on their crowns, which was believed to spit fire at the pharaoh's enemies.

12. Both men and women wore makeup in ancient Egypt.

13. Eyeliner was drawn on with kohl, a black powder.

14. Nefertiti was the wife of Pharaoh Akhenaten. She is believed to have ruled beside him for 14 years.

15. Some scholars believe Nefertiti ruled briefly as pharaoh after her husband died.

16. Nefertiti was made famous by her bust, one of the most copied works of ancient Egyptian art.

17. Shaving away all body hair—most notably the eyebrows—was part of an elaborate daily purification ritual that was practiced by the pharaoh and his priests.

18. The ancient Egyptians believed that health, good crops, victory, and prosperity depended on keeping their gods happy.

19. Egyptian religion revolved around dozens of gods, often part animal.

20. Shaving the eyebrows was also a sign of mourning, even among commoners.

21. The Greek historian Herodotus, who traveled and wrote in the fifth century B.C., said that everyone in an ancient Egyptian household would shave his or her eyebrows following the natural death of a pet cat.

22. For dogs, Herodotus reported, the household members would shave their heads and all of their body hair as well.

23. Chairs in ancient Egypt had legs shaped like animal limbs.

24. Ancient Egyptian women enjoyed equal privileges with men on many fronts, including the right to buy, sell, and inherit property; to marry and divorce; and to practice an occupation outside the home.

25. Legal rights and social privileges varied by social class rather than by gender. In other words, women and men in the same social class enjoyed fairly equal rights.

26. Queen Ahhotep, of the early 18th Dynasty, was granted Egypt's highest military decoration—the Order of the Fly—at least three times for saving Egypt during the wars of liberation against the Hyksos invaders from the north.

27. Egyptian pharaohs often married their siblings because it was believed that pharaohs were gods on Earth and thus could marry only other gods.

28. The entire Egyptian Empire depended on Nile floods. Ancient Egyptian famers built ditches and low walls to trap flood water from the Nile.

29. Ancient Egyptians brewed and drank beer. It was actually a source of nutrition, motivation, and sometimes payment for work.

30. Ancient Egyptians lined linen bandages with honey to get them to stick to skin.

31. Tutankhamun became pharaoh in 1336 B.C. at the age of nine.

32. He ruled until 1327 B.C. when he died suddenly at the age of 18.

33. Tutankhamun's tomb was discovered in 1922.

34. Graverobbers had looted the tombs of many other pharaohs, but Tut's tomb remained undisturbed for nearly 3,000 years.

35. It took archaeologists 10 years to catalog the contents of Tut's tomb.

36. The earliest step pyramid dates to around 2630 B.C.

37. At Saqqara, Egypt, the architect Imhotep built mastaba upon mastaba, fashioning a 200-foot-high pyramid as a mausoleum for Pharaoh Djoser.

38. Over the next thousand years or so, Egyptian engineers built several pyramid complexes along the Nile's west bank.

39. The most famous and popular today are those at Giza, but dozens survive.

40. We used to believe pyramids were built under the lash, but modern scholars doubt this.

41. Egyptians believed that after a pharaoh died, he would reach the heavens via sunbeams.

42. Egyptians also believed that the pyramid shape would help the pharaoh on this journey.

43. Once the pharaoh reached the heavens, he would become Osiris, the god of the dead.

44. To adequately perform the duties of Osiris, the pharaoh would need a well-preserved body and organs.

45. The Egyptians feared that if they did not prepare the pharaoh's body for the afterlife, disaster would fall upon Egypt.

46. Ancient Egyptian children played a game similar to hockey with sticks made from palm branches and pucks made of leather pouches stuffed with papyrus.

47. In 332 B.C., Alexander the Great of Greece conquered Egypt.

48. Pharaohs in the Ptolemaic Dynasty were Greeks descended from Ptolemy, one of Alexander's generals.

49. Pharaoh Ptolemy Soter began building the world's first lighthouse in the harbor of Alexandria, Egypt, in 290 B.C.

50. Cleopatra VII was the last pharaoh.

51. She and Mark Antony killed themselves.

52. Egypt became part of the Roman Empire in 30 B.C., ending the age of the pharaohs.

40 Marvelous Mummy Facts

1. Ancient Egyptians might be history's most famous embalmers, but they weren't the first.

2. In northern Chile and southern Peru, modern researchers have found hundreds of pre-Inca mummies (roughly 5000–2000 B.C.) from the Chinchorro culture.

3. The Chinchorros mummified all walks of life: rich, poor, elderly, didn't matter.

4. One of the oldest known instances of deliberate mummification comes from a rock shelter now called Uan Muhuggiag in southern Libya: the well-preserved mummy of a young boy.

5. Radiocarbon dating determined the age of the Tashwinat Mummy discovered at Uan Muhuggiag to be approximately 5,600 years old.

6. According to Egyptian lore, the god Osiris was the very first mummy.

7. The earliest known Egyptian mummy dates to around 3300 B.C.

8. As Egyptian civilization advanced, professionals formalized and refined the process of mummification.

9. *Natron*, a mixture of sodium salts abundant along the Nile, made a big difference.

10. If you extracted the guts and brains from a corpse, then dried it out it in natron for a couple of months, the remains would keep for a long time.

11. During mummification, ancient Egyptians removed all internal organs except the heart.

12. The heart was left in the body because the Egyptians believed that—as the seat of intelligence and emotions—the person would need his heart in the afterlife.

13. The canopic jars used to store a mummy's internal organs are named after the local god Canopus.

14. Canopus is also a town in the Nile delta region.

15. The entire mummification process took 40 to 70 days in ancient Egypt.

16. Many people were employed in mummification, including the embalmer, cutters, priests, scribes, servants, and other workers.

17. In ancient Egypt, chief embalmers often wore a mask of Anubis, the god of mummification and the afterlife.

18. Only the rich in ancient Egyptian society could afford to be mummified.

19. Egyptians used vast amounts of linen to mummify a body.

20. There was enough linen on one mummy from the 11th dynasty to cover three tennis courts.

21. Ancient Egyptians mummified animals, including cats, jackals, baboons, horses, birds, gerbils, fish, snakes, crocodiles, bulls, and hippos.

22. The most popular mummy in the world is likely Vladimir Lenin.

23. Millions of people have visited his mummy in Moscow, Russia.

24. Mummies have been found on every continent except Antarctica.

25. The Spirit Cave Mummy is the oldest mummy found in North America.

26. Unearthed in 1940 in Nevada, the naturally preserved 9,400-year-old was shrouded in woven reed mats and a rabbit-skin blanket.

27. The oldest well-preserved mummy in Europe is Otzi the Iceman, a natural mummy from 3300 B.C. found in the Alps.

28. Another famous mummy is Lady Dai (Xin Zhui), who died around 168 B.C.

29. Lady Dai's are some of the best preserved human remains ever discovered in China.

30. She was buried in an immense tomb at Mawangdui in China and unearthed in 1968.

31. Xin Zhui was married to Li Cang, the Marquis of Dai and Chancellor of Changsha Kingdom in ancient China.

32. The contents of Xin Zhui's tomb revealed much previously unknown information about life in the Han dynasty.

33. Perhaps the most famous mummy in history is that of Egypt's boy king Tutankhamun.

34. British archaeologist Howard Carter and his financier friend Lord George Carnarvon opened Tutankhamun's burial chamber in early 1923.

35. Two months later, Carnarvon was dead. His death is often offered as proof of the "Curse of King Tut."

36. The inner coffin of Tutankhamun is made of solid gold weighing 296 pounds.

37. Pharaoh Ramses II was the first mummy to receive a passport (needed for travel). His listed occupation is 'king.'

38. Puruchuco, a site near Lima, Peru, contained more than 2,200 mummies.

39. In Victorian England, mummy unwrapping parties were popular. The host would buy one, then the guests would unwrap it.

40. People in Victorian times also used ground up mummies for medicinal purposes.

Egyptian Mummification in 7 Steps

1. Wash and ritually purify body.

2. Remove intestines, liver, stomach, and lungs; embalm them with natron (soda ash) and place in jars.

3. Stuff body cavity with natron.

4. Remove brain through nose using a hook; throw brain away.

5. Cover body with natron and place on embalming table for 40 days.

6. Wrap 20 layers of linen around body, gluing linen strips together with resin.

7. Place mummy in protective sarcophagus, and add another layer of wrapping.

13 Military History Facts

1. The oldest military medal? Probably the Gold of Valor.

2. It was awarded by the Egyptian pharaohs around 1500 B.C.

3. The Hundred Years War lasted 116 years.

4. During his invasion of England in 1014, King Olaf's Viking ships managed to pull down London's wooden Thames River bridge. (Hence, the children's song "London Bridge Is Falling Down.")

5. The longest-running mercenary contract belongs to the world's smallest standing army—the Vatican's Swiss Guards, a 100-man company first hired by Pope Julius II in 1506.

6. Sailing into space: Pieces of Germany's High Seas Fleet, scuttled off Scotland at the end of World War I, have been used to build deep-space probes.

7. What's with General George Patton's ivory-handled revolvers? He started carrying revolvers in 1916 after he nearly blew his own leg off with the Army's newfangled automatic pistol.

8. The U.S. Marines' first land battle on foreign soil was in Derna, Libya, in 1805.

9. During this battle, 600 Marines stormed the city to rescue the crew of the USS *Philadelphia* from pirates.

10. In 1940, Brigadier General Benjamin Oliver Davis became the first African American general in the U.S. Armed Forces.

11. Britain's early 20th-century super-weapon, the battleship H.M.S. *Dreadnought*, was fitted with ultramodern weapons, so her builders naturally omitted the ancient ram from her design.

12. Her only kill was a submarine, which she sunk by ramming it.

13. Eighty percent of Soviet males born in 1923 didn't survive World War II.

53 Facts About Ancient Rome

1. Ancient Romans ruled over one of the largest and richest empires in history.

2. Rome was founded in the ninth century B.C. as a small village in central Italy. Over the centuries it expanded its territory.

3. At its largest, the Roman Empire covered almost 2 million square miles.

4. The Roman Empire was most powerful between the third century B.C. and the fifth century A.D., when their armies conquered most of Europe, North Africa, and the Middle East.

5. In its early days, Rome was ruled by kings.

6. In 509 B.C., the kingdom became a republic, ruled by a council called the Senate.

7. The highest position in the Roman Republic was the consul.

8. There were two consuls at the same time to make sure that one didn't become too powerful.

9. Pirates captured a twentysomething Julius Caesar at sea.

10. After his release, Caesar raised a fleet, hunted the pirates down, and crucified them all.

11. As its territory grew, army generals threatened to take power away from the Senate.

12. In 45 B.C., Julius Caesar took control of Rome and became its dictator.

13. Just a year later, Caesar was assassinated on March 15 (the Ides of March).

14. Caesar was stabbed 23 times by a group of senators who believed he was undermining the republic.

15. Later, his adopted son, Augustus, became the first emperor of Rome.

16. As a young man Augustus was known as Octavian.

17. When Julius Caesar was assassinated, Octavian led his armies to victory in the battle for power over Rome.

18. Octavian proclaimed himself "First Citizen" and took the name Augustus.

19. Emperors ruled Rome from 27 B.C. until the end of the empire.

20. The word emperor comes from the Latin *imperator*, meaning "military commander."

21. Emperor Augustus attempted to reform what he considered loose Roman morals.

22. In a bit of irony, Augustus then had to banish his daughter Julia for committing adultery.

23. Constantine I brought the Roman Empire together again in A.D. 324 after years of turmoil.

24. Constantine was the first Christian emperor of Rome and he issued laws protecting Christians from persecution.

25. A haruspex was a Roman priest who read messages from the gods by cutting open animals to look at their livers.

26. Before going on a long journey, Romans would bargain with the gods for safety by promising some act of devotion upon arrival.

27. If the gods finked and the trip went sour, the pious supplicant would kick the god's statue on his or her return.

28. There were only 22 letters in the Roman alphabet.

29. J was written as I, U was written as V, and W and Y did not exist.

30. Many European languages, including English, still use the Roman alphabet today.

31. Every male Roman child remained under his father's authority until that father's death. Pater could sentence filum to death;

disinherit him; block him from legal action on his own behalf; even prevent him from borrowing money.

32. The Romans were the first people to use concrete, which allowed them to build massive structures quickly and cheaply.

33. In Rome, if someone got behind on a debt, the creditor might hire a *convicium* (escort) to serenade the deadbeat with ridicule.

34. Roads built by the ancient Romans are still used today.

35. Mark Antony's wife Fulvia (77–40 B.C.) was a capable general who led eight legions in a rebellion against land confiscation, fighting to the bitter end before going into exile.

36. Ancient Romans wore soccus, the predecessors of socks.

37. A fourth-century survey of Rome counted more than 45,000 tenements, nearly 1,800 villas, 850 bathhouses, 1,300 swimming pools, and a dozen each of libraries and toilet complexes. Of course, by then, Rome had receded from its peak glories.

38. The Circus Maximus, a huge stadium built for chariot races, could accommodate over 150,000 spectators.

39. In ancient Rome, urine was as much a commodity as a waste product. The ammonia in pee was useful for bleaching togas and tunics.

40. The mad Roman emperor Caligula (reigned A.D. 37–41) led a military expedition to the North Sea, where he ordered his soldiers to gather seashells, then claimed he defeated the ocean.

41. Emperor Caligula planned to promote his horse, Incitatus, to the rank of consul—the Roman equivalent of prime minister.

42. Caligula was eventually assassinated by his bodyguards.

43. Claudius (reigned A.D. 41–54) was partially deaf and walked with a limp due to childhood illness.

44. His family ostracized him and kept Claudius out of power.

45. When Caligula was assassinated, Claudius was one of the few survivors from the emperor's family.

46. His infirmity probably saved him from being seen as a serious threat and therefore murdered.

47. Claudius was named emperor and reigned for 13 years.

48. The Bay of Neapolis (now Naples) was Rome's Padre Island or Cabo San Lucas. It was where one went for intercourse and intoxication. Not that genteel Roman society was shy about either.

49. Nero (reigned A.D. 54–68) was a bloodthirsty emperor who had his mother and wife killed. He probably didn't start the Great Fire of Rome though.

50. Emperor Hadrian (reigned A.D. 117–138) realized the Roman Empire was too large to defend and ordered his armies to withdraw to borders that were easier to protect.

51. One of these borders was Hadrian's Wall, a 73-mile line of forts and barriers that separated Roman Britain from the unconquered lands to the north.

52. The fall of the Western Roman Empire in A.D. 476 is considered the start of the "Dark Ages" (or Middle Ages) in Europe.

53. The eastern part of the empire remained a power through the Middle Ages until its fall in A.D. 1453.

25 Colosseum Facts

1. The Colosseum was one of the greatest buildings in the ancient city of Rome, at the heart of the powerful Roman Empire.

2. Construction began under Emperor Vespasian in A.D. 72 and was completed in A.D. 80 by his successor and heir, Titus.

3. Romans celebrated in A.D. 80 with 100 days of inaugural games that included the slaying of 9,000 wild animals, noonday executions, and gladiatorial brawls that usually ended in death.

4. Travertine stone was used to make up much of the exterior of the elliptical building.

5. The arena floor was made of wooden planks supported by the brick walls of the cellars underneath.

6. Sand (which was good for absorbing blood) was scattered across the arena floor.

7. Wild animals that fought in the Colosseum were kept in cellars underneath the floor.

8. When it was time for the animals to appear, they were pushed into cage elevators and winched upward.

9. There were 36 trapdoors in the floor of the Colosseum.

10. The Colosseum had 80 entrances and could hold more than 50,000 roaring Romans thirsty for a gruesome battle.

11. Seats near the front were for the rich and powerful, while regular people sat farther back.

12. The emperor had his own box.

13. There were more than 20 different types of gladiators, each with different weapons and armor.

14. Most gladiators were slaves or criminals who trained in special schools.

15. *Bestiarii* fought wild animals to entertain the public.

16. Wild beast fights usually took place just prior to fights between gladiators.

17. Exotic animals were brought to the Colosseum from all over the Roman Empire.

18. A *secutor* (or chaser) was a close-combat fighter.

19. His traditional opponent was the *retiarius*.

20. The most lightly armed gladiator, the *retiarius* (net man) had to rely on speed and agility to escape his opponent.

21. The retiarius carried a three-pronged spear (trident) and a weighted net.

22. Roman audiences loved watching the retiarius fight slower, more heavily armed opponents.

23. The *hoplomachus* was one of the most heavily armored gladiators, with a large helmet and thick coverings over his arms and legs.

24. His fighting style was based on the warriors of ancient Greece.

25. The Colosseum measures 612 feet long, 151 feet wide, and 157 feet high, making it the world's largest amphitheater.

14 Greek and Roman Gods

Greek God—Job Title—Roman God

1. Aphrodite—Goddess of love and beauty—Venus
2. Apollo—God of beauty, poetry, and music—Apollo
3. Ares—God of war—Mars
4. Artemis—Goddess of the hunt and moon—Diana
5. Athena—Goddess of war and wisdom—Minerva
6. Demeter—Goddess of agriculture—Ceres
7. Dionysus—God of wine—Bacchus
8. Hades—God of the dead and the Underworld—Pluto
9. Hephaestus—God of fire and crafts—Vulcan
10. Hera—Goddess of marriage (Queen of the gods)—Juno
11. Hermes—Messenger of the gods—Mercury
12. Hestia—Goddess of the hearth—Vesta
13. Poseidon—God of the sea, earthquakes, and horses—Neptune
14. Zeus—Supreme god (King of the gods)—Jupiter

48 Amazing Ancient Greece Facts

1. Ancient Greece is divided into three main periods: the Archaic Period (800–480 B.C.), the Classical Period (480–323 B.C.), and the Hellenistic Period (323–146 B.C.).

2. Greek poet Homer wrote the *Odyssey* and the *Iliad* during the Archaic Period.

3. Ancient Greece was ruled by city-states, which sometimes fought each other and sometimes joined forces against invaders.

4. A Greek city-state, which the Greek themselves called a *polis*, was a city and the rural lands outside its walls.

5. The two most powerful city-states were Athens and Sparta.

6. Athenians invented democracy around the fifth century B.C.

7. The word "democracy" comes from two Greek words that mean people (*demos*) and rule (*kratos*).

8. Women, children, slaves, and Greeks not born to Athenian parents couldn't vote.

9. Slaves made up approximately one-third of the population of some ancient Greek cities.

10. Sparta had a strong military culture. Boys started harsh military training at a young age.

11. Citizens of Sparta voted by pounding on their shields.

12. Ancient Greeks invented philosophy, which means love of wisdom.

13. Famous Greek philosophers include Aristotle, Socrates, and Plato.

14. Alexander the Great, one of history's most successful military commanders, expanded the ancient Greek empire surrounding the Mediterranean all the way east to India.

15. The Hellenistic Period began when Alexander the Great died and ended when Rome conquered Greece.

16. The Romans copied much of the Greek culture including their gods, architecture, language, music, and even how they ate!

17. Ancient Greeks often ate dinner while lying on their sides.

18. The first two letters of the Greek alphabet—alpha and beta—made up the word alphabet.

19. Their alphabet was the first with vowels.

20. Ancient Greeks wore draped tunics called chitons.

21. Ancient Greeks loved the theater, but only male actors were allowed to perform.

22. In Greek mythology, a sphinx had a woman's head and a lion's body.

23. Ancient Greeks believed their gods lived atop Mount Olympus.

24. They were ruled by Zeus, the king of the gods.

25. Draco was an elected Athenian leader who established a written code of laws.

26. Many punishments were particularly harsh. The adjective "draconian" is now commonly used to describe similarly unforgiving rules or laws.

27. The very first Olympic Games were held in 776 B.C. in the Greek city of Olympia.

28. The Statue of Zeus at Olympia was constructed by Phidias around 435 B.C.

29. Standing 40 feet high with a 20-foot base, the Statue of Zeus was one of the seven wonders of the world.

30. The statue depicted a seated Zeus holding a golden figure of the goddess of victory in one hand and a staff topped with an eagle in the other.

31. After the old gods were outlawed by Christian emperor Theodosius, the statue was taken as a prize to Constantinople, where it was destroyed in a fire around A.D. 462.

32. Greek hero Pheidippides ran from Athens to Sparta to request help in the Battle of Marathon against the Persians.

33. After the Greeks won, he ran 25 miles from Marathon to Athens to announce the victory. This is where the word "marathon" originated.

34. Spiked dog collars were invented in ancient Greece and were originally designed to protect the dog's throat from wolf attacks.

35. Alexander the Great never lost a battle.

36. Greek scientist Hippocrates is called the Father of Western Medicine.

37. Doctors still take the Hippocratic Oath today.

38. Coins in ancient Greece were decorated with images of bees.

39. Ancient Greeks didn't call their land Greece. Its official name was Hellenic Republic, but Greeks have called it Hellas or Hellada.

40. The English word "Greece" comes from the Latin word *Graecia*, meaning "the land of the Greeks."

41. The Greeks used a trick wooden horse to finally win the war against the Trojans.

42. Ancient Greek warriors were well organized and heavily armored.

43. They fought in a rectangular group called a phalanx, covering themselves with shields for protection.

44. Jury trials in ancient Greece had as many as 500 jurors.

45. For breakfast, Greeks ate bread dipped in wine.

46. Ancient Greeks drank wine, usually mixed with water, with meals.

47. Sparta was the only Greek city-state to mandate public education for girls.

48. Most of those splendid Greek statues you see in off-white marble used to be painted in realistic colors. We know because tiny bits of paint remain embedded in nooks and crannies.

The 9 Muses

The nine muses were Greek goddesses (the daughters of Zeus and Mnemosyne) who ruled over the arts and sciences and offered help and inspiration to mortals.

✳   ✳   ✳   ✳

1. Calliope—Muse of epic poetry

2. Clio—Muse of history

3. Erato—Muse of love poetry

4. Euterpe—Muse of music

5. Melpomene—Muse of tragedy

6. Polyhymnia—Muse of sacred poetry or mime

7. Terpsichore—Muse of dance

8. Thalia—Muse of comedy

9. Urania—Muse of astronomy

37 Facts About the Maya

1. The Maya or Mayan peoples made their home in modern-day Mexico and Central America.

2. Mayan culture was well established by 1000 B.C. and lasted until A.D. 1697.

3. The Maya civilization consisted of a large number of cities.

4. The cities shared a common culture and religion, but each city had its own noble ruler.

5. Mayan kings frequently fought with each other over tribute (gifts) and prisoners to sacrifice to the gods.

6. The ancient Maya developed a picture-based written language around 300 B.C.

7. Mayan weapons were made from volcanic rock called obsidian.

8. Ancient Maya nobles pierced their skin with spines from stingrays as part of a blood sacrifice made to the gods.

9. The Maya cultivated stingless bees for honey.

10. There was never a single Mayan empire, but rather a widespread, interconnected civilization.

11. The Maya believed their rulers could communicate with the gods through ritual bloodletting and human sacrifice.

12. Self-sacrifice was also common among the Maya.

13. Crossed eyes, flat foreheads, and big noses were attractive features to the Maya.

14. Mothers tried to flatten their baby's foreheads with boards and purposely make their children cross-eyed since these were features of nobility.

15. Mayan noblewomen filed their teeth into sharp points.

16. Mayans grew a lot of maize (yellow corn), which was a staple in their diet, along with avocados, tomatoes, chili peppers, beans, papayas, pineapples, and cacao.

17. Corn was made into a kind of porridge called *atole*.

18. The Maya were among the first cultures to use saunas.

19. Mayans played a ballgame on a court in which the losers were sacrificed to the gods.

20. Mayan kings were thought to become gods when they died.

21. The Maya were expert mathematicians and astronomers.

22. Mayans followed a 52-year Calendar Round, which resulted in two calendar cycles: the Haab and the Tzolkin.

23. The Haab was made up of 365 days organized into 18 months of 20 days, with 5 unlucky days added at the end.

24. In the Tzolkin or Sacred Round, 20 day names were combined with 13 numbers to give a year 260 days.

25. The two cycles reach the same point after 52 years.

26. For periods longer than 52 years, the Maya used a separate system called the Long Count.

27. The Maya were the first civilization to use the number zero as a place holder, but they were not the first to use it in mathematics.

28. Jade was the most precious material to the Maya.

29. It was associated with water, the life-giving fluid, and with the color of the corn plant, their staple food.

30. Both men and women wore jewelry, including ear spools, neck-laces, and bracelets.

31. The Maya used human hair for sutures.

32. Although some Mayan cities continued to thrive until the 16th century, the Mayan civilization began to decline after A.D. 800.

33. During the 9th century A.D., cities in the central Maya region began to collapse, likely due to a combination of causes, including endemic inter warfare, overpopulation, severe environmental degradation, and drought.

34. During this period, known as the Terminal Classic, the northern cities of Chichén Itzá and Uxmal showed increased activity.

35. Many cities such as Tikal (in modern-day Guatemala) were swallowed up by the rain forest and not rediscovered until modern times.

36. Tikal was one of the biggest Mayan cities, with as many as 60,000 people living there.

37. The last surviving Mayan city was Tayasal, which existed until 1696 when the Spanish conquered it.

11 Ancient Cities: Then and Now

1. **Memphis (now the ruins of Memphis, Egypt):** By 3100 B.C., this Pharaonic capital bustled with an estimated 30,000 people. Today it has none—but modern Cairo, 12 miles north, is home to an estimated 21,750,000 people.

2. **Ur (now the ruins of Ur, Iraq):** Sumer's great ancient city once stood near the Euphrates with a peak population of 65,000 around 2030 B.C. Ur now has a population of zero.

3. **Alexandria (now El-Iskandariyah, Egypt):** Built on an ancient Egyptian village site near the Nile Delta's west end, Alexander the Great's city once held a tremendous library. In its heyday, it may have held 250,000 people; today an estimated 5.4 million people call it home.

4. **Babylon (now the ruins of Babylon, Iraq):** Babylon may have twice been the largest city in the world, in about 1700 B.C. and 500 B.C.—perhaps with up to 200,000 people in the latter case. Now, it's windblown dust and faded splendor.

5. **Athens (Greece):** In classical times, this powerful city-state stood miles from the coast but was never a big place—something like 30,000 residents during the 300s B.C. It now reaches the sea with about 3,154,000 residents.

6. **Rome (Italy):** With the rise of its empire, ancient Rome became a city of more than 500,000 and the center of Western civilization. Though that mantle moved on to other cities, Rome now has around 4.3 million people in the metro area.

7. **Xi'an (China):** This longtime dynastic capital, famed for its terra-cotta warriors but home to numerous other antiquities, reached 400,000 people by A.D. 637. Its 8.5 million people make it as important a city now as then.

8. **Constantinople (now Istanbul, Turkey):** Emperor Constantine the Great made this city, first colonized by Greeks in the 1200s B.C., his eastern imperial Roman capital with 300,000 people. As Byzantium, it bobbed and wove through the tides of faith and conquest. Today, it is Turkey's largest city with more than 15.5 million people.

9. **Baghdad (Iraq):** Founded around A.D. 762, this center of Islamic culture and faith was perhaps the first city to house more than 1,000,000 people. Today, more than 7.5 million people call Baghdad home.

10. **Tenochtitlán (now Mexico City, Mexico):** Founded in A.D. 1325, this island-built Aztec capital had more than 200,000 inhabitants within a century. Most of the surrounding lake has been drained over the years. Today, more than 22 million souls call Mexico City home.

11. **Carthage (now the ruins of Carthage, Tunisia):** Phoenician seafarers from the Levant founded this great trade city in 814 B.C. Before the Romans obliterated it in 146 B.C., its

population may have reached 700,000. Today, it sits in empty silence ten miles from modern Tunis—population 2.4 million.

22 Facts About Petra

1. The ancient city of Petra is a World Heritage Site and a Jordanian national treasure.

2. Located within the Kingdom of Jordan, Petra is some 80 miles south of Amman in the Naqab (Negev) Desert, about 15 miles east of the Israeli border.

3. Petra was a key link in the trade chain connecting Egypt, Babylon, Arabia, and the Mediterranean.

4. In 600 B.C., the narrow red sandstone canyon of Petra housed a settlement of Edomites: seminomadic Semites said to descend from the biblical Esau.

5. With the rise of the incense trade, Arab traders began pitching tents at what would become Petra. We know them as the Nabataeans.

6. The Nabataeans showed up speaking early Arabic in a region where Aramaic was the business-speak.

7. The newcomers thus first wrote their Arabic in a variant of the Aramaic script. But Petra's trade focus meant a need to adopt Aramaic as well, so Nabataeans did.

8. By the end (about 250 years before the rise of Islam), Nabataean "Arabaic" had evolved into classical (Koranic) Arabic.

9. The Nabataeans of Petra weren't expansionists, but defended their homeland with shrewd diplomacy and obstinate vigor. Despite great wealth, they had few slaves.

10. Petra's Nabataeans showed a pronounced democratic streak, despite monarchical government. Empires rose and fell around them; business was business.

11. The core commodity was incense from Arabia, but many raw materials and luxuries of antiquity also passed through Petra—

notably bitumen (natural asphalt), useful in waterproofing and possibly in embalming.

12. Nabataean women held a respected position in society, including property and inheritance rights.

13. While no major ancient Near Eastern culture was truly egalitarian, the women of Petra participated in its luxuriant prosperity.

14. Most Nabataeans were pagan, worshipping benevolent fertility and sun deities.

15. Jews were welcome at Petra, as were Christians in its later days.

16. In 70 B.C., Petra was home to about 20,000 people.

17. Ornate homes and public buildings rivaling Athenian and Roman artistry were carved into the high red sandstone walls of the canyon.

18. Petra's last king, Rabbel II, willed his realm to Rome.

19. When he died in A.D. 106, Nabataea became the Roman province of Arabia Petraea. Again the Nabataeans adjusted—and kept up the trade.

20. In the 2nd and 3rd centuries A.D., the caravans began using Palmyra (in modern Syria) as an alternate route, starting a slow decline at Petra.

21. An earthquake in 363 delivered the knockout punch: damage to the intricate water system sustaining the city.

22. By about A.D. 400, Petra was an Arabian ghost town.

45 Cool Facts About Celts

1. Ancient Celts believed the head was the seat of the soul.

2. The Celts were fierce warriors from central Europe.

3. By 200 B.C. their civilization stretched across much of northern and western Europe.

4. The Celts did not have a single empire like the Romans, but

instead lived in separate tribes with similar languages, religion, and customs.

5. Celts spoke related languages that survive today as Scottish Gaelic, Welsh, Irish, and Breton.

6. The Celts belonged to tribes ruled by kings, queens, and chieftains.

7. Classical Celtic culture emerged in central Europe around modern Austria, Bavaria, and Switzerland.

8. The earliest major Celtic settlement, dating from 1200 B.C., was found in Hallstatt, Upper Austria.

9. Celtic armies attacked Rome in 387 B.C. and invaded Greece in 279 B.C.

10. Some of these invaders then crossed into Galatia (modern-day Turkey) and built settlements there.

11. The Celts were expert metalworkers in iron, bronze, silver, and gold.

12. Iron was used for tools and weapons.

13. The Celts made metal rings called torcs that they wore around their necks.

14. The bravest Celtic warriors went into battle wearing nothing except a torc.

15. The Celts invented chainmail (around 300 B.C.) and helmets later used by Roman legionaries.

16. The ancient Celts used an early form of hair gel made from vegetable oil and resin from trees.

17. The ancient Celts traded goods with people in China.

18. The Celts believed in many gods and goddesses.

19. They sacrificed valuable objects and animals—and sometimes people—to the gods.

20. Ancient Celts threw precious objects into rivers and lakes, which they believed to be entrances to the world of the gods.

21. Celtic women had more rights than Greek or Roman women.

22. In A.D. 60, Queen Boudicca and her Iceni tribe rose in revolt against the Romans.

23. The walls of ancient Celtic homes were made of woven willow branches smeared with mud and animal dung.

24. The religious leaders of the Celts in Britain, Ireland, and Gaul (France) were called druids.

25. As the priestly class of Celtic society, the druids served as the Celts' spiritual leaders.

26. Druids were repositories of knowledge about the world and the universe, as well as authorities on Celtic history, law, religion, and culture. In short, they were the preservers of the Celtic way of life.

27. To become a druid, one had to endure as many as 20 years of oral education and memorization.

28. Celtic women could become druids.

29. They could also buy or inherit property, assume leadership, wage war, and divorce men.

30. In terms of power, the druids took a backseat to no one.

31. Even the Celtic chieftains, well-versed in power politics, recognized the overarching authority of the druids.

32. Celtic society had well-defined power and social structures and territories and property rights.

33. The druids were deemed the ultimate arbiters in all matters relating to such.

34. If there was a legal or financial dispute between two parties, it was unequivocally settled in special druid-presided courts.

35. Armed conflicts were immediately ended by druid rulings. Their word was final.

36. There were two forces to which even the druids had to succumb—the Romans and Christianity.

37. With the Roman invasion of Britain in A.D. 43, Emperor Claudius decreed that druidism throughout the Roman Empire was to be outlawed.

38. The Romans destroyed the last vestiges of official druidism in Britain with the annihilation of the druid stronghold of Anglesey in A.D. 61.

39. Surviving druids fled to unconquered Ireland and Scotland, only to become completely marginalized by the influence of Christianity within a few centuries.

40. In Scotland, the Celts made small artificial islands called crannogs in lakes and rivers, and built houses on them.

41. Most Celts lived in round houses with one central fire for heating, boiling water, and cooking.

42. Celtic warriors fought with iron swords and daggers, on foot or horseback.

43. The Romans and Greeks wrote about the Celts as swaggering barbarians who indulged in head-hunting and human sacrifices.

44. The Celts held large banquets to celebrate victories in battle. These lasted several days.

45. No one knows what the Celts called themselves. The term "Celt" is believed to come from Greek *Keltoi* or Latin *Celtae*.

33 Knights Templar Facts

1. The Crusades, Christendom's quest to recover and hold the Holy Land, saw the rise of several influential military orders, including the Knights Templar.

2. On July 15, 1099, the First Crusade stormed Jerusalem and slaughtered everyone in sight—Jews, Muslims, and Christians.

3. This unleashed a wave of pilgrimage, as European Christians flocked to now-accessible Palestine and its holy sites.

4. Though Jerusalem's loss was a blow to Islam, it was a bonanza for thieves, as it brought a steady stream of naive pilgrims to rob.

5. French knight Hugues de Payen, with eight chivalrous comrades, swore to guard the travelers.

6. In 1119, they gathered at the Church of the Holy Sepulchre and pledged their lives to poverty, chastity, and obedience before King Baldwin II of Jerusalem.

7. The Order of Poor Knights of the Temple of Solomon took up headquarters in said Temple.

8. The Templars did their work well, and in 1127 King Baldwin sent a Templar embassy to Europe to secure a marriage that would ensure the royal succession in Jerusalem.

9. Not only did they succeed, influential nobles showered the Order with money and real estate, the foundation of its future wealth.

10. With this growth came a formal code of rules.

11. Templars could not desert the battlefield or leave a castle by stealth.

12. They had to wear white habits—except for sergeants and squires, who could wear black.

13. They had to tonsure (shave) their crowns and wear beards.

14. They were required to dine in communal silence, broken only by Scriptural readings.

15. Templars had to be chaste, except for married men joining with their wives' consent.

16. With offices in Europe to manage the Order's growing assets, the Templars returned to Palestine to join in the Kingdom's ongoing defense.

17. In 1139, Pope Innocent II decreed the Order answerable only to the Holy See.

18. Now the Order was entitled to accept tithes!

19. By the mid-1100s, the Templars had become a church within a church, a nation within a nation, and a major banking concern.

20. Templar keeps were well-defended depositories, and the Order

became financiers to the crowned heads of Europe—even to the papacy.

21. Their reputation for meticulous bookkeeping and secure transactions underpinned Europe's financial markets, even as their soldiers kept fighting for the faith in the Holy Land.

22. In 1187, Saladin the Kurd retook Jerusalem, martyring 230 captured Templars.

23. Factional fighting between Christians sped the collapse.

24. In 1291, the last Crusader outpost fell to the Mamelukes of Egypt. The Templars' troubles had just begun.

25. King Philip IV of France owed the Order a lot of money, and in 1307, Philip ordered all Templars arrested.

26. They stood accused of devil worship, sodomy, and greed.

27. Hideous torture produced piles of confessions.

28. The Order was looted and officially dissolved.

29. In March 1314, Jacques de Molay, the last Grand Master of the Knights Templar, was burned at the stake.

30. Many Templar assets passed to the Knights Hospitallers.

31. The Order survived in Portugal as the Order of Christ, where it exists to this day in form similar to British knightly orders.

32. A Templar fleet escaped from La Rochelle and vanished; it may have reached Scotland.

33. Swiss folktales suggest that some Templars took their loot and expertise to Switzerland, possibly laying the groundwork for what would one day become the Swiss banking industry.

18 Facts About Picts

1. The Picts inhabited ancient Scotland before the Scots.

2. Some scholars believe that the Picts were descendants of the Caledonians or other Iron Age tribes who invaded Britain.

3. No one knows what the Picts called themselves; the origin of their name comes from other sources and probably derives from the Pictish custom of tattooing or painting their bodies.

4. The Irish called them *Cruithni*, meaning "the people of the designs."

5. The Romans called them *Picti*, Latin for "painted people."

6. However, the Romans probably used the term for all the untamed peoples living north of Hadrian's Wall.

7. The Picts themselves left no written records.

8. All descriptions of their history and culture come from second-hand accounts.

9. The earliest of these is a Roman account from A.D. 297 stating that the Picti and the Hiberni (Irish) were already well-established enemies of the Britons to the south.

10. Before the arrival of the Romans, the Picts spent most of their time fighting amongst themselves.

11. The threat posed by the Roman conquest of Britain forced the squabbling Pict kingdoms to come together and eventually evolve into the nation-state of Pictland.

12. The united Picts were strong enough not only to resist conquest by the Romans, but also to launch periodic raids on Roman-occupied Britain.

13. Having defied the Romans, the Picts later succumbed to a more benevolent invasion launched by Irish Christian missionaries.

14. Arriving in Pictland in the late 6th century, they succeeded in converting the polytheistic Pict elite within two decades.

15. Much of the written history of the Picts comes from the Irish Christian annals.

16. If not for the writings of the Romans and the Irish missionaries, we might not have knowledge of the Picts today.

17. Despite the existence of an established Pict state, Pictland disappeared with the changing of its name to the Kingdom of Alba in

A.D. 843, a move signifying the rise of the Gaels as the dominant people in Scotland.

18. By the 11th century, virtually all vestiges of the Picts had vanished.

35 Facts About the Maori

1. The Maori are a group of people who settled on the islands of New Zealand between A.D. 800 and 1200.

2. They had traveled across the South Pacific from the islands of Polynesia.

3. Maori tradition says they came from an island called Hawaiki, the mother island of the east Polynesians.

4. The Maori called the islands of New Zealand Aotearoa, which means "Land of the Long White Cloud."

5. Early Maori settlers of New Zealand hunted the moa, large flightless birds, to extinction.

6. Maori built canoes (*Waka*), which ranged in size from small one-man boats to large double-hulled war canoes called *Waka taua*, which were as long as 130 feet.

7. Traditional Maori clothing included skirts made from flax, a common plant found on New Zealand, and elaborate cloaks for warmth and to indicate status.

8. Te Reo Maori is the language of the Maori.

9. Today it is an official language of New Zealand.

10. Maori worship more than 70 gods, each strongly linked with nature.

11. According to legend, all Maori gods descended from Papatuanuku (Earth Mother) and Ranginui (Sky Father).

12. The Maori creation story describes the world being formed by the violent separation of Ranginui and Papatuanuku by their children.

13. Many Maori carvings and artworks graphically depict this struggle.

14. Maori society was divided into a number of large tribes known as *iwis*.

15. Each *iwi* consisted of a number of smaller tribes called *hapus*, with each led by an *ariki*, or chief.

16. Each *hapu* was further divided into *whanaus*, or families.

17. The families came together as a *hapu* for rituals and activities in a common meetinghouse known as a *wharenui*.

18. Elaborately carved meetinghouses were the pride of the tribe.

19. Tiki motifs often decorated meetinghouses.

20. A feast centering around a pit in the ground was a cultural tradition of the Maori.

21. Food would be put in ovens called *hangi*.

22. Heated stones would cook the meat and vegetables.

23. Maori held eggs in high esteem and would place them in the hands of the dead before burial.

24. Traditional arts such as carving, weaving, *kapa haka* (group performance), *whaikorero* (oratory), and *moko* (tattoo) are common Maori practices still used today.

25. Maori society became more warlike as their population grew.

26. Maori headhunters removed the flesh from the skulls of their enemies, then smoked and dried it.

27. This process preserved distinctive tribal tattoos, which meant that the deceased could be identified.

28. The *haka* is an ancient traditional war cry, dance, or challenge of the Maori.

29. A *haka* was often performed on a battlefield prior to engaging the enemy or by warriors prior to heading off to battle.

30. Not all *hakas* center around warfare.

31. Today, a *haka* is used to greet important people visiting a *marae* (communal area), at national festivals, sporting events, and other important occasions.

32. Before European settlers arrived in 1840, the Maori were New Zealand's sole human residents.

33. Today there are more than 500,000 Maori, most of whom live on New Zealand's North Island.

34. Maori are well represented in New Zealand's rugby and netball teams.

35. The New Zealand national rugby union team frequently perform a *haka*, a traditional Maori challenge, before events.

32 Boston Tea Party Facts

1. Contrary to popular belief, the Boston Tea Party was more about wealthy smugglers protecting their economic interests than about protesting "taxation without representation."

2. This tale begins with the 1765 Stamp Act.

3. Britain wanted help paying to defend the colonies. This would happen through tax stamps, similar to modern postage stamps, required on various documents, printed materials, goods, etc.

4. Mobs of colonists tarred and feathered government officials, burned them in effigy, and torched their homes and possessions.

5. Within months, the British gave up on the Stamp Act fiasco.

6. Next the British tried the Townshend Act (1767), imposing customs duties on imported goods, including tea.

7. Enterprising Dutch smugglers snuck shiploads of tea past British customs officials.

8. Seeing opportunity, clever colonial businessmen bought and distributed smuggled tea.

9. Colonials boycotted legally imported tea, often refusing to let it be unloaded from ships.

10. Tea smuggling grew rampant and was a lucrative business venture for American colonists, such as John Hancock and Samuel Adams.

11. The British realized the Townshend Act wasn't working.

12. In response, Parliament passed the 1773 Tea Act, which granted the British East India Company a monopoly on tea sales in the American colonies.

13. The direct sale of tea by the British East India Company to the American colonies undercut the business of colonial merchants.

14. Smuggled tea became more expensive than British East India Company tea.

15. By the time of the Boston Tea Party, American colonists drank an estimated 1.2 million pounds of tea each year.

16. On the night of December 16, 1773, three ships carrying cargoes of British East India Company tea were moored in Boston Harbor: the *Beaver*, the *Dartmouth*, and the *Eleanor*.

17. Members of a protest group called the Sons of Liberty dressed as Mohawk Indian warriors, boarded the ships, dragged the cargo up from the holds, and dumped the tea into Boston Harbor on December 16.

18. By the end of the night, 340 chests (approximately 46 tons) of tea had been dumped overboard, and tea leaves washed up on Boston shores for weeks.

19. Afterward, the Sons of Liberty wandered home, proud of their patriotic accomplishment.

20. Similar tea "parties" occurred in other colonial ports.

21. The colonies had successfully impugned King George III and maintained a healthy business climate for smugglers.

22. For weeks after the Boston Tea Party, the 92,000 pounds of tea dumped into the harbor caused it to smell.

23. The British shut down Boston Harbor until all of the 340 chests of British East India Company tea were paid for.

24. Benjamin Franklin offered to pay for the tea that was dumped if Britain reopened the harbor.

25. Britain refused, so Franklin never paid.

26. Despite the popular misconception that the ships were British, all three ships were in fact built in America and owned by Americans.

27. The *Beaver* and the *Dartmouth* were built and owned by an affluent Nantucket Quaker family.

28. The *Eleanor* was one of several ships owned by Boston merchant John Rowe.

29. The tea the Sons of Liberty dumped into Boston Harbor was from China, not India.

30. Famous Sons of Liberty who participated included John Adams, Paul Revere, John Hancock, and Samuel Adams.

31. The actual location of the Boston Tea Party, Griffin's Wharf, is thought to be at the corner of Congress and Purchase Streets.

32. Once underwater, it is now a busy intersection.

36 Amazing Angkor Wat Facts

1. Angkor Wat and other ancient temples in present-day Cambodia were erected between the 9th and 12th centuries.

2. While Europe languished in the Dark Ages, the Khmer Empire of Indochina was reaching its zenith.

3. The earliest records of the Khmer people date to the middle of the 6th century.

4. They migrated from southern China and settled in what is now Cambodia.

5. The early Khmer retained many Indian influences—they were Hindus, and their architecture evolved from Indian building methods.

6. In the early 9th century, King Jayavarman II laid claim to an independent kingdom called Kambuja.

7. He established his capital some 190 miles north of Phnom Penh, the modern Cambodian capital.

8. Jayavarman II also introduced the cult of *devaraja*, which claimed that the Khmer king was a representative of Shiva, the Hindu god of destruction and rebirth.

9. As such, in addition to the temples built to honor the Hindu gods, temples were also constructed to serve as tombs for kings.

10. The Khmer built more than 100 stone temples spread out over about 40 miles.

11. The temples were made from laterite (a material similar to clay that forms in tropical climates) and sandstone.

12. The sandstone provided an open canvas for the statues and reliefs celebrating the Hindu gods.

13. During the first half of the 12th century, Kambuja's King Suryavarman II decided to raise an enormous temple dedicated to the Hindu god Vishnu, a religious monument that would subdue the surrounding jungle and illustrate the power of the Khmer king.

14. His masterpiece—the largest temple complex in the world—would be known to history by its Sanskrit name, "Angkor Wat," or "City of Temple."

15. Pilgrims visiting Angkor Wat in the 12th century would enter the temple complex by crossing a square, 600-foot-wide moat that ran around the temple grounds.

16. Visitors would tread the moat's causeway to the main gateway.

17. From there, they would follow a spiritual journey representing the path from the outside world through the Hindu universe and into Mount Meru, the home of the gods.

18. Visitors would pass a giant, eight-armed statue of Vishnu as they entered the western *gopura*, or gatehouse, known as the "Entrance of the Elephants."

19. They would then follow a stone walkway decorated with *nagas* (mythical serpents) past sunken pools and column-studded buildings once believed to house sacred temple documents.

20. At the end of the stone walkway, a pilgrim would step up to a platform surrounded with galleries featuring six-foot-high bas-reliefs of gods and kings.

21. One depicts the Churning of the Ocean of Milk, a Hindu story in which gods and demons churn a serpent in an ocean of milk to extract the elixir of life.

22. Another illustrates the epic battle of monkey warriors against demons whose sovereign had kidnapped Sita, the beautiful wife of Rama (the Hindu deity of chivalry and virtue).

23. Others depict the gruesome fates awaiting the wicked in the afterlife.

24. A visitor to Suryavarman's kingdom would next ascend the dangerously steep steps to the temple's second level, an enclosed area boasting a courtyard decorated with hundreds of dancing *apsaras*, female images ornamented with jewelry and elaborately dressed hair.

25. For kings and high priests, the journey would continue with a climb up more steep steps to a 126-foot-high central temple, the pinnacle of Khmer society.

26. Spreading out some 145 feet on each side, the square temple includes a courtyard cornered by four high conical towers shaped to look like lotus buds.

27. The center of the temple is dominated by a fifth conical tower soaring 180 feet above the main causeway.

28. Inside it holds a golden statue of the Khmer patron, Vishnu, riding a half-man, half-bird creature in the image of King Suryavarman.

29. Angkor Wat was constructed with an estimated five million tons of sandstone, all transported from a quarry 25 miles away, without the aid of machinery.

30. Angkor Wat is the largest religious monument in the world, spread across 400 acres.

31. In the late 13th century, Angkor Wat shifted from a Hindu to Buddhist temple and remains a place of Buddhist worship today.

32. Today, Angkor Wat is a World Heritage Site and Cambodia's top tourist destination.

33. Angkor Wat is featured on the Cambodian flag.

34. Cambodia is one of the only nations to feature a building on its flag.

35. One of the first Westerners recorded at the site was Portuguese monk Antonio da Madalena in 1586.

36. However, it was when French explorer Hanri Mouhot discovered it in the mid-18th century and wrote about it in his book, *Travels in Siam, Cambodia, Laos and Annam*, that it gained international fame.

27 Incredible Facts About the Inca

1. The Inca civilization in South America started out as a highland tribe in present-day Peru.

2. The term "Inka" means ruler or lord in Quechua, the language of the Incas.

3. In the 12th century, Cuzco (also spelled Cusco) was established as a city-state and later the capital of the Inca Empire.

4. The founder of the Inca civilization and its first ruler was named Manco Cápac.

5. The civilization remained a relatively small tribe until the Inca began their conquests in the early 15th century under Emperor Pachacuti (also called Pachacutec).

6. Within one hundred years, the Inca Empire had expanded from modern-day Peru into parts of what are now Ecuador, Bolivia, Argentina, and Chile.

7. At its height, the Inca Empire included an Andean population of about 12 million people.

8. The Inca Empire was ruled by the ancestors of the original Inca people, who established the city of Cuzco.

9. At the top of Inca society was the emperor, called the Sapa Inca.

10. The Inca worshipped many different gods, but their main deity was the sun god Inti.

11. The Inca were skilled at trepanation, the practice of cutting or drilling a hole in the skull.

12. A study found that the odds of surviving trepanation were far better in the ancient Inca Empire than in the American Civil War.

13. The Inca developed a way of recording things on a system of knotted strings called a *quipu*.

14. The Inca were master stonemasons, creating large buildings and walls so precisely engineered that no mortar was needed.

15. They were also adept at road building.

16. An elaborate road system spanning 25,000 miles was constructed to connect the vast Inca Empire.

17. Some of the greatest achievements of the Inca are still in use today.

18. They invented the technique of freeze drying, as well as the rope suspension bridge.

19. Incan architecture was built to resist earthquakes.

20. The collapse of the Inca Empire came in 1532, when conquistador Francisco Pizarro conquered Peru for Spain.

21. The most notable victim of Spanish conquest was the Incan Emperor Atahualpa.

22. When the Inca and Spanish met in battle in 1532, the Inca had greater numbers.

23. But the superior steel weapons and armor of the Spanish were decisive.

24. Though he first presented himself in friendship, Pizarro soon kidnapped the Incan ruler.

25. Atahualpa tried to secure his release by offering a ransom of 24 tons of gold and silver.

26. The Spanish happily accepted the ransom, then murdered Atahualpa anyway.

27. Atahualpa was the 13th and last Inca emperor.

16 Machu Picchu Facts

1. Machu Picchu exists as one of the great legacies of the Inca.

2. The ancient stone city in the Andes is in southern Peru.

3. Most archaeologists believe that Machu Picchu was built as an estate or retreat for Inca Emperor Pachacuti around A.D. 1460.

4. Construction likely continued until the Inca Empire was conquered by the Spanish in the mid-1500s.

5. Machu Picchu was one of the few major Inca sites to escape Spanish capture.

6. It was rediscovered in 1911 by an American explorer named Hiram Bingham.

7. Machu Picchu means "old mountain" or "old peak" in the Quechua language of the Inca.

8. It refers to the lower of the two peaks that stand on each side of the city.

9. Today, Machu Picchu is a UNESCO World Heritage Site.

10. There are around 140 buildings in the city, built with stones fitted together tightly without the use of mortar.

11. Like most of the Inca Empire, Machu Picchu was built without the use of the wheel, draft animals, knowledge of iron or steel, or even a system of writing.

12. One of the sacred structures found in Machu Picchu is the Intihuatana.

13. The Inca believed this stone structure helped to hold the sun in place and keep it on its correct path.

14. Machu Picchu is located about 50 miles from Cuzco, the capital city of the Inca Empire.

15. The Inca built a stone road from Cuzco to Machu Picchu.

16. Many people still hike this trail today on their way to see Machu Picchu.

45 Facts About the Pilgrims

1. When the Pilgrims began their voyage to the New World, they didn't expect to sail on the *Mayflower*, nor did they plan to land at Plymouth Rock.

2. The story of the Pilgrims begins back in 1606—14 years before they set sail on the *Mayflower*.

3. A band of worshippers from Scrooby Manor wishing to separate from the Church of England escaped to Holland.

4. Although they worshipped freely, they feared their children were becoming more Dutch than English.

5. Meanwhile, English noblemen were seeking brave, industrious people to sail to America and establish colonies in Virginia (which extended far beyond modern Virginia).

6. They offered the Separatists a contract for land at the mouth of the Hudson River, near present-day New York City.

7. Led by William Brewster and William Bradford, the Separatists accepted the offer and even bought their own boat: the *Speedwell*.

8. In July 1620, they sailed to England to meet 52 more passengers who rode in their own ship, the *Mayflower*.

9. The Separatists, who called themselves "Saints," referred to these new people as "Strangers."

10. After two disastrous starts on the leaky *Speedwell*, the Saints joined the Strangers on the *Mayflower* on September 6.

11. The *Mayflower* was just 30 yards long—about the length of three school buses.

12. The 50 Saints rode in the "tween" deck, an area between the two decks that was actually the gun deck.

13. Its ceilings were just five feet high.

14. Crammed into close quarters were 52 Strangers, 50 Saints, 30 crewmen, 2 dogs, barley, oats, shovels, hammers, tools, beer, cheese, cooking pots, and chamber pots.

15. Storms pushed the *Mayflower* off course.

16. After 65 days on the high seas, they realized they were nowhere near the Hudson River.

17. Instead, they sighted the finger of Cape Cod—more than 220 miles away from their destination.

18. Though they were far from the land contracted for the English colony, the settlers saw their arrival in the New World as an opportunity to build a better life.

19. In November 1620, 41 free men (Saints and Strangers alike) signed the Mayflower Compact.

20. They agreed to work together for the good of the colony and to elect leaders to create a "civil body politic."

21. After anchoring in a harbor (which is now Provincetown), the Saints formed three expeditions to locate a suitable place to live.

22. One expedition ventured 30 miles west to a place called "Plimouth," which had been mapped several years earlier by explorer John Smith.

23. The settlers first noticed a giant rock, probably weighing 200 tons, near the shore.

24. The land nearby had already been cleared. Likely, more than a thousand Native people had lived there before being wiped out by an epidemic. Some remaining bones were still visible.

25. In December 1620, the group decided to make Plymouth its settlement.

26. According to legend, each passenger stepped on Plymouth Rock upon landing.

27. If this actually happened, leader William Bradford did not record it.

28. Yet the first steps on land—no matter where they took place—remain important.

29. Helped by Native people, the Saints and Strangers would live and work together to form one of the first British settlements in North America.

30. Many of the *Mayflower*'s passengers got sick and died during the first winter.

31. At one point there were only six people well enough to continue working.

32. By springtime, only 47 of the original 102 settlers were still alive.

33. The colony likely would not have survived without the help of the Wampanoag peoples.

34. A Wampanoag man named Squanto taught the Pilgrims how to plant corn and where to hunt and fish.

35. The "First Thanksgiving" in 1621 was little more than a harvest feast.

36. Abraham Lincoln proclaimed Thanksgiving a national holiday in 1863.

37. Much of what we know of the pilgrims comes from William Bradford's detailed journal, now known as *Of Plymouth Plantation*.

38. In it, he detailed the history of the Plymouth colony between 1630 and 1647.

39. Bradford was one of the first to sign the Mayflower Compact and served as the Plymouth colony's governor intermittently between 1621 and 1657.

40. The journey to America left the Pilgrims deeply in debt to English merchants who had financed the trip.

41. The early fruits of Pilgrim labor were sent back to England to repay the debt.

42. The first historical references to "Plymouth Rock" appeared more than 100 years after the Pilgrims' landing.

43. Pieces of Plymouth Rock can be seen today at Pilgrim Hall Museum, as well as at the Smithsonian.

44. According to estimates, Plymouth Rock is now only about $1/3$ its original size.

45. Over the years, the rock has been broken during attempts to move it, and people have chipped off pieces as souvenirs.

19 U.S. Founding Facts

1. Because the colonists inadvertently chose a low tide to execute the Boston Tea Party, nearly 350 crates of tea piled up in the shallow water.

2. The colonists had to jump overboard and actually smash open the tea crates in order to make sure it was ruined.

3. The first major battle of the Revolutionary War was the Battle of Bunker Hill, which took place in 1775.

4. It followed the smaller opening battles at Lexington and Concord.

5. At the gallows before being hung by the British in 1776, American patriot Nathan Hale said, "I only regret that I have but one life to lose for my country."

6. Morocco was the first country to recognize the United States as a sovereign nation, in 1777.

7. Benjamin Franklin considered the bald eagle a "bird of bad

moral character" and resented its being chosen to represent the United States of America.

8. Thomas Jefferson thought the Constitution should be rewritten every generation.

9. The proud American motto "E pluribus unum"—out of many, one—was originally used by Virgil to describe salad dressing.

10. During a fierce battle with the British in 1779, John Paul Jones, America's first naval hero, yelled, "I have not yet begun to fight!"

11. The final major battle of the Revolutionary War was the Battle of Yorktown in October 1781.

12. It ended with the surrender of the British.

13. New York City served as the nation's first capital, followed by Philadelphia.

14. The U.S. capital moved to Washington, D.C., in 1800.

15. The original Library of Congress was burned down by the British in 1814 along with the Capitol.

16. To replace it, the Congress bought Thomas Jefferson's personal book collection, which consisted of approximately 6,500 volumes.

17. The White House hasn't always been known as such.

18. This presidential domicile has also been called the "President's Palace," the "President's House," and the "Executive Mansion."

19. In 1901, President Theodore Roosevelt officially gave the White House its current name.

57 Civil War Facts to Consider

1. Abraham Lincoln won the presidential election of 1860 without one Southern state.

2. News of his victory prompted a secession movement across the South.

3. By the time Lincoln took office, seven Southern states had formed the Confederate States of America.

4. Four others soon joined.

5. America's bloodiest war began at Fort Sumter in Charleston, South Carolina, on April 12, 1861.

6. By war's end, the dead would number more than 620,000.

7. Fort Sumter was preceded by attacks on other forts and military installations in Confederate territory.

8. The first gun fired in defense of the Union was in Pensacola, Florida, on January 8, 1861, when state troops tried to occupy federal forts.

9. On January 9, 1861, Mississippi followed South Carolina to become the second state to secede from the Union.

10. The Confederacy consisted of 11 states and had a population of 9 million, which included 3.5 million slaves.

11. Without the states that seceded, the Union was made up of 25 states (including the border states of Missouri, Kentucky, Maryland, Delaware, and later West Virginia) and various organized territories in the West.

12. The Union's population was 22 million.

13. During the Civil War, the westernmost region represented by a delegate to the Confederate Congress was the Territory of Arizona.

14. Tennessee was the last of 11 states to secede from the Union in 1861 but the first state to be readmitted when it rejoined in July 1866.

15. Robert E. Lee was offered the chance to lead the largest army in the Union in 1861, but he chose to remain loyal to his home state of Virginia.

16. More Civil War battles were fought in Virginia than in any other state.

17. The Tennessee or Virginia quickstep was the name Union troops fighting in the South gave to diarrhea.

18. The exact number of Confederate enlisted men during the war years is unknown. Estimates range from as many as 1.4 million to as few as 600,000.

19. Some Confederate volunteers wore homespun attire dyed a yellowish-brown using, among other things, walnut shells.

20. Their uniform color led to their nickname of Butternuts.

21. Uniforms of Union troops weren't standardized until after the First Battle of Bull Run.

22. Typical Union gear included a musket, bayonet, cartridge box with 40 rounds, belt, cap pouch, haversack, canteen, knapsack, ground blanket, shelter half (or "dog tent"), winter greatcoat, tin cup and plate, and leggings.

23. Although the first national draft law allowed men to hire a substitute or purchase an exemption for $300, it also stimulated volunteering.

24. Postage service was abolished between the North and South in 1861.

25. At the time, the basic postal rate was 3 cents, with adjustments for distance.

26. On March 9, 1862, the USS *Monitor* and the CSS *Virginia* engaged in a four-hour close-range battle resulting in a draw.

27. This naval warfare milestone marked the first engagement between two ironclad warships.

28. In July 1862, David Glasgow Farragut was promoted to rear admiral, the first officer to hold that rank in the history of the U.S. Navy.

29. Ulysses S. Grant led the Army of Tennessee in the early stages of the war.

30. Grant claimed early victories at Fort Henry and Fort Donelson, earning the nickname "Unconditional Surrender."

31. After winning major victories at Shiloh and Vicksburg, Lincoln promoted Grant to lead the entire Union Army.

32. Rose O'Neal Greenhow was a renowned Confederate spy.

33. She passed along information on Washington's defenses and Union troop movements to Confederate officials.

34. Clara Barton was a famous nurse to Union troops. She also founded the American Red Cross.

35. The Battle of Antietam was fought on September 17, 1862, and was the bloodiest single day of fighting in American history.

36. Union casualties at the Battle of Antietam were 12,400, including 2,100 killed; Confederate casualties were 10,320, including 1,550 killed.

37. Contrary to popular belief, the Emancipation Proclamation didn't free all the slaves in America.

38. Lincoln wrote the edict in September 1862. The language of the document was clear: Any slave that was still held in the states that had seceded from the Union was "forever free" as of January 1, 1863.

39. Significantly, the Emancipation Proclamation did not include border states in which slaves were still held, such as Kentucky or Missouri, because Lincoln didn't want to stoke rebellion there.

40. More than 186,000 African Americans joined the U.S. Armed Forces by the end of the war.

41. Of these, an estimated 93,500 Black soldiers were former slaves.

42. Civil War soldiers devised makeshift "dog tags" before battle—handkerchiefs or pieces of paper with the soldiers' names and addresses on them that they pinned to their uniforms.

43. About two-thirds of the deaths in the Civil War were due to disease.

44. Union General William Tecumseh Sherman gained notoriety

after his role in the Battle of Shiloh, but he is most famous for his "March to the Sea."

45. Sherman's troops destroyed everything in their path during the 90-day march through Georgia.

46. The Civil War marked the first time troops were deployed en masse via the railroads in a large-scale war.

47. Confederate General Thomas J. "Stonewall" Jackson was accidentally shot by one of his own men at the Battle of Chancellorsville.

48. His left arm was amputated, then retrieved the next day and buried at a nearby plantation cemetery. Stonewall Jackson died days later.

49. A tombstone marking the grave of Stonewall Jackson's amputated arm still stands today in Fredericksburg, Virginia.

50. Lieutenant Thomas Custer, who fought for the Union, was the only soldier to win the Medal of Honor twice during the Civil War.

51. More than a decade later, he would meet his death at Little Big Horn under the leadership of his brother General George Armstrong Custer.

52. Confederate General Henry Heth graduated dead last in his West Point class.

53. Heth accidentally started the Battle of Gettysburg. He sent two brigades ahead on reconnaissance, and they encountered and engaged Union troops.

54. Lincoln's famous Gettysburg Address was only 269 words long.

55. Contrary to popular belief, Lincoln did not write the Gettysburg Address on the back of an envelope while riding the train.

56. On April 9, 1865, Confederate General Robert E. Lee surrendered to Ulysses S. Grant at Appomattox, Virginia.

57. Grant offered generous terms, allowing the defeated army to return home with their small arms and giving them rations.

15 Facts About the History of Underwear

1. The earliest and most simple undergarment was the loincloth—a long strip of material passed between the legs and around the waist.

2. King Tutankhamun was buried with 145 loincloths.

3. The style didn't go out with the Egyptians. Loincloths are still worn in many Asian and African cultures.

4. Men in the Middle Ages wore loose, trouser-like undergarments called braies, which one stepped into and tied around the waist and legs about mid-calf.

5. To facilitate urination, braies were fitted with a codpiece, a flap that buttoned or tied closed.

6. Medieval women wore a close-fitting undergarment called a chemise.

7. Corsets began to appear in the 18th century.

8. Early versions were designed to flatten a woman's bustline, but by the late 1800s, corsets were reconstructed to give women an exaggerated hourglass shape.

9. Bras were invented in 1913 when American socialite Mary Phelps-Jacob tied two handkerchiefs together with ribbon.

10. Phelps-Jacob patented the idea a year later.

11. Maidenform introduced modern cup sizes in 1928.

12. Around 1920, as women became more involved in sports such as tennis and bicycling, loose, comfortable bloomers replaced corsets as the undergarment of choice.

13. The constricting corset soon fell out of favor altogether.

14. The thong made its first public U.S. appearance at the 1939 World's Fair, when New York Mayor Fiorello LaGuardia required nude dancers to cover themselves, if only barely.

15. Thongs gained popularity as swimwear in Brazil in the 1970s and are now a form of underwear.

41 Women's Suffrage Facts

1. Margaret Brent, a landowner in Maryland, was the very first woman in the United States to call for voting rights.

2. In 1647, Brent insisted on two votes in the colonial assembly— one for herself and one for the man for whom she held power of attorney. The governor rejected her request.

3. Abigail Adams asked her husband John in 1776 to "remember the ladies" in the new code of laws he was drafting.

4. "If particular care and attention is not paid to the ladies, we are determined to foment a rebellion, and will not hold ourselves bound by any laws in which we have no voice or representation," Abigail Adams wrote.

5. Judith Sargent Murray wrote the essay "On the Equality of the Sexes" in 1790, arguing that women were just as capable of intellectual accomplishment as men and that an education would liberate women from economic dependence.

6. Between 1818 and 1820, Fanny Wright, a feminist from Scotland, lectured throughout the United States on such topics as voting rights for women, birth control, and equality between the sexes in education and marriage laws.

7. Before the Civil War, the women's suffrage and abolition movements were closely linked.

8. However, female delegates, including Lucretia Mott and Elizabeth Cady Stanton, were barred from participating in the 1840 World Anti-Slavery Convention in London because of their gender. This prompted Stanton and Mott to organize a women's convention.

9. In 1848, Stanton (1815–1902) and Mott (1793–1880) held the first national convention to discuss women's rights, including the right to vote, in Seneca Falls, New York.

10. Stanton presented her *Declaration of Sentiments*, the first formal

action by women in the United States to advocate civil rights and suffrage.

11. Modeled after the Declaration of Independence, Stanton's document proclaimed that "all men and women are created equal" and resolved that women would claim the rights of citizenship denied to them by men.

12. The *Declaration of Sentiments* was adopted officially at the Seneca Falls Convention in July 1848 and signed by 68 women and 32 men.

13. Two groups formed at the end of the 1860s: the National Woman Suffrage Association (NWSA) and the American Woman Suffrage Association (AWSA).

14. The NWSA, led by Susan B. Anthony and Stanton, worked to change voting laws on the federal level by way of an amendment to the U.S. Constitution.

15. The AWSA, led by Lucy Stone and Julia Ward Howe, worked to change the laws on the state level.

16. The two groups united in 1890 and as the National American Woman Suffrage Association (NAWSA).

17. Victoria Woodhull (1838–1927) became the first woman to run for U.S. president in 1872.

18. While women were not able to vote, there were no laws prohibiting them from running for office.

19. Amelia Jenks Bloomer (1818–94) was an active member of the suffrage movement in the United States throughout her life and helped popularize the Turkish style of pantaloons in the 1850s that took on her name (bloomers).

20. New Zealand granted its female citizens suffrage in 1893, making it the first nation or territory to formally allow women to vote in national elections.

21. Wyoming became the first state to grant women suffrage when it was admitted to the Union in 1890.

22. In 1913, suffragists organized a parade down Pennsylvania Avenue in Washington, D.C.

23. The parade was the first major suffrage spectacle organized by the National American Woman Suffrage Association (NAWSA).

24. In 1916, Alice Paul formed the National Woman's Party (NWP).

25. Suffragists staged a months-long vigil outside the White House in 1917.

26. Tolerated at first, the picketers drew increasing criticism after the United States entered World War I.

27. In the fight for suffrage, women were arrested and imprisoned for such "offenses" as holding open-air meetings, obstructing traffic, and silently picketing the White House.

28. While in prison, women were subjected to violence and intimidation, unsanitary living conditions, and were often placed in solitary confinement.

29. When some of the suffragists instituted a hunger strike, they were painfully force-fed.

30. The resulting publicity helped put pressure on Congress to consider a suffrage amendment.

31. When Congress ratified the 19th Amendment in 1920, women were finally given the right to vote in all U.S. government elections. It had been a long time coming.

32. Although legally entitled to vote, Black women were effectively denied voting rights in numerous Southern states until 1965.

33. Azerbaijan was the first Muslim-majority country to enfranchise women in 1918.

34. In 1921, a monument honoring Stanton, Anthony, and Mott was unveiled at the U.S. Capitol, where it remains today.

35. Alice Paul drafted the text of the Equal Rights Amendment in 1923.

36. The 19th Amendment gave women the right to vote, but not full equality, so Alice Paul and the National Woman's Party took up the fight for equal rights.

37. Albanian women voted for the first time in the 1945 election.

38. In 1947, on its independence from the United Kingdom, India granted equal voting rights to all men and women.

39. Women in Switzerland gained the right to vote in federal elections after a referendum in February 1971.

40. In Liechtenstein, women were finally given the right to vote by the women's suffrage referendum of 1984. Three prior referendums had failed to secure women's right to vote.

41. In 1963 Betty Friedan wrote the book *The Feminine Mystique*, which analyzed women's limited roles in American society and galvanized the modern-day feminist movement.

89 World War II Facts

1. Adolf Hitler and his Nazi Party swept to power in 1933.

2. Hitler rebuilt the army and air force, and expanded Germany's borders.

3. Most notoriously, the Germans undertook the systematic brutalization and mass murder of Jews, Romany, homosexuals, and others deemed "undesirable."

4. In an ominous prelude to World War II, Japanese forces invaded eastern China in 1937, starting a war there that would continue until 1945.

5. In September 1939, Germany invaded Poland and France. Britain declared war on Germany. World War II had begun.

6. The 12-day conquest of Yugoslavia cost Germany only 151 lives. Partisan fighting after the surrender would claim far more.

7. Poland was the first to discover the underlying clues to the German ENIGMA encryption machine.

8. In 1939 the Polish handed the ball off to the British; it then became the ULTRA project.

9. This handsome present may have been Poland's greatest gift to Allied victory.

10. The main success of the Nazi Blitzkrieg or "lightening war" that raced across Europe was due to tank units supported from the air by dive-bombers, such as the Junkers Ju87 (Stuka).

11. The Stukas were fitted with sirens, which sounded like screaming to terrify the population.

12. The Blitz, Germany's terror-bombing campaign against Britain in 1940–41, was a horrifying experience.

13. More than 43,000 civilians died during the bombing; one million homes were destroyed or damaged.

14. The Blitz created dangerous morale problems in Britain.

15. To galvanize the people, Churchill lied and said that a German invasion might be imminent. The fib worked, and the British rallied.

16. After the Japanese attack on Pearl Harbor on December 7, 1941, the United States joined the global war that had been raging for nearly two years.

17. At the time of the Pearl Harbor attack, there were 96 ships anchored there.

18. During the attack, 18 were sunk or seriously damaged, including eight battleships.

19. Three hundred and fifty aircraft were destroyed or damaged.

20. There were 2,402 American men killed and 1,280 injured at Pearl Harbor.

21. Japanese pilots attacking Pearl Harbor had studied tourist postcards (provided by spy Takeoyoshi Yoshikawa) to learn where ships were normally positioned in the harbor.

22. Knowing that America would enter the war after the attack on Pearl Harbor, Winston Churchill said he went to bed that night and "slept the sleep of the saved and thankful."

23. The Soviet Union joined the Allies when Germany invaded in June 1941.

24. The Allied powers included Britain, the U.S., and the Soviet Union, among others.

25. The Axis powers were Germany, Japan, and Italy.

26. Before the war, Japanese pilots were selected at age 14 from the best academic and most physically fit students.

27. Their five-year training program was the world's most arduous.

28. A U.S. draftee had to be at least 5 feet tall and weigh 105 pounds; he also had to have at least 12 teeth.

29. The two most common physical attributes that led to a draftee being rejected were flat feet and venereal disease.

30. Denmark was the only European country to cut back on its military after war broke out.

31. Beginning in the fall of 1942, each American family was limited to three gallons of gasoline per week and two new pairs of shoes per year.

32. The famous U.S. bazooka antitank weapon fired a 60-mm projectile.

33. Germany's counterpart, the less common Panzerschreck, packed an 88-mm wallop, but had a much shorter effective range.

34. The Nazis paid one of their most successful spies, the valet to the British ambassador in Turkey, in counterfeit currency.

35. In 1943, Vichy France's coerced tribute accounted for roughly 25 percent of Nazi Germany's gross national product.

36. U.S. riflemen often used condoms to keep mud and dirt out of the barrels of their rifles.

37. The longest battle of World War II was the Battle of the Atlantic, which was fought 1939–1945.

38. From 1942, U.S. Marines in the Pacific used the Navajo language as their secret code.

39. The language didn't have the vocabulary for existing WWII technology, so existing words had to be given new meanings.

40. Around 400 Navajo Indians (Code Talkers) were trained to use the code, and the Japanese never cracked it.

41. The 1942–43 Battle and Siege of Stalingrad resulted in more Russian deaths (military and civilian) than the U.S. and Britain sustained (combined) in all of World War II.

42. Four of every five German soldiers killed in the war died on the Eastern Front.

43. Auschwitz-Birkenau was one of six Nazi extermination camps; the others were Belzec, Chelmno, Majdanek, Sobibor, and Treblinka. All were located in Poland.

44. Concentration camps, as distinct from extermination camps, were sited throughout Germany and across Nazi-occupied Europe.

45. Auschwitz was built near the Polish town of Oswiecim, on dank, marshy ground that had been considered unsuited for development.

46. Hamburgers were dubbed "Liberty Steaks" in the U.S. during the war to avoid the German-sounding name.

47. The first bomb dropped on Berlin by the Allies killed an elephant in the Berlin Zoo.

48. A Soviet soldier's daily vodka ration was 100 grams, or about 3.5 ounces.

49. The desperate house-to-house fighting in Stalingrad led the German soldiers to call it a "Rat's War."

50. Nearly 180 Londoners died during a raid on March 3, 1943—but not from bombs. A woman tripped entering an underground station; the crowd rushing for cover crushed her and many others, suffocating them.

51. The first air-reconnaissance photos of the Auschwitz complex were taken by American aircraft in April 1944.

52. London and Washington had been aware of the camp's existence as early as 1943.

53. The Allies mounted no concerted bombing against Auschwitz, nor attempted to otherwise liberate it, because the camp was deemed to be of no military value, and thus unsuited for expenditure of Allied resources.

54. The first national European capital captured by the Allies during the war was Rome on June 4, 1944—two days before the Normandy invasion.

55. D-Day—June 6, 1944—was the launching date for Operation Overlord, the Allied invasion of Nazi-occupied Western Europe.

56. In one of the most complex operations in military history, U.S., British, and Canadian forces landed simultaneously on five separate beachheads in Normandy, France.

57. For the D-Day invasion, British vehicles carried huge spools of coconut-husk matting to unroll on the beach, which helped tanks avoid bogging down in the soft ground.

58. About 12,000 Allied aircraft supported the Normandy invasion.

59. D-Day owed its success in part to weather reports from lonely, icy stations in Arctic Greenland and Spitsbergen and Jan Mayen in Norway.

60. The invasion needed to occur on June 5, 6, or 7 to take advantage of the moon and the tides.

61. D-Day was postponed a day due to rain and gale-force winds, which would have been disastrous for the Allies.

62. During one phase of the battle for Iwo Jima, U.S. forces expended 10,000 gallons of flamethrower fuel every day.

63. Iwo Jima proved an effective haven for more than 2,200 damaged U.S. bombers that made emergency landings on the island's airstrip.

64. The Sullivans, a family from Waterloo, Iowa, lost five sons in U.S. Navy service in the Pacific (Madison, 23; Joseph, 24; Francis, 27; George, 28; and Albert, 29).

65. Their sister, Genevieve, promptly took up her fallen brothers' sword: She enlisted in the Navy's Women Accepted for Voluntary Emergency Service (WAVES).

66. When U.S. Marines landed on Guam, they were greeted by a sign that read, "Welcome Marines."

67. The sign had been left by U.S. Navy frogmen who had earlier scouted the area for mines.

68. After the Dunkirk evacuation, the British were short of everything, rifles included.

69. The home guard armed itself with shotguns, pitchforks, clubs, broom handles, axes, and even golf clubs.

70. Nazi propaganda minister Joseph Goebbels believed so strongly in the power of patriotic music that Germany's troops were issued a patriotic songbook as part of their standard kit.

71. Contrary to most accounts, Goebbels did not have a "club foot." He had suffered from a bone marrow affliction as a child; the surgery shortened and weakened his left leg.

72. Panzer is simply German for "tank."

73. The Panther tank that appeared in 1943 was a panzer, so was the Panzer Ib, so was the King Tiger, and so forth.

74. Betty Grable was the number-one GI pinup girl; Rita Hayworth was a close second.

75. Italy went to great lengths to shield its antiquities from wartime damage: Statues were surrounded by sandbags; monuments were shelled in masonry and sandbagged scaffolding.

76. World War II was the first major conflict in which penicillin was used. It drastically reduced battlefield fatalities.

77. Navy submarines carried no doctors; a pharmacist's mate had to handle medical emergencies.

78. Among many lifesaving accomplishments during the war, Navy pharmacist's mates tallied 11 emergency appendectomies.

79. Prior to taking office as president in April 1945, Harry Truman knew nothing of the Manhattan Project, which produced the first atomic bombs.

80. By 1945 Hitler could not have read these words. His eyes had grown so bad that all his reports had to be prepared on special Führer typewriters, which used a typeface triple the normal size.

81. Greek defenders weren't very impressed with the Italian invaders. One Greek bomber pilot dropped chamber pots on Italian troops after releasing his conventional bomb load.

82. In the first week of May 1945, following Hitler's suicide on April 30, the Nazi regime collapsed.

83. Berlin fell to the Soviets, and Axis armies in Italy gave up. On May 7, 1945, Germany surrendered, and the war for Europe was over.

84. The Department of Defense estimates between 15 and 20 percent of U.S. combat deaths—some 21,000—were due to friendly fire.

85. The Enola Gay became well known for dropping the first atomic bomb on Hiroshima, but few people know the name of the B-29 that bombed Nagasaki.

86. It was Bock's Car, named after the plane's usual commander, Frederick Bock.

87. World War II ended on September 2, 1945, when Japan signed a surrender agreement following nuclear attacks on Hiroshima and Nagasaki.

88. Japan's Lt. Hiroo Onoda didn't surrender until 1974, having fought a guerilla war on the Philippine island of Lubang since early 1945.

89. For 29 years, he refused to believe Allied stories about World War II ending and only gave up the fight after his old unit commander was flown to the Philippines and ordered him to lay down his arms.

51 Facts About Inventions

1. The first can opener was invented almost 50 years after the first tin can.

2. John Harington, godson to Queen Elizabeth I, invented a flushing toilet in the 1590s.

3. He installed one for the queen, but she was not impressed.

4. In 1608, Dutch eyeglass maker Hans Lippershey discovered he could make far-away objects look larger by fitting two lenses in a long tube.

5. Italian scientist Galileo Galilei improved Lippershey's telescope in 1609 and used it to look at the stars.

6. In 1668, English scientist Isaac Newton developed the reflecting telescope, which used mirrors rather than lenses.

7. The first bicycle was invented in 1817 by German nobleman Baron von Drais.

8. His wooden-framed, iron-wheeled bicycle was pushed along by the rider's feet.

9. The first successful pedal-driven bicycle was invented by Pierre Lallement in 1864.

10. The first pedal-driven bicycles became known as "boneshakers" because they shook over every bump in the road.

11. The electric light bulb was invented by Henry Woodward and Mathew Evans in 1874.

12. These Canadians paved the way for Thomas Edison's later improvements to the bulb.

13. When Thomas Edison died in 1931, he held a world-record 1,093 U.S. patents.

14. Alexander Graham Bell invented the telephone in 1876.

15. Otto Lilienthal invented the hang glider in the late 19th century.

16. Unfortunately, in 1896, Lilienthal plunged more than 50 feet during one of his test runs and died shortly after.

17. Traffic lights were in use before cars were invented.

18. Mary Anderson invented a swinging-arm device with a rubber blade that the driver operated by using a lever.

19. In 1903, Anderson received a patent for what became known as the windshield wiper.

20. Inventor of vaccines and pasteurization, Louis Pasteur was the head of the science program at a French school during the mid 1800s.

21. Movie actress Hedy Lamarr's invention was a matter of national security.

22. Lamarr and composer George Anthiel developed a secret communications system to help the Allies in World War II— their method of manipulating radio frequencies was used to create unbreakable codes.

23. The invention proved invaluable again two decades later when it was used aboard naval vessels during the Cuban Missile Crisis.

24. The "spread spectrum" technology that Lamarr helped to pioneer became the key component in the creation of cellular phones, fax machines, and other wireless devices.

25. John Harvey Kellogg and W.K. Kellogg, who invented Corn Flakes, ran a medical facility called the Battle Creek Sanitarium in the late 1800s.

26. The anti-gravity flight suit, invented by Dr. Wilbur Franks in 1941, enabled combat pilots to withstand G-force pressure and extreme acceleration.

27. The Hula-Hoop's origins date back some 3,000 years, when children in ancient Egypt made hoops out of grapevines and twirled them around their waists.

28. Wham-O introduced the plastic version of the toy in 1958, and 20 million were sold in the first six months.

29. In the late 1950s, Ruth Handler set out to create a grown-up, three-dimensional doll.

30. Handler named her creation after her daughter, and the Barbie doll was introduced in 1959.

31. Handler was one of the founders of the toy giant Mattel.

32. Leo Fender, who invented the first mass-produced electric guitar, never learned to play the guitar.

33. *Chindogu* is a Japanese term that refers to the strange art of creating seemingly useful yet ultimately useless inventions—useless because no one would want to be seen with one in public.

34. An example is the hay-fever hat, which consists of a roll of toilet paper secured to the head by a chinstrap.

35. Harold von Braunhut came up with the Sea Monkey, a simple three-step kit that allowed youngsters to breed their own aquatic creatures.

36. The wee serpents are actually a unique species of brine shrimp.

37. Women came up with ideas and specifications for such useful items as life rafts (Maria Beasley), circular saws (Tabitha Babbitt), medical syringes (Letitia Geer), and underwater lamps and telescopes (Sarah Mather).

38. In 1822 Englishman Charles Babbage began designing complex computing machines, which he called "engines."

39. He never actually built one though.

40. The world's first working computer was developed during World War II and used to make secret codes for Germany.

41. The top-secret Z3 was invented by German engineer Konrad Zuse in 1941.

42. British scientists also built a secret computer during World War II.

43. Called Colossus, it helped break German secret codes and win the war.

44. Giuliana Tesoro was a prolific inventor in the textile industry.

45. Flame-resistant fibers and permanent-press properties are among her many contributions.

46. The Tesoro Corporation holds more than 125 of her textile-related patents.

47. Ermal Fraze invented the pop-top aluminum can in 1959.

48. In 1963, he received U.S. patent number 3,349,949 for the design.

49. John Landis Mason invented Mason jars.

50. Mason also invented screw-top salt shakers. Before that, you just poured salt onto your food willy-nilly.

51. The man who created the Nobel Peace Prize, Alfred Nobel, is also responsible for a destructive invention: dynamite.

35 Miscellaneous Facts

1. Talk about staying power—China's Zhou Dynasty (approximately 1100 B.C. to 221 B.C.) was the longest-running dynasty in history.

2. Early medieval English coins were only legal tender for a few

years, after which the authorities melted them down and gave out slightly fewer new ones. Today we call this "tax."

3. Mongol emperor Kublai Khan kept 5,000 astrologers busy working for him.

4. Those who attended Genghis Khan's burial (A.D. 1227)—all 2,900 of them—were executed.

5. Evidently the Mongols were fairly sure someone would try to plunder the grave.

6. During the Black Death of the 1300s, fanatical groups called Flagellants wandered Germany and France whipping themselves bloody.

7. They believed God would call off the plague if they kept it up long enough.

8. Although many sources claim that the popular children's rhyme "Ring Around the Rosy" is about the Black Death, five centuries passed between the plague and the first mentions of the rhyme.

9. One of history's great unknown female generals was the Kahinah Dahiyan (A.D. 680), a seer and leader of a Jewish tribe in the Atlas Mountains (now Algeria) during the Islamic explosion.

10. She led the region's diverse tribes in battle for years until the Arab armies finally overwhelmed her.

11. Sati, the Indian practice of widows immolating themselves on their husbands' funeral pyres, is also recorded among some Slavic peoples in the post-Roman era before A.D. 1000.

12. The average 17th-century American woman gave birth to 13 children.

13. Frankish soldiers learned to throw the francisque (battle-axe) at the enemy's shield to shatter it for an advantage in close combat.

14. A millennium later in World War II, the sordid Vichy French régime superimposed the Franciscan Axe on its short-lived version of the famous Tricolor flag.

15. The "Pig War" between the United States and Great Britain was fought over possession of the San Juan Islands off the coasts of Washington state and British Columbia.

16. The 1859 confrontation was set off by the shooting of a pig.

17. Egyptian pharaoh Thutmose III (probably reigned 1479–1425 B.C.) resented his powerful stepmom, Hatshepsut.

18. After she died, Thutmose vandalized her statues and dumped them in a quarry.

19. Then he set out to chisel every instance of her name off all stone inscriptions. Despite his best efforts, we still know Hatshepsut existed.

20. In terms of per capita death and destruction, King Philip's War (1675–76) between the Wampanoags and English settlers was the worst in U.S. history.

21. Of 90 Puritan towns, 52 were attacked and 12 were completely destroyed during King Philip's War.

22. The world's longest war is commonly considered to be that which existed between the Netherlands and Isles of Scilly off the Cornish Coast.

23. War was declared by the Netherlands during the confusing years of the English Civil War (1642–51), and the proclamation was forgotten until 1985. Officially, the war lasted 335 years.

24. Before marrying Mark Antony, Cleopatra was married to two of her brothers. But not at the same time!

25. She married one brother, Ptolemy XIII, when he was 11.

26. When he died, Cleopatra married her other brother, Ptolemy XIV. He died not long after they were married.

27. Salem witch trivia: If there wasn't witchery, why did several young girls suddenly act so bewitched? The answer may lie in moldy rye grain, a constant problem in the Colonies. Rye mold contains a hallucinogen called ergot.

28. Herodotus reports that the Scythians, a fierce nomadic people from the Caspian Sea region of Central Asia, made drinking cups by gilding the interiors of human skulls. This died out when the Vikings took over.

29. Vikings didn't just raid England, France, and Ireland.

30. Besides sailing to America and colonizing Greenland and Iceland, they assailed Portugal, Spain, the French Riviera, Italy, and Arctic Russia.

31. Swedish Vikings traded all the way down to Byzantium.

32. Arab ambassador Ibn Fadlan on the Swedish Rus (Vikings, c. 923): "They are the filthiest of God's creatures. They do not wash after discharging their natural functions, neither do they wash their hands after meals. They are as stray donkeys."

33. Ten of the first twelve U.S. presidents owned or had owned slaves.

34. One of the U.S. presidents was not a U.S. citizen at his time of death: John Tyler, a Virginia native, died on January 18, 1862, a citizen of the Southern Confederacy.

35. The 1893 World's Fair, hosted by Chicago, introduced the Ferris wheel for the first time.

Chapter 4

Food and Drink

42 Cheesy Facts

1. Cheddar cheese production involves "cheddaring"—the repeated cutting and piling of curds to create a firm cheese.

2. During World War II and for several years after, nearly all British cheese production was devoted to cheddar.

3. Fresh cheddar curds—the natural shape of the cheese before it's pressed into a block and aged—are called "squeaky curds."

4. Cheddar cheese is naturally white or pale yellow.

5. People started dying cheese orange back in the 17th century to prevent seasonal color variations.

6. These days, much of it is dyed orange with seeds from the annatto plant.

7. Early cheese makers used carrot juice and marigold petals.

8. Before 1850, nearly all cheese produced in the United States was cheddar.

9. Traditional English cheddar is produced in wheels and aged in cloth for a minimum of six months.

10. Someone who sells cheese professionally at a cheese shop or specialty food store is called a cheesemonger.

11. Age is the only thing that makes mild cheddar taste different from sharp cheddar.

12. Mild cheddar is usually aged for a few months.

13. The sharpest cheddars might be aged two years or more.

14. Archaeological surveys show that cheese was being made from the milk of cows and goats in Mesopotamia before 6000 B.C.

15. Travelers from Asia are thought to have brought the art of cheese making to Europe, where the process was adapted and improved in European monasteries.

16. American cheese is a cheddar that undergoes additional processing.

17. Processed American cheese was developed in 1915 by J. L. Kraft as an alternative to traditional cheeses that had a short shelf life.

18. A one-ounce serving of cheese is about the size of four dice.

19. The Pilgrims brought cheese onboard the *Mayflower* in 1620.

20. Philadelphia cream cheese is named after a village in upstate New York, not the Pennsylvania city.

21. Cheese contains trace amounts of naturally occurring morphine, which comes from the cow's liver.

22. The holes in Swiss cheese were once seen as a sign of imperfection and something cheesemakers tried to avoid.

23. Cheddar has been around since at least the 1100s.

24. A purchase of more than 10,000 pounds of cheddar is listed in the financial records of King Henry II, dated 1170.

25. Cheddar is the most popular kind of cheese in the world.

26. Legend has it that cheddar was first created when a woman left a pail of milk in England's Cheddar Gorge caves.

27. By the time she returned, the milk had turned to cheese.

28. Queen Victoria received a 1,000-pound wheel of cheddar as a wedding present in 1840.

29. The crunchy bits you sometimes get inside of aged cheese are amino acid crystals.

30. Wisconsin produces about 2.6 billion pounds of cheese each year.

31. A cheddar wheel displayed at the Toronto Industrial Exposition in the 1800s inspired the poem "Ode on the Mammoth Cheese Weighing over 7,000 Pounds," by James McIntyre.

32. British explorer Robert Falcon Scott brought 3,500 pounds of cheddar cheese with him on his ill-fated expedition to the South Pole in 1901.

33. Founded in 1882, the Crowley Cheese Factory in Healdville, Vermont, is the nation's oldest cheesemaker still in operation.

34. In 1964, the Wisconsin Cheese Foundation created a 34,665-pound cheddar wheel for the 1964 World's Fair in New York City.

35. President Andrew Johnson once served a 1,400-pound block of cheddar at a White House party.

36. Wisconsin-based Simon's Specialty Cheese made a 40,060-pound cheddar wheel named the "Belle of Wisconsin" in 1988.

37. June is National Dairy Month, and the last week in June is National Cheese Week.

38. The Cheese Days celebration in Monroe, Wisconsin, has been held every other year since 1914.

39. Highlights include a 400-pound wheel of Swiss cheese and the world's largest cheese fondue.

40. A cheesemaker in Quebec created a 57,518-pound round of cheddar cheese in 1995, claiming it to be the world's largest.

41. In 2010, chef Tanys Pullin sculpted a 1,100-pound crown out of cheddar cheese in honor of Queen Elizabeth II's coronation anniversary. It was the largest cheese sculpture in history at the time.

42. Stilton blue cheese frequently causes odd, vivid dreams.

10 Smelliest Cheeses

1. Limburger

2. Epoisses

3. Stinking Bishop

4. Stilton

5. Taleggio

6. Roquefort

7. Munster d'Alsace

8. Camembert

9. Pont l'Eveque

10. Serra da Estrela

28 Pretzel Facts

1. According to one origin story, Italian monks invented soft pretzels in A.D. 610 to motivate catechism students.

2. The monks called the treat *pretiola*, or "little rewards."

3. The twists resemble arms folded in prayer, and the three holes might represent the Father, Son, and Holy Spirit.

4. Pretzels are made of a simple mixture of water, flour, and salt.

5. In Germany, there are stories that pretzels were the invention of desperate bakers held hostage by local dignitaries.

6. Bakers in Austria have their own coat of arms: two lions holding a pretzel.

7. It was granted in 1510, after monks baking pretzels in a basement heard invaders digging tunnels under Vienna's city walls and helped defeat the invasion.

8. Pretzels show up in medieval religious art, including depictions of the Last Supper and in a prayer book created in 1440.

9. In 2015, archaeologists found a 250-year-old pretzel in Bavaria, perhaps the oldest ever discovered in Europe.

10. Hard pretzels were invented in the 1600s in Pennsylvania.

11. Legend has it that a baker's apprentice overcooked a batch of pretzels, accidentally creating this tasty snack.

12. In 1861, Julius Sturgis created the first commercial pretzel bakery in Lititz, Pennsylvania.

13. The Bavarian pretzel, or *Bayerische breze*, is on the European Union's protected origins list.

14. Only pretzels made in the state of Bavaria can be sold as Bavarian pretzels in the E.U.

15. The average American eats about two pounds of pretzels in a year.

16. Philadelphians eat about six times that amount.

17. Pretzels are a symbol of good luck and have been used in New Year's celebrations in Germany.

18. This food is also a symbol of eternal love. They've even been included in wedding ceremonies—the possible origin of the phrase "tying the knot."

19. The high school in Freeport, Illinois, nicknamed "Pretzel City USA," has a pretzel as a mascot.

20. The world's largest pretzel was baked in El Salvador in 2015.

21. It measured 29 feet 3 inches long by 13 feet 3 inches wide. It weighed 1,728 pounds.

22. President George W. Bush choked on a pretzel while watching a football game in 2002 and temporarily lost consciousness.

23. Americans buy more than $550 million worth of pretzels each year. About 80 percent are made in Pennsylvania.

24. In the United States, April 26 is National Pretzel Day.

25. Every pretzel in the world was made by hand until 1935, when the first automated pretzel machine was introduced in Pennsylvania.

26. Unsalted pretzels are nicknamed "baldies."

27. Pretzels have long been a beloved bar snack.

28. In fact, Prohibition in the 1920s hit the pretzel business hard as bars were legally required to close their doors.

44 Amazing Alcohol Facts

1. The name whiskey is the result of a mispronunciation of the Gaelic *uisge beatha*, which literally means "water of life" and was commonly used to refer to distilled liquor.

2. The oldest evidence of alcoholic beverages ever found dates as far back as 7000 B.C. in China.

3. The workers who built Egypt's Great Pyramids may have received their payment in beer.

4. It takes about six minutes for alcohol to start affecting the brain.

5. The Code of Hammurabi, ancient Babylon's set of laws, stipulated a daily ration of beer for citizens.

6. Gin and tonics originated as a way to make anti-malarial medication more palatable.

7. Sloe gin isn't gin at all. It's a liqueur made with sloe berries (blackthorn bush berries).

8. About 600 grapes go into one bottle of red wine.

9. As they age, white wine gets darker and red wine gets lighter.

10. It can take 40 years for vintage Port to reach full maturity.

11. Every bottle of champagne has an estimated 49 million bubbles.

12. The world's oldest existing bottle of wine was buried in A.D. 350 in Germany.

13. It was found again in 1867 and is now on display at a museum.

14. California produces more wine than any country in the world, with the exception of France, Italy, and Spain.

15. Kentucky produces all but five percent of the world's bourbon.

16. Tequila has no worm in it.

17. Mezcal does, although it's not even a worm, it's a moth larva.

18. Early thermometers often contained brandy instead of mercury.

19. Despite the perceived warming effects, alcohol actually lowers your body temperature.

20. A particularly drunken Christmas party at West Point Academy in 1826 resulted in a riot and 19 people being expelled.

21. The incident became known as the Eggnog Riot.

22. During Prohibition in the United States, doctors could still prescribe whiskey to patients.

23. This loophole helped Walgreens grow exponentially in those years.

24. Prohibition in the U.S. created a boom for Canadian distillers, and rye (the Canadian name for Canadian whiskey) became part of the national identity.

25. Rye is generally sweeter than bourbon and retains a worldwide following among liquor connoisseurs.

26. The exact origin of absinthe is unknown, but the strong alcoholic liqueur was probably first commercially produced around 1797.

27. It takes its name from one of its ingredients, *Artemisia absinthium*, which is the botanical name for the bitter herb known as wormwood.

28. Green in color due to the presence of chlorophyll, absinthe became an immensely popular drink in France by the 1850s.

29. The United States banned absinthe in 1912, years before Prohibition, and it remained banned until 2007.

30. Most Americans consider sake a Japanese rice wine, but it's actually more akin to beer.

31. Sake also may have originated in China, not Japan.

32. Contrary to urban legend, Jägermeister does not contain elk blood.

33. The founder was an avid hunter, and the name Jägermeister literally translates to "hunt master."

34. What's the difference between brandy and cognac? All cognac is brandy, but not all brandy is cognac.

35. Brandy refers to any distilled spirit made from fermented fruit juice.

36. Cognac is a type of brandy from the Cognac region of France.

37. Strict rules govern where the grapes must be grown, and how the spirit is then produced.

38. Tarantula brandy is a popular drink in Cambodia.

39. The 1808 Rum Rebellion in Australia kicked out New South Wales governor William Bligh when he got in the way of the colony's rum business, among other things.

40. The main difference between scotch and whiskey is geographic. Ingredients and spellings can differ, too.

41. Scotch is whisky made in Scotland (and spelled without an "e"), while bourbon is whiskey made in the U.S.

42. Scotch is made mostly from malted barley, while bourbon is distilled from corn.

43. If you ask for a whisky in England, you'll get scotch. But in Ireland, you'll get Irish whiskey (they spell it with an "e").

44. After a Tennessee whiskey, like Jack Daniel's, for example, is distilled, it's filtered through sugar-maple charcoal.

31 Mush-ruminations

1. France was the first country to cultivate mushrooms, in the mid-17th century.

2. From there, the practice spread to England and made its way to the United States in the 19th century.

3. In 1891, New Yorker William Falconer published *Mushrooms: How to Grow Them—A Practical Treatise on Mushroom Culture for Profit and Pleasure*, the first book on the subject.

4. In North America alone, there are an estimated 10,000 species of mushrooms, only 250 of which are known to be edible.

5. A mushroom is a fungus (from the Greek word *sphongos*, meaning "sponge").

6. A fungus differs from a plant in that it has no chlorophyll, produces spores instead of seeds, and survives by feeding off other organic matter.

7. Mushrooms are related to yeast, mold, and mildew, which are also members of the "fungus" class.

8. There are approximately 1.5 million species of fungi, compared with 250,000 species of flowering plants.

9. An expert in mushrooms and other fungi is called a mycologist—from the Greek word *mykes*, meaning "fungus."

10. A mycophile is a person whose hobby is to hunt edible wild mushrooms.

11. Ancient Egyptians believed mushrooms were the plant of immortality.

12. Pharaohs decreed them a royal food and forbade commoners to even touch them.

13. White agaricus (a.k.a. "button") mushrooms are by far the most popular, accounting for more than 90 percent of mushrooms bought in the United States each year.

14. Brown agaricus mushrooms include cremini and portobellos, though they're really the same thing: Portobellos are just mature cremini.

15. Cultivated mushrooms are agaricus mushrooms grown on farms.

16. Exotics are any farmed mushroom other than agaricus (think shiitake, maitake, oyster).

17. Wild mushrooms are harvested wherever they grow naturally—in forests, near riverbanks, even in your backyard.

18. Many edible mushrooms have poisonous look-alikes in the wild.

19. For example, the dangerous "yellow stainer" closely resembles the popular white agaricus mushroom.

20. "Toadstool" is the term often used to refer to poisonous fungi.

21. In the wild, mushroom spores are spread by wind.

22. On mushroom farms, spores are collected in a laboratory and then used to inoculate grains to create "spawn," a mushroom farmer's equivalent of seeds.

23. A mature mushroom will drop as many as 16 billion spores.

24. Mushroom spores are so tiny that 2,500 arranged end-to-end would measure only an inch in length.

25. Mushroom farmers plant the spawn in trays of pasteurized compost, a growing medium consisting of straw, corncobs, nitrogen supplements, and other organic matter.

26. The process of cultivating mushrooms—from preparing the compost in which they grow to shipping the crop to markets—takes about four months.

27. The small town of Kennett Square, Pennsylvania, calls itself the Mushroom Capital of the World—producing more than 51 percent of the nation's supply.

28. September is National Mushroom Month.

29. One serving of button mushrooms (about five) has only 20 calories and no fat.

30. Mushrooms provide such key nutrients as B vitamins, copper, selenium, and potassium.

31. Some experts say the taste of mushrooms belongs to a "fifth flavor"—beyond sweet, sour, salty, and bitter—known as umami, from the Japanese word meaning "delicious."

38 Revolting Food Facts

1. Baby mice wine, which is made by preserving newborn mice in a bottle of rice wine, is a traditional health tonic from Korea said to aid the rejuvenation of one's vital organs.

2. Ever get a hankering for soft-boiled duck embryos? Balut are duck eggs that have been incubated for 15 to 20 days (a duckling takes 28 days to hatch) and then boiled.

3. The egg is then consumed—both the runny yolk and the beaky, feathery, veiny duck fetus. Balut are eaten in the Philippines, Cambodia, and Vietnam.

4. Black pudding, eaten in Britain and Ireland, is congealed pig

blood that's been cooked with oatmeal and formed into a small disk. It tastes like a thick, rich, beef pound cake.

5. The Sardinian delicacy casu marzu is a hard sheep's milk cheese infested with *Piophila casei*, the "cheese fly."

6. The larvae produce enzymes that break down the cheese into a tangy goo, which Sardinians dive into and enjoy, larvae and all.

7. Surströmming, primarily a seasonal dish in northern Sweden, is rotten fermented herring. Even the Swedes rarely open a can of it indoors.

8. History shows that people will make alcohol from any ingredients available. That includes bananas, mashed with one's bare feet and buried in a cask. The result is pombe, an east African form of beer.

9. Cobra heart delivers precisely what it promises: a beating cobra heart, sometimes accompanied by a cobra kidney and chased by a slug of cobra blood.

10. Preparations of the Vietnamese delicacy involve a large blade and a live cobra.

11. Escamoles are the eggs, or larvae, of the giant venomous black *Liometopum* ant.

12. This savory Mexican chow has the consistency of cottage cheese and a surprisingly buttery and nutty flavor.

13. The key ingredient of bird's nest soup is the saliva-rich nest of the cave swiftlet, a swallow that lives on cave walls in Southeast Asia.

14. Hákarl, an Icelandic dish dating back to the Vikings, is putrefied shark meat.

15. Traditionally, it has been prepared by burying a side of shark in gravel for three months or more.

16. Nowadays, hákarl might be boiled in several changes of water or soaked in a large vat filled with brine and then cured in the open air for two months.

17. This is done to purge the shark meat of urine and trimethyl-amine oxide.

18. Durian is a football-size fruit with spines.

19. It smells like unwashed socks but tastes sweet. Imagine eating vanilla pudding while trying not to inhale.

20. Lutefisk is a traditional Scandinavian dish made by steeping pieces of cod in lye solution.

21. The result is translucent and gelatinous, stinks to high heaven, and corrodes metal kitchenware.

22. Enjoy lutefisk covered with pork drippings, white sauce, or melted butter, with potatoes and Norwegian flatbread on the side.

23. Kava is the social lubricant of many island nations. Take a pepper shrub root (*piper methysticum*), and get someone to chew or grind it into a pulp. Mix with water and enjoy.

24. Pacha is a sheep's head stewed, boiled, or otherwise slow-cooked for five to six hours together with the sheep's intestines, stomach, and feet. Yum!

25. Vegemite, made from leftover brewers' yeast extract, looks like chocolate spread, smells like B vitamins, and tastes overwhelmingly salty.

26. Australians love Vegemite on sandwiches or baked in meatloaf.

27. When you're a Mongolian nomad and there are no taverns, you're happy to settle for fermented mare's milk, airag.

28. It takes only a couple of days to ferment and turns out lightly carbonated.

29. Spiders are popular fare in parts of Cambodia, especially in the town of Skuon.

30. Tarantulas are sold on the streets and are said to be very good fried with salt, pepper, and garlic.

31. If you find yourself hungry as you hustle through London, grab a jellied eel from a street vendor.

32. It tastes like pickled herring with a note of vinegar, salt, and pimiento, all packed in gelatin.

33. Drink enough scotch, and you'll eventually get so hungry you'll eat haggis—sheep innards mixed with oatmeal and boiled in the sheep's stomach.

34. Some tribespeople in Papua New Guinea consider Sago beetle grubs delicious.

35. A specialty of Newfoundland, seal flipper pie is made from the chewy cartilage-rich flippers of seals, usually cooked in fatback with root vegetables and sealed in a flaky pastry crust or topped with dumplings.

36. Eating improperly prepared pufferfish can result in sudden death. As such, Japan requires extensive training and apprenticeship, as well as special licensing, for the chefs preparing this highly sought-after delicacy.

37. Menudo is basically cow-stomach soup.

38. If you can tolerate the slimy, rubbery tripe chunks, the soup itself tastes fine. It's often served in Mexico for breakfast to cure a hangover.

51 Popcorn Facts

1. Popcorn's scientific name is *zea mays everta.*

2. It is the only type of corn that will pop.

3. People have been enjoying popcorn for thousands of years.

4. The first evidence of popcorn has been radiocarbon-dated as 6,700 years old (c. 4700 B.C.), based on macrofossil cobs unearthed between 2007 and 2011 at archaeological sites on the northern coast of Peru.

5. It is believed that the Wampanoag American Indian tribe brought popcorn to the Pilgrims for the first Thanksgiving in Plymouth, Massachusetts.

6. Traditionally, American Indian tribes flavored popcorn with dried herbs and spices, possibly even chili.

7. They also made popcorn into soup and beer, and made popcorn headdresses and corsages.

8. Some American Indian tribes believed that a spirit lived inside each kernel of popcorn.

9. The spirits wouldn't usually bother humans, but if their home was heated, they would jump around, getting angrier and angrier, until eventually they would burst out with a pop.

10. Christopher Columbus allegedly introduced popcorn to the Europeans in the late 15th century.

11. Charles Cretors invented the first commercial popcorn machine in Chicago in 1885.

12. The business he founded still manufactures popcorn machines and other specialty equipment.

13. American vendors began selling popcorn at carnivals in the late 19th century.

14. Movie theater owners initially feared that popcorn would distract their patrons.

15. It took a few years for them to realize that popcorn could be a way to increase revenues, and popcorn has been served in movie theaters since 1912.

16. Nowadays, many movie theaters make a greater profit from popcorn than they do from ticket sales.

17. Popcorn also makes moviegoers thirsty and more likely to buy expensive drinks.

18. Each popcorn kernel contains a small amount of moisture.

19. As the kernel is heated, this water turns to steam.

20. The popcorn kernel's shell is not water-permeable, so the steam cannot escape and pressure builds up until the kernel finally explodes, turning inside out.

21. On average, a kernel will pop when it reaches a temperature of 347°F (175°C).

22. Unpopped kernels are called "old maids" or "spinsters."

23. There are two possible explanations for old maids.

24. The first is that they didn't contain sufficient moisture to create an explosion.

25. The second is that their outer coating (the hull) was damaged, so that steam escaped gradually, rather than with a pop.

26. Good popcorn should produce less than two percent old maids.

27. Ideally, the moisture content of popcorn should be around 13.5 percent, as this results in the fewest old maids.

28. Popcorn is naturally high in fiber, low in calories, and sodium-, sugar-, and fat-free, although oil is often added during preparation, and butter, sugar, and salt are popular toppings.

29. Americans consume 17 billion quarts of popped popcorn each year.

30. That's enough to fill the Empire State Building 18 times!

31. Nebraska produces more popcorn than any other state in the country—around 250 million pounds per year.

32. That's about a quarter of all the popcorn produced annually in the United States.

33. The employees at The Popcorn Factory in Lake Forest, Illinois, made a 3,423-pound popcorn ball in 2006.

34. It was the world's largest popcorn ball at the time.

35. You can visit the World's Largest Popcorn Ball in Sac City, Iowa.

36. Built in 2016, the popcorn ball weighs 9,370 pounds and stands taller than eight feet.

37. There are at least five contenders claiming to be the "Popcorn Capital of the World" due to the importance of popcorn to their local economies: Van Buren, Indiana; Marion, Ohio; Ridgway, Illinois; Schaller, Iowa; and North Loup, Nebraska.

38. Popped popcorn comes in two basic shapes: snowflake and mushroom.

39. Movie theaters prefer snowflake because it's bigger.

40. Confections such as caramel corn use mushroom because it won't crumble.

41. In 1948, ears of popcorn (the variety of corn grown for this particular purpose) were discovered in the Bat Cave in New Mexico.

42. They were around 5,600 years old.

43. Aztec Indians of the 16th century used garlands of popped maize as a decoration in ceremonial dances.

44. They were a symbol of goodwill and peace.

45. Colonial women poured sugar and cream on popcorn and served it for breakfast—likely the first "puffed" cereal!

46. Some colonists popped corn in a type of cage that revolved on an axle and was positioned over a fire.

47. During the Great Depression, popcorn was ubiquitous on city streets because it was an inexpensive way to stave off hunger.

48. According to the Popcorn Institute, popcorn is high in carbohydrates and has more protein and iron than potato chips, pretzels, and soda crackers.

49. If you made a trail of popcorn from New York City to Los Angeles, you would need more than 352,028,160 popped kernels!

50. The world's longest popcorn string measured 1,200 feet.

51. It was created by Gary Kohs, Laura Scaccia, and Dave Vandenbossche at Drake Park in Marine City, Michigan, in October 2016.

34 Fortune Cookie Facts

1. To most Americans, fortune cookies are synonymous with Chinese food, but they didn't originate in China.

2. It's extremely difficult to find fortune cookies in China.

3. This is because the fortune cookie actually traces its origins back to Japan, not China.

4. Fortune cookies are a mainstay in the United States, but they are also served in Britain, Italy, France, and Mexico.

5. More than three billion fortune cookies are made each year, the vast majority of them in the United States.

6. The recipe for fortune cookies is surprisingly simple.

7. All you need is flour, sugar, vanilla, and sesame oil.

8. Most manufacturers add other ingredients. Some use vegetable shortening or butter instead of sesame oil. Starch, eggs, and food coloring are also regularly used.

9. No one knows for sure who first introduced the fortune cookie to the United States, but two entrepreneurs in California are given credit.

10. According to one legend, Japanese immigrant Makoto Hagiwara, a landscape designer responsible for Golden Gate Park's Japanese Tea Garden, introduced the first U.S. fortune cookie in 1914 in San Francisco.

11. A second legend credits Chinese immigrant David Jung, a founder of a noodle company, with introducing the cookie in 1918 in Los Angeles.

12. According to the story, Jung was concerned about the number of poor people living on the streets, so he passed out free fortune cookies to them.

13. Each cookie contained an inspirational verse written for Jung by a Presbyterian minister.

14. In 1983, San Francisco's Court of Historical Review held a mock trial to determine whether Hagiwara or Jung should get credit for bringing fortune cookies to U.S. diners.

15. Not surprisingly, the judge ruled for San Francisco and Hagiwara.

16. A piece of evidence that surfaced during the trial was a fortune saying, "S.F. judge who rules for L.A. not very smart cookie."

17. As far back as the 19th century, a cookie very similar in appearance to the modern fortune cookie was made in Kyoto, Japan.

18. A woodblock image from 1878 shows what seems to be a Japanese street vendor grilling fortune cookies.

19. The Japanese version of the cookie is a bit larger and darker than the American fortune cookie.

20. Their batter contains miso paste and sesame rather than vanilla and butter.

21. The fortunes were never put inside the cookies either.

22. Instead, they were tucked into the fold of the fortune cookie on the outside.

23. This kind of cookie is called *tsujiura senbei*.

24. Fortune cookies became common in Chinese restaurants after World War II.

24. Edward Louie, owner of the Lotus Fortune Cookie Company in San Francisco, invented a machine in 1974 that could insert the fortune and fold the cookie.

25. In 1980, Yong Lee created an upgraded, fully-automated fortune-cookie machine.

26. Wonton Food, Inc., in New York, is the largest producer of fortune cookies in the United States.

27. The factory churns out 4.5 million cookies per day.

28. The company boasts a database of 15,000 possible fortunes.

29. Company officials say that only about 25 percent of these fortunes are used at any given time.

30. Some fortune cookies don't contain fortunes at all. Crack one open, and you'll often find lucky lottery numbers or a philosophical message.

31. Fortune cookies today come in a wide variety of flavors.

32. Some are covered in chocolate or caramel. Many bakeries also sell fortune cookies decorated for Christmas, Valentine's Day, and other holidays.

33. Unless you really love them, you won't gain too much weight eating fortune cookies.

34. The average fortune cookie contains about 30 calories and no fat.

29 Wacky Jelly Belly Flavors

1. Pickle
2. Black Pepper
3. Booger
4. Dirt
5. Earthworm
6. Earwax
7. Sausage
8. Rotten Egg
9. Soap
10. Vomit
11. Sardine
12. Grass
13. Skunk Spray
14. Bacon
15. Baby Wipes
16. Pencil Shavings
17. Toothpaste
18. Moldy Cheese
19. Buttered Popcorn
20. Dr. Pepper
21. Jalapeño
22. Margarita
23. Spinach
24. Cappuccino
25. Peanut Butter
26. Café Latte
27. 7UP
28. Pomegranate
29. Baked Beans

13 Vodka Facts

1. This colorless alcohol hails from Russia, where its original name is *zhiznennaia voda*, or "water of life."

2. Vodka may be made with vegetables (such as potatoes or beets) or grains (barley, wheat, rye, or corn).

3. Grain vodka is considered higher quality.

4. Early vodkas were crudely manufactured and tasted pretty bad.

5. It was common to mix vodka with herbs, spices, or honey to mask the harsh, offensive taste.

6. Today's distillation processes create clean-tasting vodkas, many of which are enhanced with vanilla or fruit flavors.

7. In 1540, Ivan the Terrible stopped fighting long enough to establish the country's first vodka monopoly.

8. In the late 17th century, Peter the Great explored improved methods of distillation and means of export.

9. During the reign of Peter the Great, it was customary that foreign ambassadors visiting Russia consume a liter and a half of vodka.

10. Lightweight ambassadors began to enlist stand-ins to drink so that the official could discuss important matters with a clear head.

11. Since vodka is an alcohol, consider using it to clean razors—the liquid disinfects the blade and prevents rusting.

12. Vodka can help remove bandages.

13. Saturating the bandage with vodka will dissolve the adhesive, making removal of the bandage painless.

34 Chocolatey Facts

1. Archaeologists have found cocoa residue in pottery unearthed in Honduras from 1400 B.C.

2. Four hundred cocoa beans go into each pound of chocolate.

3. A cacao tree is between four and five years old before it grows its first beans.

4. It takes about a year for a single cacao tree to produce enough beans to make 10 Hershey bars.

5. Cacao trees can live for hundreds of years, but they produce quality cocoa beans for only 25 years.

6. The Maya used cocoa beans as currency.

7. Chocolate helped researchers decode the Maya's written language.

8. The word *ka-ka-w*, or cacao, was written on containers of chocolate found buried with Maya dignitaries.

9. The cacao tree's scientific name is *Theobroma cacao*, or "food of the gods."

10. The cacao tree is in the same family as okra and cotton.

11. Cacao trees are native to Central and South America, but most cacao now comes from West Africa, primarily Côte d'Ivoire.

12. Chocolate was only enjoyed as a beverage for centuries.

13. The Aztec mixed cacao seeds with chilies to make a frothy, spicy drink.

14. Aztec Emperor Montezuma is said to have drunk 50 goblets of chocolate, flavored with chili peppers, every day.

15. Explorer Hernán Cortés brought the Aztec drink back to Spain.

16. French and Spanish nobles added cinnamon and cane sugar to sweeten the bitter beverage.

17. In 1847, Joseph Fry & Sons in England made the first "eating" chocolate bar.

18. The ancient Olmec of present-day Mexico and Central America are believed to be the first people to use the cacao plant, as early as 1000 B.C.

19. A single fruit of the cacao tree contains between 20 and 60 cocoa seeds.

20. The "blood" running down the drain in *Psycho*'s shower scene was actually chocolate syrup.

21. If you're feeling down after watching a sad film, eating chocolate can improve your mood.

22. German chocolate cake isn't German.

23. It was named after an American named Sam German.

24. The familiar Hershey's chocolate bar was first sold in 1900.

25. The first Hershey's Kisses were introduced in 1907.

26. White chocolate technically isn't chocolate.

27. It's made with only cocoa butter, not cocoa powder, as real chocolate bars are.

28. People in Switzerland consume more chocolate than anyone else in the world, with an average resident going through nearly 20 pounds each year.

29. Chocolate-covered bacon is a popular snack at state fairs.

30. In 2007, a man used chocolate to help him make friends with some guards at a Belgian bank.

31. After gaining the guards' trust, the man calmly walked out of the bank with $28 million worth of stolen diamonds.

32. A study published in 2016 found people were more likely to make purchases if a store smelled like chocolate.

33. Some researchers have found that people rate the quality of a chocolate higher if they eat it while looking at a nice painting.

34. The Baby Ruth candy bar was not named for baseball slugger Babe Ruth but actually for President Grover Cleveland's daughter Ruth.

10 Food Origins

1. **Pizza:** Pizza originated in the South Italian region around Naples, where flat yeast-based bread topped with tomatoes was a local specialty. In 1889, baker Raffaele Esposito made pizzas for the visiting Italian King Umberto I and Queen Margherita. The queen's favorite pizza featured basil leaves over mozzarella cheese and tomatoes—a dish soon known as a Margherita pizza.

2. Sandwich: This lunchtime favorite likely dates back to the ancient Hebrews, who may have put meat and herbs between unleavened bread during Passover. But an 18th-century British noble, John Montagu, 4th Earl of Sandwich, gave it its name. The earl was a keen card player and commonly ate meat between pieces of bread to keep from getting the cards greasy.

3. Caesar Salad: Caesar salad isn't named after Julius or Augustus. It's the creation of Italian-born Mexican chef Caesar Cardini, who according to one story, whipped up the salad when faced with a shortage of ingredients for a Fourth of July celebration in 1924. Another tale attests that Cardini made it for a gourmet contest in Tijuana. Either way, the salad is worthy of his name!

4. Ranch Dressing: Ranch dressing actually did start out on a ranch—a dude ranch in Santa Barbara, California. Opened in 1954 by Steve and Gayle Henson, the "Hidden Valley Ranch" served a special house dressing that was so popular visitors came just to buy it.

5. Taco: The taco is one of the first "fusion" foods—as much a mix of cultures as the Mexican people today. The native Nahuatl people ate fish served in flat corn bread, but it was 16th-century Spanish explorers who gave the bread the name *tortilla* and began filling it with beef and chicken as well.

6. Frozen Dinners: Swanson's TV Brand Frozen Dinner was a popular prime-time meal in 1953, so it may come as a surprise that frozen dinners predated Swanson's by nearly a decade. William L. Maxson devised a prepackaged frozen meal for airplanes in 1944. Still, clever marketing and better distribution ensured that the TV Dinner from Swanson became the nationally recognized leader in frozen food.

7. Steak Tartare: The nomadic Tartar warrior tribe was known for eating raw meat, which was usually pressed underneath the saddle of a horse, but the raw meat dish today, which is usually chopped beef or horsemeat with a liberal amount of seasoning and spices, may take its name from the Italian word *tartari*, which means raw steak.

8. **Pasty:** The forerunner of the potpie was originally cooked up as a lunch meal by Cornish miners, who were unable to return to the surface to eat. The pastry crust allowed the miners the ability to eat the contents, which typically included meat, vegetables, and gravy, and then discard the shell.

9. **Thousand Island Dressing:** It turns out thousand island dressing owes its name to the "Thousand Islands" region in upstate New York, which actually boasts 1,800 small islands. Back in the early 1900s, actress May Irwin was served this scrumptious salad dressing while at a dinner party in the Thousand Islands area. She named the dressing and spread the word.

10. **Waffles:** During the Middle Ages, a thin crisp cake was baked between wafer irons. Oftentimes, the irons included designs that helped advertise the kitchen that produced the waffle. Early waffles were made of barley and oats, but by the 18th century, the ingredients changed to the modern version of leaven flour.

12 Quotes About Food

1. *"He was a bold man that first ate an oyster."* –JONATHAN SWIFT

2. *"So long as you have food in your mouth, you have solved all questions for the time being."* –FRANZ KAFKA

3. *"If you are ever at a loss to support a flagging conversation, introduce the subject of eating."* –LEIGH HUNT

4. *"Our minds are like our stomachs; they are whetted by the change of their food, and variety supplies both with fresh appetite."* –QUINTILIAN

5. *"One should eat to live, not live to eat."* –MOLIÈRE

6. *"The secret of success in life is to eat what you like and let the food fight it out inside."* –MARK TWAIN

7. *"My weaknesses have always been food and men— in that order."* –DOLLY PARTON

8. *"There is no sincerer love than the love of food."* –GEORGE BERNARD SHAW

9. *"Food is symbolic of love when words are inadequate."* –ALAN D. WOLFELT

10. *"All happiness depends on a leisurely breakfast."* –JOHN GUNTHER

11. *"An Englishman teaching an American about food is like the blind leading the one-eyed."* –A. J. Liebling

12. *"Anything is good if it's made of chocolate."* –Jo Brand

21 Canadian Foods

1. Arctic char is the northernmost freshwater fish in North America, caught commercially since the 1940s. Char is a little like salmon in color and texture, but its unique flavor elevates it to a delicacy.

2. Back bacon, also called Canadian bacon in the U.S., has less fat than other kinds of bacon. Its taste and texture are similar to ham. Peameal bacon is cured back bacon that's coated with ground yellow peas.

3. Bakeapples are also referred to as baked-apple berries, chicoute, and cloudberries. Found mostly in the provinces of Nova Scotia and Newfoundland, they taste—surprise—like a baked apple. These can be eaten raw or used in pies and jams.

4. To ward off the cold, Newfoundlanders have long boiled up bangbelly—a pudding of flour, rice, raisins, pork, spices, molasses, and sometimes seal fat. The result is comparable to bread pudding and is commonly served at Christmastime.

5. No aquatic rodents are harmed in the making of beavertails, an Ottawa specialty. These fried, flat pastries, shaped like a beaver's tail, are similar to that carnival staple called elephant ears. Top with sugar, cinnamon, fruit, even cream cheese and salmon.

6. Similar to the Bloody Mary, the Bloody Caesar cocktail is popular all over Canada. It's made of vodka, tomato-clam juice, Worcestershire sauce, and hot sauce, served on the rocks in a glass rimmed with celery salt.

7. Along with so many other aspects of their culture, the Scots brought butter tart to Canada. These are little pecan pies without the pecans, perhaps with chocolate chips, raisins, or nuts. You haven't had Canadian cuisine until you've had a butter tart.

8. Cipaille, a layered, spiced-meat-and-potato pie, is most popular in Quebec. Look for it on menus as "sea pie" in Ontario, not for any aquatic additives but because that's exactly how the French word is pronounced in English.

9. Break your morning butter-and-jam routine and have some cretons instead! This Quebecois tradition is a seasoned pork-and-onion pâté often spread on toast.

10. Dulse is a tasty, nutritious, protein-packed seaweed that washes up on the shores of Atlantic Canada. It's used in cooking much the same ways one uses onions: chopped, sautéed, and added to everything from omelets to bread dough.

11. Figgy duff is a traditional bag pudding from Newfoundland. Common ingredients include breadcrumbs, raisins, brown sugar, molasses, butter, flour, and spices.

12. Jigg's dinner is a traditional meal from Newfoundland that incorporates salt beef, cabbage, boiled potatoes, carrots, turnips, and homemade pease pudding.

13. Many consider the Prince Edward Island delicacies called malpeques the world's tastiest oysters, harvested with great care by workers who rake them out of the mud by hand. If you can find them, the "pride of P.E.I." will cost you dearly.

14. The Nanaimo bar is a chocolate bar layered with nuts, butter-cream, and sometimes peanut butter or coconut. Nanaimo bars are well liked throughout Canada and in bordering U.S. regions (especially around Seattle).

15. Close to 90 percent of Canada's maple syrup comes from Quebec, and Canada is the world's largest producer of this sweet, sticky pancake topping.

16. Moosehunters are what Canadians call molasses cookies.

17. Nougabricot is a Quebecois preserve consisting of apricots, almonds, and pistachios.

18. Canada's large waves of Slavic immigration have brought pierogi to the True North. They're small dumplings with a variety of

fillings, including cheese, meat, potatoes, mushrooms, cabbage, and more. Top with sour cream and onions.

19. Persians are sort of a cross between a large cinnamon bun and a doughnut, topped with strawberry icing; unique to Thunder Bay, Ontario.

20. Acadia, the Cajuns' ancestral homeland, loves its buckwheat pancakes. But the greenish-yellow griddle cakes known as ployes contain no milk or eggs, so they're not actually pancakes. Eat ployes with berries, whipped cream, cretons, or maple syrup.

21. Dump gravy and cheese curds on French fries: *Voilà la poutine*! Quebec is the homeland of poutine, but you can get it all over the nation.

40 Coffee Facts

1. Coffee is one of the most popular beverages in the world, with more than 400 billion cups consumed each year.

2. Coffee is brewed from roasted beans of the plant species *Coffea*.

3. Coffee beans are technically fruit pits, not beans.

4. Coffee is believed to originate in Ethiopia.

5. According to legend, a goat herder named Kaldi noticed his goats became energetic after eating the berries from a certain tree.

6. He reported his findings to the abbot of a local monastery, who made a drink with the berries.

7. Coffee cultivation and trade began on the Arabian Peninsula.

8. By the 16th century, coffee houses were common in Persia, Egypt, Syria, and Turkey.

9. Brazil is the top coffee producing country, accounting for roughly 40 percent of the world's coffee supply.

10. The second-largest coffee producer is Vietnam, responsible for about 20 percent of global supply.

11. Around 25 million farmers worldwide depend on coffee crops for their economic livelihood.

12. After oil, coffee is the world's most valuable traded commodity.

13. Coffee contains caffeine, the stimulant that increases your metabolism.

14. Caffeine is the most popular drug in the world, and 90 percent of people in the U.S. consume it in some form every day.

15. It takes a day to fully eliminate caffeine from your system.

16. The effects of caffeine reach their peak around 30 to 60 minutes after consumption.

17. There are two main types of coffee: Arabica and Robusta.

18. Dark-roast coffee actually has less caffeine than coffee that's been lightly roasted.

19. The word coffee comes from Kaffa, a region in Ethiopia where coffee beans may have been discovered.

20. As early as the ninth century, people in the Ethiopian highlands were making a stout drink from ground coffee beans boiled in water.

21. By the 17th century, coffee had made its way to Europe and was becoming popular across the continent.

22. Dutch merchants shipped live coffee plants from the Yemeni port of Mocha to India and Indonesia, where they were grown on plantations to supply beans to Europe.

23. Some Europeans feared the new beverage, calling it the "bitter invention of Satan."

24. Pope Clement VIII (1536–1605) gave coffee papal approval after trying a cup.

25. By the middle of the 17th century, there were over 300 coffee houses in London.

26. In 1674, the Women's Petition Against Coffee claimed the

beverage was turning British men into "useless corpses" and proposed a ban on it for anyone under the age of 60.

27. Scandinavia boasts the highest per-capita coffee consumption in the world.

28. On average, people in Finland drink more than four cups of coffee a day.

29. The traditional Finnish way of brewing coffee is a variation on Turkish coffee where water and coffee grounds are brought just barely to a boil repeatedly.

30. Espresso contains three times more caffeine per ounce than standard brewed coffee.

31. The first webcam was set up to watch a coffeepot so computer scientists at Cambridge University could monitor the coffee situation without leaving their desks.

32. The world's most expensive coffee, Kopi Luwak, sells for $100 to $600 per pound.

33. The pricy coffee beans are collected from Indonesian civet droppings.

34. The civet (a small weasel-like critter) eats coffee cherries but can't digest the beans, so they pass through the animal's digestive tract and are handpicked by locals.

35. Germany banned coffee pods in all government buildings.

36. The first "instant coffee" appeared in Britain in 1771.

37. It was called a "coffee compound" and had a patent granted by the British government.

38. In 1971, a group of Seattle-based entrepreneurs opened a coffee shop called Starbucks.

39. Americans consume an average of 9.7 pounds of coffee per year, making the U.S. only the 25th biggest consumer of coffee worldwide on a per-person basis.

40. The average person in the U.S. consumes about three cups of coffee per day.

20 Top Coffee Consumers

Country—Pounds per Person per Year

1. Finland—26.45 lbs.
2. Norway—21.82 lbs.
3. Iceland—19.84 lbs.
4. Denmark—19.18 lbs.
5. Netherlands—18.52 lbs.
6. Sweden—18 lbs.
7. Switzerland—17.42 lbs.
8. Belgium—15 lbs.
9. Luxembourg—14.33 lbs.
10. Canada—14.33 lbs.
11. Bosnia and Herzegovina—13.67 lbs.
12. Austria—13.45 lbs.
13. Italy—13 lbs.
14. Brazil (tied)—12.79 lbs.
15. Slovenia (tied)—12.79 lbs.
16. Germany—12.13 lbs.
17. Greece (tied)—11.9 lbs.
18. France (tied)—11.9 lbs.
19. Croatia—11.24 lbs.
20. Cyprus—10.8 lbs.

17 Apple Facts

1. Apples are a member of the rose family.

2. There are approximately 7,500 varieties of apples grown around the world.

3. The crabapple is the only apple native to North America.

4. Apples will ripen six times faster if you leave them at room temperature rather than refrigerate them.

5. A peck of apples weighs 10.5 pounds.

6. A bushel of apples weighs 42 pounds, and will yield 20 to 24 quarts of applesauce.

7. Apple trees produce fruit four to five years after they're planted.

8. The science of apple growing is called pomology.

9. It takes about 36 apples to create one gallon of apple cider.

10. About 25 percent of an apple's volume is air; that's why they float.

11. Around two pounds of apples are needed to make one 9-inch pie.

12. The largest apple picked weighed three pounds.

13. A medium apple is about 80 calories.

14. A large-sized apple has about 130 calories.

15. Early American apple orchards produced very few apples because there were no honeybees.

16. Historical records indicate that colonies of honeybees were first shipped from England and landed in the colony of Virginia early in 1622.

17. Archeologists have found evidence that humans have been enjoying apples since at least 6500 B.C.

20 Pasta Translations

1. Tortellini = twists
2. Vermicelli = worms
3. Spaghetti = strings or twine
4. Farfalle = butterflies
5. Fettuccine = little ribbons
6. Fusilli = springs or rifles
7. Linguine = little tongues
8. Manicotti = muffs or sleeves
9. Mostaccioli = mustaches
10. Penne = pens or quills
11. Rotelle = little wheels
12. Ziti = bridegrooms
13. Campanelle = little bells
14. Ditali = thimbles
15. Conchiglie = shells
16. Lasagna = flat sheets
17. Barbina = little beards
18. Ditalini = small fingers
19. Reginette = little queens
20. Lumaconi = snails

48 Fast Food Facts

1. Consumption statistics show that Americans eat more fast food than any other country.

2. The U.S. is home to more than 200,000 fast food restaurants.

3. France ranks second in fast food consumption.

4. KFC's iconic founder, Colonel Sanders, was never in the military.

5. Kentucky Governor Ruby Laffoon named Harland Sanders an honorary colonel.

6. Kentucky Fried Chicken was the brainchild of Harland Sanders, who opened his first restaurant during the Great Depression in a gas station in Corbin, Kentucky.

7. In the 1930s, Sanders developed his secret recipe of 11 herbs and spices.

8. McDonald's was the most-valuable fast food brand in 2022, followed by Starbucks.

9. McDonald's was founded in the early 1940s by Dick and Mac McDonald as a barbecue drive-in.

10. The early McD's even offered carhop service.

11. French fries weren't introduced at Mickey D's until 1949.

12. Until then, potato chips had been offered instead.

13. McDonald's has its own university (of sorts).

14. Hamburger University opened in 1961.

15. There graduates receive "Bachelor of Hamburgerology" degrees.

16. Ronald McDonald first appeared in 1966 when McDonald's aired its first television commercial.

17. The Hamburglar, Grimace, and Mayor McCheese joined him five years later.

18. Insta-Burger King was founded in Jacksonville, Florida, in 1953.

19. James McLamore and David Edgerton purchased the company in 1954 and renamed it "Burger King."

20. The restaurant was based on an assembly line production system inspired by a visit to the McDonald brothers' hamburger stand.

21. Today, Burger King has more than 17,000 locations worldwide.

22. Australia is the only country in which Burger King does not operate under its own name.

23. Burger King is known as Hungry Jack's in Australia.

24. Burger King says there are 221,184 possible ways you could order its Whopper hamburger.

25. Taco Bell comes from a family name: Glen Bell started working on the chain in the late 1940s in San Bernardino—the same place where McDonald's was born.

26. His first venture was a hot dog stand called Bell's Drive-In.

27. By the early '50s, Bell started adding Mexican items onto the menu, eventually opening a secondary restaurant called Taco Tia.

28. The first actual Taco Bell didn't open until 1962.

29. In 1958, brothers Dan and Frank Carney of Wichita, Kansas, founded Pizza Hut.

30. With more than 18,000 restaurants, Pizza Hut is today the world's largest pizza chain in terms of locations.

31. Brothers Tom and James Monaghan purchased a pizza store in Ypsilanti, Michigan, called DomiNick's for $500 in 1960.

32. A year later, Tom became the restaurant's sole owner when James traded his share of the business for a Volkswagen Beetle.

33. Tom renamed the store Domino's Pizza and it soon became one of the world's leading pizza chains.

34. Today, Domino's is the world's top pizza chain by revenue.

35. Arby's entered the restaurant world in 1964.

36. The first location opened in Boardman, Ohio, featuring the roast beef sandwiches that are still the chain's signature item today.

37. The name Arby's actually represents the initials "R" and "B."

38. The letters stand for "Raffel Brothers," in homage to founders

Leroy and Forrest Raffel, although the company says many suspect it also stood for "roast beef."

39. Subway was founded in 1965 by 17-year-old college freshman Fred DeLuca and family friend Dr. Peter Buck.

40. If all the sandwiches made by Subway in a year were placed end to end, they would wrap around the world an estimated six times.

41. Subway's Italian B.M.T. is named after the Brooklyn Manhattan Transit.

42. Wendy's joined the fast food mix in 1969 when founder Dave Thomas opened his first restaurant in Columbus, Ohio.

43. Wendy's was named for Dave Thomas's daughter, Melinda Lou "Wendy" Thomas.

44. McDonald's Filet-O-Fish was originally developed specifically for Catholic customers abstaining from eating meat on Fridays.

45. Steve Ellis founded Chipotle in 1993 after graduating from culinary school in order to fund his dream of opening a fine-dining restaurant.

46. According to the book *Eat This, Not That*, Chick-fil-A is the healthiest overall fast food chain.

47. Chick-fil-A was the first fast-food chain in the United States to offer a menu completely free of trans fat.

48. Shaquille O'Neal owns some 155 Five Guys restaurants in North America.

15 Food Phrases

1. **There's No Such Thing as a Free Lunch:** In the 1840s, bars in the United States offered anyone buying a drink a "free lunch." It was really just a bunch of salty snacks that made customers so thirsty, they kept buying drinks.

2. **Egg on Your Face:** During slapstick comedies in the Victorian theater, actors made the fall guy look foolish by breaking eggs on his forehead.

3. **The Big Cheese:** In 1802, a cheesemaker delivered a 1,235-pound wheel of cheese to President Thomas Jefferson. Citizens referred to both the wheel and its important recipient as the "big cheese."

4. **In a Nutshell:** This saying, which indicates a lot of information conveyed succinctly, is so old that Cicero used it. He said that Homer's *Iliad* was penned in such small handwriting that all 24 books could fit "in a nutshell."

5. **Cool as a Cucumber:** Even on a warm day, a field cucumber stays about 20 degrees cooler than the outside air. Though scientists didn't prove this until 1970, the saying has been around since the early 18th century.

6. **Bring Home the Bacon:** The Dunmow Flitch Trials, an English tradition that started in 1104, challenged married couples to go one year without arguing. The winners took home a "flitch" (a side) of bacon.

7. **Spill the Beans:** In ancient Greece, the system for voting new members into a private club involved secretly placing colored beans into opaque jars. Prospective members never knew who voted for or against them—unless the beans were spilled.

8. **Gone to Pot:** Dating to pre-Elizabethan England, this phrase refers to hardened pieces of meat that were on the verge of spoiling—and good only for the stew pot.

9. **Easy as Pie:** Making a pie from scratch isn't easy; the phrase is a contraction of the late-19th-century phrase "easy as eating pie."

10. **Cook Your Goose:** In 1560, a town attacked by the Mad King of Sweden, Eric XIV, hung up a goose—a symbol of stupidity—in protest. The furious king threatened, "I'll cook your goose!"

11. **In a Pickle:** From the old Dutch phrase "de pikel zitten," which means to sit in a salt solution used for preserving pickles—sure to be an uncomfortable situation.

12. **With a Grain of Salt:** To take something "with a grain of salt" is to consider the subject with skepticism or suspicion. Salt was

once believed to have healing properties, and to eat or drink something with a grain of salt was to practice preventive medicine against potential poisoning or illness.

13. **Happy as a Clam:** The original phrase was "happy as a clam at high tide." Because clam diggers are able to gather clams only at low tide, the clams are much safer (and happier) when the tide is high and the water is too deep to wade into.

14. **Take the Cake:** The phrase originated at cakewalk contests, where individuals would parade and prance in a circle to the audience's delight. The person with the most imaginative swagger would take home first prize, which was always a cake.

15. **A Baker's Dozen:** It was once common that English medieval bakers would cut corners and dupe customers by making loaves that contained more air pockets than bread. By 1266, authorities enacted a law that required bakers to sell their bread by weight. To avoid paying the heavy penalties, bakers started adding an extra loaf for every dozen: hence the number 13.

33 Odd Beer Names

1. Pet Rock and a Moon Boot
2. Moose Drool
3. Tactical Nuclear Penguin
4. Barbarian Streisand
5. Tart Side of the Moon
6. Citra Ass Down
7. Pathological Lager
8. Hoptimus Prime
9. 668 the Neighbor of the Beast
10. Duck Duck Gooze
11. Leafer Madness
12. Little Sumpin' Sumpin'
13. Beasts of Bourbon
14. Pour Decisions
15. Purple Monkey Dishwasher
16. Smooth Hoperator
17. Ryes Against the Machine
18. For Richer or Porter
19. My Wife's Bitter
20. Brew Free or Die
21. Punk'in Drublic
22. Chug Norris
23. Peter, Pale and Mary
24. Haulin' Oats Stout

25. L.L. Cool Haze

26. Audrey Hopburn

27. Bockslider

28. Fraggle Bock

29. Pickleodeon

30. Monk in the Trunk

31. San Quentin Breakout Stout

32. Aroma Borealis Herbal Cream Ale

33. Tongue Buckler

27 Surprising Food Facts

1. Twist-ties (and plastic tabs) on bread sold in stores are color-coded to tell the date the bread was baked.

2. While poppy seeds do contain morphine, eating a poppy-seed muffin is unlikely to result in a positive drug-test.

3. The legal threshold for a positive drug-test result was raised in 1998 from 300 nanograms per milliliter to 2,000 nanograms per milliliter.

4. The shelf life of a Twinkie is about 25 days.

5. Jell-O is indeed made from animal parts.

6. Jell-O's gelatin is processed animal collagen (made from the bones, skin, and tendons of animals, primarily cows or pigs).

7. Contrary to popular belief, the tomato is a fruit, not a vegetable.

8. Until the 19th century, most Americans believed that the "love apple" (tomato) was poisonous.

9. To prove that tomatoes were perfectly safe, in 1820, Robert Gibbon stood on the courthouse steps in Salem, New Jersey, and ate an entire basket of tomatoes in front of the townspeople.

10. 7-Up soda originally counted lithium among its ingredients. "It's an UP thing," indeed.

11. Spicy foods don't cause ulcers. The bacteria *Helicobacter pylori* causes nearly all ulcers, except those triggered by medications such as aspirin.

12. Spicy foods can exacerbate ulcers, which may cause people to mistakenly blame the spicy food.

13. Ears of corn always have an even number of rows of kernels because the female flowers of corn kernels occur in pairs.

14. For every tasty corn kernel in the ear, there is a corresponding piece of silk—one of which will inevitably get caught between your teeth.

15. Despite its alleged sleep-promoting effects, drinking alcohol before bed may actually disrupt sleep and increase wakefulness.

16. The adhesive on a lickable U.S. postage stamp contains only one-tenth of a calorie.

17. Cockroaches like eating the adhesive on stamps—a ringing endorsement for buying the self-stick kind.

18. It takes 340 to 350 squirts from a cow's udders to make a gallon of milk.

19. Cooking carrots can make them more nutritious.

20. Boiling carrots increased carotenoids (antioxidants) by 14 percent.

21. Onions belong to the lily flower family. Also in this very pungent family: garlic, leeks, and chives.

22. It's a myth that celery takes more energy to digest than it provides in calories, according to the Mayo Clinic.

23. Head cheese isn't a cheese product at all—it's a jellied loaf of sausage.

24. Head cheese is made with meaty bits from the head of a cow, pig, or sheep.

25. It might include other edible animal parts, including feet, tongues, and hearts.

26. The holes in Swiss cheese are actually popped bubbles of carbon dioxide gas.

27. During the curing process, a special strain of bacteria called

Propionibacter shermani eats away at the lactic acid in the cheese curd, tooting carbon dioxide gas all the while.

25 Top Beer Consumers

Country—Liters per Person per Year

1. Czech Republic—181.7
2. Austria—107.8
3. Romania—100.3
4. Germany—99
5. Poland—97.7
6. Namibia—95.5
7. Ireland—92.9
8. Spain—88.8
9. Croatia—85.5
10. Latvia—81.4
11. Estonia—80.5
12. Slovenia—80
13. Netherlands—79.3
14. Bulgaria—78.7
15. Panama—78.3
16. Slovakia—76.1
17. Australia—75.1
18. Lithuania—74.4
19. Hungary—73.7
20. United States—72.7
21. Finland—72
22. Mexico—70.5
23. United Kingdom—70.3
24. Bosnia and Herzegovina—68.6
25. Gabon—67

17 Facts About Pizza

1. Antica Pizzeria Port'Alba, in Naples, Italy, is believed to be the world's first pizzeria.

2. First established as a stand in 1738, the restaurant opened in 1830.

3. Italian pizza was originally square-shaped (today, we call this "Sicilian cut").

4. It also used Romano cheese before switching to mozzarella.

5. The first pizzeria in the United States was opened by Gennaro Lombardi in 1895 in New York City.

6. Pepperoni is the most popular pizza topping in the U.S.

7. Americans consume more than 251 million pounds of pepperoni annually.

8. The Hawaiian pizza, topped with pineapple and ham, was invented in 1962 by restaurateur Sam Panopoulos in Canada.

9. Held annually in Las Vegas, the Pizza Expo is the largest pizza-only trade show.

10. Chicago deep dish pizza puts the sauce on top of the cheese.

11. October is officially designated National Pizza Month in the United States.

12. Philadelphia's Pizza Brain is half restaurant, half museum.

13. It offers a large collection of pizza-related items and pizza varieties from around the world.

14. YouTuber Airrack and Pizza Hut teamed up in January 2023 to make the world's largest pizza.

15. It covered a colossal 13,990 square feet and was made using 13,653 pounds of dough, 4,948 pounds of marinara sauce, over 8,800 pounds of cheese, and about 630,496 pieces of pepperoni.

16. Approximately three billion pizzas are sold in the United States every year, plus an additional one billion frozen pizzas.

17. The Celentano Brothers invented the frozen pizza in 1957.

Favorite Pizza Toppings from 10 Countries

1. **India**—pickled ginger, minced mutton, and paneer (a form of cottage cheese)

2. **Russia**—mockba (a combination of sardines, tuna, mackerel, salmon, and onions), red herring

3. **Brazil**—green peas

4. **Japan**—eel, squid, and Mayo Jaga (mayonnaise, potato, bacon)

5. France—flambé (bacon, onion, fresh cream)

6. Pakistan—curry

7. Australia—shrimp, pineapple, barbecue sauce

8. Costa Rica—coconut

9. Netherlands—"Double Dutch"—double meat, double cheese, double onion

10. United States—pepperoni, mushrooms, sausage, green pepper, onion, and extra cheese

18 Top Wine-Consuming Countries

Country—Gallons per Person per Year

1. Portugal—13.71

2. Italy—12.31

3. France—12.15

4. Switzerland—9.43

5. Austria—7.90

6. Australia—7.34

7. Argentina—7.29

8. Germany—7.26

9. Sweden—7.13

10. Netherlands—6.39

11. Spain—6.31

12. Romania—6.21

13. Czech Republic—6.13

14. United Kingdom—5.86

15. Canada—3.67

16. United States—3.22

17. Russia—2.27

18. South Africa—1.95

Chapter 5

Sports and Games

25 Early Sports & Games Facts

1. The ancient Egyptians loved playing a checkerboard game called senets.

2. Players moved pieces around a board determined by tossing numbered throwing sticks.

3. Chinese acrobatics were performed at least as far back as the Han Dynasty, around 200 B.C.

4. Central American Mayans developed their own team sports similar to lacrosse, football, and soccer.

5. Ancient Egypt had its own version of the Olympics, featuring gymnastics, javelin, running, swimming, and other events.

6. Native American tribes had many versions of, and names for, the modern sport of lacrosse.

7. The Cherokee called it "Little Brother of War," because it was good training for combat.

8. The Roman game of quoits, where a ring is tossed at a stake in the ground, is the forerunner of modern horseshoes.

9. An ancient festival devoted to cheese rolling is still held each May in Gloucestershire, England.

10. The festival dates back hundreds of years and involves pushing and shoving a large wheel of ripe Gloucestershire cheese downhill in a race to the bottom.

11. Skipping and jumping are natural movements of the body, and

the inclusion of a rope in these activities dates back at least as far as A.D. 1600, when Egyptian children jumped over vines.

12. Early Dutch settlers brought the game of jump rope to North America.

13. It flourished and evolved from a simple motion into the often-elaborate form prevalent today: double Dutch.

14. With two people turning two ropes simultaneously, a third, and then fourth, person jumps in, often reciting rhymes.

15. Ice skates were used as far back as the Bronze Age, although early skates were likely used for winter travel, not for fun.

16. Something like today's Hula-Hoop existed some 3,000 years ago, when children in ancient Egypt made hoops out of grape-vines and twirled them around their waists.

17. The first historical mention of the yo-yo dates to Greece in 500 B.C., but it was a man named Pedro Flores who brought the yo-yo to the United States from the Philippines in 1928.

18. The original name of the game volleyball was "mintonette."

19. It was created in 1895 when a YMCA gym teacher borrowed from basketball, tennis, and handball to create a new game.

20. Playing cards have been a tradition in China for millennia.

21. The markings on cards today date back to 14th-century France.

22. The four suits represent the major classes of society at that time.

23. Hearts translate to nobility and the church; the spear-tip shape of spades stands in for the military; clubs are clover, meant to represent rural peasantry; and diamonds are similar to the tiles then associated with retail shops and signified the middle class.

24. The classic territorial-capture board game Go originated in China some 4,000 years ago.

25. Legend has it that an emperor invented it to sharpen his son's thinking.

17 Ancient Olympic Games Facts

1. The Games took place at Olympia, a religious sanctuary.

2. The nearest town was a small one called Elis that was about 40 miles away.

3. Reportedly the Games began in 776 B.C.

4. At the time, Greece consisted of city-states, and Rome was not yet an empire.

5. Legend has that the King of Elis held the Games for the first time after an Oracle told him it would appease Zeus and end a plague.

6. Elis continued to host the Games as a multi-day event every few years.

7. Over time, training facilities and arenas were built to accommodate the events, which grew larger over time.

8. Footraces, equestrian, and combat events all took place.

9. Any male athlete could try out.

10. The Games were winner-take-all—no silver or bronze medals.

11. Married women were not permitted to compete nor attend.

12. One exception to this rule held that female racing-chariot owners were allowed.

13. A priestess of Demeter was also permitted to attend.

14. *Pankratists*—freestyle fighters—and other athletes including charioteers and boxers, literally risked their health and lives.

15. Injuries and deaths were not uncommon.

16. The audience flocked to the Games despite a lack of sanitation and the risk of sunstroke.

17. The Games were banned by Roman Emperor Theodosius I in A.D. 393 to promote Christianity.

32 Facts About Golf

1. King James II of Scotland banned golf in 1457 because golfers spent too much time on the game instead of improving their archery skills.

2. The first golf course in England was the Royal Blackheath Golf Club, founded in 1608.

3. The first golf course constructed in the United States was Oakhurst Links Golf Club in White Sulfur Springs, West Virginia, in 1884.

4. It was restored in 1994, but then closed to repair flood damage from 2016.

5. The oldest continuously existing golf club in the United States is St. Andrews Golf Club in New York.

6. It was formed by the Apple Tree Gang in 1888.

7. The first golf balls were made of wood.

8. Next were leather balls filled with goose feathers, followed by rubber balls, gutta-percha balls (made of a leathery substance from tropical trees), and then modern balls.

9. The first sudden-death playoff in a major championship was in 1979, when Fuzzy Zoeller beat Tom Watson and Ed Sneed in the Masters.

10. The longest sudden-death playoff in PGA Tour history was an 11-hole playoff between Cary Middlecoff and Lloyd Mangrum in the 1949 Motor City Open.

11. They were declared co-winners.

12. Beth Daniel is the oldest winner of the LPGA Tour.

13. She was 46 years, 8 months, and 29 days old when she won the 2003 BMO Financial Group Canadian Women's Open.

14. Most golf courses in Japan have two putting greens on every hole—one for summer and another for winter.

15. Tiger Woods has won eleven PGA Player of the Year Awards as of the 2022 season, the most won by any PGA Tour player.

16. Tom Watson is second, with six awards.

17. The youngest player to win a major championship was Tom Morris Jr. (known as "Young" Tom), who was 17 years old when he won the 1868 British Open.

18. Sam Snead and Tiger Woods top the list of PGA Tour winners with 82 victories each as of 2022.

19. They are followed by Jack Nicklaus with 73 wins, Ben Hogan with 64, and Arnold Palmer with 62.

20. To make the LPGA Hall of Fame, players must be active on the tour for ten years, have won an LPGA major championship, and secured a significant number of trophies, among other qualifications.

21. The shortest hole played in a major championship is the 106-yard, par-3 seventh hole at Pebble Beach Golf Links in California.

22. The longest hole in the United States is the 841-yard, par-6 12th hole at Meadows Farms Golf Club Course in Locust Grove, Virginia.

23. In 1899, Dr. George Grant, a New Jersey dentist, invented and patented the first wooden tee.

24. Before tees were invented, golfers elevated balls on a tiny wet-sand mound.

25. The oldest player to win a major championship was 48-year-old Julius Boros at the 1968 PGA Championship.

26. The youngest golfer to shoot a hole-in-one was five-year-old Coby Orr.

27. It happened in Littleton, Colorado, in 1975.

28. The chances of making two holes-in-one in a single round of golf are 1 in 67 million.

29. There are more than 11,000 golf courses in North America.

30. Playing on a downhill hole and with the help of a tailwind, Bob Mitera sank a 444-yard hole-in-one on the Miracle Hill Country Club course in Omaha, Nebraska.

31. When you score two shots under par, it's called an eagle.

32. When you shoot three under par on a single hole, it's called an albatross.

14 Great Sports Nicknames

1. One of hockey's tougher pugilists, 6'5" Stu "The Grim Reaper" Grimson earned more than 2,100 penalty minutes during his 729-game NHL career.

2. Basketball player "Pistol Pete" Maravich used his eerie peripheral vision to pull off hotdog passes and circus shots like one of the Harlem Globetrotters.

3. During Wayne Gretzky's heyday with the great Edmonton teams, Dave "Cementhead" Semenko had one job: Keep Wayne safe.

4. Football player Dick "Night Train" Lane reportedly got the odd nickname from associating with fellow Hall-of-Famer Tom Fears, who constantly played the record *Night Train* on his phonograph.

5. Basketball player Damon Stoudamire stood only 5'10". Between this and a tattoo of "Mighty Mouse" on his arm, the nickname "Mighty Mouse" was inevitable.

6. Football player for the Chicago Bears and the Philadelphia Eagles, William "The Refrigerator" Perry got attention for Chicago coach Mike Ditka's willingness to use the 326-pound "Fridge" at fullback on goal-line plays.

7. The last pitcher legally allowed to throw the spitball under the grandfather clause when baseball outlawed ball-doctoring, Burleigh "Ol' Stubblebeard" Grimes always showed up to the diamond with a faceful of scruffy whiskers.

8. Basketball player Darrell "Dr. Dunkenstein" Griffith of the Utah Jazz was only 6'4" but could jump as though grafted to a pogo stick.

9. Baseball player Mike "The Human Rain Delay" Hargrove got his nickname by fooling around in the batter's box: He would adjust his helmet, adjust his batting glove, pull on his sleeves, and wipe his hands on his pants before every pitch.

10. Pepper "The Wild Horse of the Osage" Martin was a player for the St. Louis Cardinals in the 1930s and '40s. His rather wild, free-spirited base-running got him the nickname, though it also probably referred to his love of practical jokes. "Pepper" was a nickname too: he was born Johnny Leonard Roosevelt Martin.

11. Basketball star Charles Barkley's nicknames included "The Round Mound of Rebound."

12. Norwegian biathlete Ole Einar Bjorndalen is known as the "King of Biathlon," which makes sense. He's also known as "The Cannibal," leading to a number of headlines about him being "hungry for success" and (after a defeat) "toothless."

13. With the last name "Crawley," it was probably inevitable that British Cricketeer John Crawley would be dubbed "Creepy."

14. Speaking of inevitable: Sprinter Usain Bolt, of course, is known as "Lightning Bolt."

17 Facts About Curling

1. Curling began in Scotland in the 1500s, played with smooth river-worn stones.

2. The game is also called "The Roaring Game" because of the noise the rocks make on the ice.

3. The Royal Montreal Curling Club began in 1807.

4. In 1927, Canada held its first national curling championship.

5. Curling was played as a demonstration sport in the Olympic Games in 1932, 1988, and 1992, and officially added in 1998.

6. A curling tournament in called a bonspiel.

7. A "cashspiel" is slang for a tournament with large prizes.

8. The standard curling rink measures 146 feet by 15 feet.

9. At each end are 12-foot-wide concentric rings called houses, the center of which is the button.

10. There are four curlers on a team.

11. Each shoves two rocks in an effort to get as close to the button as possible.

12. The players with brooms aren't trying to keep the ice clear of crud.

13. The team skip (captain) determines strategy and advises the players using the brooms in the fine art of sweeping to guide the stone with surprising precision.

14. One particular arrangement of rocks in the house is called a Christmas Tree for its shape.

15. The hog line is a blue line in roughly the same place as a hockey blue line.

16. One must let go of the rock before crossing the near hog line—and the rock must cross the far hog line—or it's hogged (removed from play).

17. Good sportsmanship is prized, and players are even expected to call their own fouls.

13 Sports Salaries Facts

1. In the National Basketball Association's first season, 1946–47, the top-paid player was Detroit's Tom King, who made $16,500.

2. King also acted as the team's publicity manager and business director.

3. Players who ran the floor in the first professional basketball league, which was formed in 1898, were paid $2.50 for home games and $1.25 for road matches.

4. As a young man, legendary coach Vince Lombardi coached football, basketball, and baseball at Cecilia High School in Englewood, New Jersey, while also teaching Latin, algebra, physics, and chemistry.

5. He did all of this for $1,700 a year.

6. Hank Greenberg was the first baseball player to earn $100,000 a year in 1947.

7. In 1930, Babe Ruth signed a contract that would pay him $80,000 a year—a shocking sum that earned him more than then-President Herbert Hoover.

8. Ruth famously quoted, "Why not? I had a better year than he did."

9. In 1975, the late Brazilian soccer legend Pelé joined the North American Soccer League team the New York Cosmos for a record salary of $1.4 million per year.

10. In 2012, then-undefeated boxer Floyd Mayweather was ranked by *Forbes* as the world's highest-paid athlete.

11. He competed in two pay-per-view fights, earning $85 million in the process.

12. Earl Anthony was the first man to accumulate $1 million in career earnings as a professional bowler.

13. In 2017, *Forbes* produced a list of the 100 highest-paid athletes. The only woman on the list was Serena Williams, who made about $27 million that year.

24 Early Football Facts

1. The University of Toronto played the first documented game of something rugby-like that was potentially the first football game in the world in 1861.

2. In 1869, Princeton traveled to Rutgers for a rousing game of "soccer football."

3. The field was 120 yards long by 75 yards wide, about 25 percent longer and wider than the modern field.

4. It played more like soccer than modern football, and with 25 players on a side, the field was a crowded place. Rutgers prevailed 6–4.

5. In 1892, desperate to beat the Pittsburgh Athletic Club team, Allegheny Athletic Association leaders created the professional football player by hiring William "Pudge" Heffelfinger to play for their team.

6. Heffelfinger played a pivotal role in AAA's 4–0 victory.

7. The first Rose Bowl was played in 1902.

8. Michigan, having scored its regular schedule 501–0, drubbed Stanford 49–0.

9. In 1902, Charles Follis joined the Shelby Blues, a team in Ohio, becoming the first Black professional football player.

10. Follis later went to play baseball for the Cuban Giants, the first Black professional baseball club.

11. One of Follis's football teammates, Branch Rickey, also went on to a career in baseball. Rickey later signed Jackie Robinson to the Dodgers.

12. In 1905, disgusted at the mortality rate among college football players, Teddy Roosevelt told the Ivy League schools: "Fix this blood sport, or I'll ban it."

13. A number of changes followed, including a ban on mass plays.

14. The mortality rate went from 19 deaths in 1905 to 11 in the following year.

15. In 1911, in the first championship of the National Football League (NFL), the Pittsburgh Steelers defeated the Philadelphia Athletics by a score of 11 to 0.

16. In 1912, the Rules Committee determined that a touchdown is worth six points.

17. In 1921, fans heard the first commercially sponsored radio broadcast of a game, with University of Pittsburgh beating West Virginia 21–13.

18. In 1922, the American Professional Football Association became the National Football League (NFL).

19. The NFL didn't begin recording statistics until 1932.

20. The modern football took its current shape in the mid-1930s, after a couple of decades of gradual evolution from the egglike rugby ball.

21. In 1939, the Brooklyn Dodgers–Philadelphia Eagles game was the first to be beamed into the few New York homes that could afford TV sets.

22. The first coast-to-coast TV broadcast of an NFL game was in 1951 as the Los Angeles Rams faced the Cleveland Browns in the league championship game.

23. The American Football League (AFL), the NFL's rival, began play in 1960.

24. The two merged in 1970.

12 College Football Facts

1. Legendary Notre Dame football coach Knute Rockne won more than 88 percent of his games, with a 105–12–5 career.

2. Ohio State running back Archie Griffin was the first player to win the Heisman Trophy twice, in 1974 and 1975.

3. The first freshman to win the Heisman was Texas A&M quarterback Johnny Manziel in 2013.

4. Lafayette and Lehigh in Eastern Pennsylvania have been famous football rivals since 1884.

5. Professional basketball player Charlie Ward played football in college for Florida State.

6. The University of Nebraska–Lincoln sold out a game against Missouri in 1962—and just kept selling out.

7. The Cornhuskers didn't have an unsold seat for 375 games.

8. The Michigan Wolves have won more Division I games than any other team.

9. "Dotting the i" in the "Script Ohio" is one of the most

memorable traditions among marching bands nationwide and is considered to be a high honor.

10. Toledo won the first overtime game in major college football history in 1995, defeating Nevada in the Las Vegas Bowl.

11. The Colorado Buffaloes have been running onto the field behind their mascot "Ralphie," a real live buffalo, since 1967.

12. There have been six Ralphies, and despite the name, they have all been female.

29 Tennis Facts

1. The first reliable accounts of tennis come from tales of 11th-century French monks who played a game called *jeu de paume* ("palm game," that is, handball) off the walls or over a stretched rope.

2. The main item separating tennis from handball—a racket— evolved within these French monasteries.

3. Rackets had been used before that in ancient Greece, in a game called *sphairistike* and then in *tchigan*, played in Persia.

4. Henry VII popularized the game in England.

5. In 1529 Henry VII built a tennis court on the grounds of his Hampton Court palace.

6. It was rebuilt in 1625 and is now the oldest tennis court in existence.

7. Thirty cases of golf clubs and tennis rackets for A. G. Spalding's sporting goods company were being shipped by transatlantic mail on the *Titanic* when it sank.

8. That wasn't the only *Titanic* tennis tie: Known for his victories as a singles and doubles tennis player at the U.S. Open and Wimbledon during the 1910s and '20s, R. Norris Williams also gained fame as a survivor of the *Titanic* disaster.

9. Six competitors from four nations participated in the women's singles event at the 1900 Olympics, the second in the modern era.

10. The first female Olympic champion was Charlotte Cooper of Great Britain, who won the tennis singles and the mixed doubles.

11. For his habit of indulging temper tantrums, John McEnroe was once dubbed by *The New York Times* "the worst advertisement for our system of values since Al Capone."

12. Wimbledon drama! In 1995, tennis player Jeff Tarango disagreed vehemently with the chair umpire.

13. His wife Benedictine came onto court and slapped the official. The two then stormed off the court.

14. Drama in tennis is nothing new. While playing a game of tennis in 1606, the artist Caravaggio and his young opponent got into an argument.

15. Caravaggio stabbed and killed his opponent and immediately had to flee Rome.

16. Catgut (the name of the product which tennis racket strings are made of) is made from the intestines of mostly livestock animals, but not cats.

17. In 2002, *Sports Illustrated* ran a profile of a 17-year-old from Uzbekistan, Simonya Popova.

18. She was everything a rising tennis star would want to be, except for the fact that she didn't exist: The article was a spoof.

19. Between 1927 and 1932, four French tennis players who were dubbed "the Four Musketeers" won six consecutive Davis Cups, as well as numerous Wimbledon and French championships.

20. In 1973, Billie Jean King defeated Bobby Riggs in three straight sets in a tennis match billed as the "Battle of the Sexes."

21. Sir Elton John wrote a song for Billie Jean King, titled "Philadelphia Freedom," that was released in 1975.

22. According to most tennis historians, modern tennis dates back to the early 1870s, when the delightfully named Major Walter Clopton Wingfield devised a lawn game for the entertainment of party guests on his English country estate.

23. Wingfield (whose bust graces the Wimbledon Tennis Museum) based his game on an older form of tennis that long had been popular in France and England, called "real tennis."

24. Wingfield opted to borrow the counting system from earlier versions of tennis—in French, scoring mimicked the quarter-hours of the clock: 15–30–45.

25. For some unknown reason, 45 became 40.

26. A number of historians argue that Wingfield borrowed the terms for his new game from the older French version.

27. Hence, *l'oeuf* (meaning "egg") turned into "love" and *deux le jeu* ("to two the game") became "deuce."

28. Actor Brad Pitt played tennis in high school at Kickapoo High in Springfield, Missouri.

29. In 1957, tennis great Althea Gibson won Wimbledon, the first African American player to do so.

25 Facts About Jim Thorpe

1. Generally considered one of the finest athletes of all time, James Thorpe played baseball and football at a professional level.

2. He also competed in the Olympics as a runner in the decathlon and pentathlon events.

3. Born in 1887 or 1888 in what is now Oklahoma, Thorpe was Native American, a member of the Sac and Fox Nation.

4. His Native American name translates to "Bright Path."

5. As a teenager, he was coached in football by legendary Glenn "Pop" Warner.

6. Thorpe also competed in track and field events, baseball, and lacrosse as a teenager.

7. He led his team to victory in 1912 in the national collegiate championship.

8. He competed in track and field events in the 1912 Summer Olympics in Stockholm, Sweden.

9. Thorpe won gold in the pentathlon and decathlon.

10. He won eight of the 15 individual events that comprised the two competitions.

11. Thorpe was honored on his returned home with a ticker-tape parade on Broadway.

12. After the Olympics, Thorpe continued to play various sports.

13. In 1913, a story broke that Thorpe had played semi-professional baseball in 1909 and 1910.

14. Since he had not been an amateur, he was stripped of his Olympic medals.

15. Because the reports did not come out until a year later, Thorpe's disqualification was actually against the rulebook for the 1912 Olympics, which stated that protests had to be made within 30 days.

16. Because of this, and because it wasn't uncommon for college players to earn money playing during the summer (often using an alias), Thorpe's disqualification was perceived as unfair.

17. Thorpe's medals were reinstated in a ceremony in 1983 after a long campaign by people who felt he had been treated unfairly and that racism might have been part of the disqualification.

18. His children received his reinstated medals on his behalf.

19. Thorpe played professional baseball in both the major and minor leagues.

20. He also played for several different NFL teams.

21. He was part of a traveling basketball team that barnstormed the country in 1927 and 1928.

22. Thorpe appeared in several films, playing a football coach in *Always Kickin'* and an extra in Westerns.

23. Thorpe was the subject of a 1951 movie starring Burt Lancaster: *Jim Thorpe – All American*.

24. Thorpe struggled with financial issues and alcoholism later in life.

25. He died in 1953 at the age of 65.

21 Baseball Firsts

1. In 1882, Paul Hines became the first ballplayer to wear sunglasses on the field. They weren't corrective; he just didn't like having the sun in his eyes.

2. On July 4, 1939, Lou Gehrig was the first major-league player to have his number (4) retired.

3. On May 23, 1901, Nap Lajoie was the first player in baseball history to be intentionally walked with the bases loaded.

4. During a spring training game on March 7, 1941, Pee Wee Reese and Joe Medwick of the Brooklyn Dodgers were the first major-league players to wear plastic batting helmets.

5. The first night game in World Series history was Game 4 of the 1971 series, when Pittsburgh hosted Baltimore.

6. The cork center was added to the official baseball in 1910. Before that, the core of a baseball was made of rubber.

7. In 1957, Warren Spahn became the first left-handed pitcher to win a Cy Young Award.

8. In 1953, respected and innovative National League umpire Bill Klem was the first ump elected to baseball's Hall of Fame.

9. On May 17, 1939, the first televised baseball game took place. It was a college game between Princeton and Columbia broadcast on W2XBS. Princeton won 2–1.

10. The first All-Star Game broadcast on television took place on July 11, 1950.

11. On April 18, 1956, Ed Rommel was the first umpire to wear glasses in a regular-season game.

12. Hall of Fame player Frank Robinson was the first Black manager in the majors, but the Blue Jays' Cito Gaston was the first Black manager whose team won the World Series.

13. The first "Babe" in baseball was Babe Adams, who pitched from 1906 to 1926.

14. The first major-leaguer to hit a home run with the lights on was Babe Herman of the Reds in July 1935.

15. Babe Ruth hit his first career home run at the Polo Grounds against the New York Yankees on May 6, 1915.

16. The American League's first Most Valuable Player (MVP) Award was given to St. Louis Browns first baseman George Sisler in 1922. The second was given to Babe Ruth.

17. In 1928, the Hollywood Stars of the Pacific League became the first team to travel by air.

18. In 1981, Rollie Fingers of the Milwaukee Brewers became the first relief pitcher to win the American League MVP Award.

19. On April 23, 1952, Hoyt Wilhelm won his first game and hit his first home run, which became the only home run he ever hit in 1,070 games.

20. In 1960, Bill Veeck of the White Sox became the first owner to put players' names on the back of their uniforms.

21. In 1992, the Toronto Blue Jays became the first MLB team from a country other than the United States to win the World Series.

20 World Series Facts

1. The modern World Series was first played in 1903.

2. Before 1903, a number of 19th-century competitions were held between various leagues. These series were sometimes called the "World's Series."

3. In the 1903 series, the Boston Americans (the predecessors to the Red Sox) of the American League beat the Pittsburgh Pirates of the National League.

4. The World Series was not held in 1904 due to a boycott.

5. Other than that, the competition has been held each year except 1994, when there was a players' strike.

6. An earthquake interrupted Game 3 of the 1989 World Series between the San Francisco Giants and the Oakland Athletics.

7. Candlestick Park had to be evacuated and the game postponed.

8. The first trio of brothers ever to win World Series titles were the Molina brothers.

9. Bengie, the oldest, and Jose won championships as Los Angeles Angels teammates in 2002, and younger brother Yadier joined the club with the St. Louis Cardinals in 2006 and '11—all as catchers.

10. Sparky Anderson was the first manager to win World Series titles in both leagues.

11. He captured his first two championships with the Cincinnati Reds in the National League,.

12. Anderson then took the Detroit Tigers of the American League to the top in 1984.

13. With 18 World Series home runs, Mickey Mantle holds the record for most career home runs in World Series play.

14. He topped Babe Ruth's previous record by three.

15. Reggie Jackson earned the nickname "Mr. October" in part by hitting home runs on four straight swings of the bat in the 1977 World Series.

16. Yogi Berra played in the World Series a record 14 times.

17. In 2000, Derek Jeter became the first in Major League history named MVP of both the All-Star Game and World Series in the same year.

18. "Shoeless" Joe Jackson went 12-for-32 (.375) during the 1919 World Series—the one for which he and seven Chicago White Sox teammates were banned for life for their roles in "fixing" games.

19. The New York Yankees have made it to the World Series 40 times, winning 27 times.

20. The Seattle Mariners have never played in a World Series.

43 Basketball Facts

1. In 1936, basketball was added as an Olympic sport.

2. In 2004, the University of Connecticut won both the men's and the women's NCAA basketball championships.

3. It is the only school to ever do the double dip.

4. The silhouette of a dribbling basketball player on the National Basketball Association's logo is an image of former Los Angeles Lakers great Jerry West.

5. There are numerous rules on how to properly dribble a basketball, but bouncing the ball with such force that it bounds over the head of the ball handler is not illegal.

6. When Dr. James Naismith first drafted the rules for the game that eventually became known as basketball, the dribble wasn't an accepted method of moving the ball.

7. In the game's infancy, the ball was advanced from teammate to teammate through passing.

8. Naismith played the game only twice because he felt he committed too many fouls.

9. He believed this was because his extensive experience in wrestling and football made physical contact come naturally to him.

10. Although the term *dunk* was commonly used to describe the action of propelling a basketball through the hoop from above the rim, the phrase *slam dunk* was coined by the late Los Angeles Lakers announcer Francis "Chick" Hearn.

11. The colorful commentator also originated the terms *air ball* (ball that misses the entire backboard), *charity stripe* (foul line), and *finger roll* (rolling the ball off the fingertips).

12. George Mikan, a 6'10" pioneer in pro basketball who played for the Chicago American Gears and the Minneapolis Lakers through the 1940s and '50s, was probably the first man to use the dunk as an offensive weapon.

13. In 2002, Lisa Leslie became the first woman in the WNBA to dunk.

14. Wilt Chamberlain holds a number of NBA records, including the unapproachable mark of 100 points in a game.

15. Chamberlain also holds the record for 55 rebounds in a game.

16. During his 100-point game, Chamberlain converted 28 of his 32 free throw attempts.

17. He was a 51-percent career shooter from the free throw line, but hit an .875 rate in his signature game.

18. In 1993, Michael Jordan broke the hearts of fans by retiring from the Chicago Bulls.

19. He returned to basketball 17 months later.

20. The film *Hoosiers* was inspired by the Cinderella story of Milan High School—a school that, despite having only 73 male students, won the Indiana State Championship in 1954.

21. "Be tall, bask," is an anagram of basketball.

22. Before 1937, the basketball referee tossed a jump ball after every basket.

23. In 1895, the first American college basketball game was played between the Minnesota State School of Agriculture and Hamline College.

24. At the time, peach baskets were still used for hoops.

25. Minnesota State won the game, 9 to 3.

26. Michael Jordan launched a 17-foot winning shot in the 1982 NCAA Title Game that led North Carolina to victory over Georgetown.

27. Jordan was a freshman at the time.

28. In 1946, the Basketball Association of America (BAA) was founded.

29. The BAA merged with the National Basketball League in 1949 to form the National Basketball Association (NBA).

30. In 2013, Rick Pitino became the first coach to lead two different schools to men's NCAA Division I championships.

31. UCLA had an incredible 88-game winning streak in 1974, broken by Notre Dame.

32. Lewis Alcindor, before he became known as Kareem Abdul-Jabbar and a professional player, was honored as Most Outstanding Player in the NCAA Final Four a record three times, in 1967, '68, and '69.

33. Legendary coach Pat Summit won eight national titles during her tenure at Tennessee from 1974 to 2012.

34. Under Don Haskins in 1966, Texas Western College upset Kentucky to become the first NCAA men's basketball champions with an all-Black starting lineup.

35. Before they were famous, Magic Johnson and Larry Bird battled for the 1979 NCAA championship.

36. Michigan State's Magic Johnson prevailed over Indiana State's Larry Bird in what was then the highest-rated college basketball game in television history.

37. Michael Jordan led the Chicago Bulls to six NBA championships in the 1990s.

38. Jordan was MVP of all six of his NBA Finals appearances, a record.

39. LeBron James, as of 2022, has been named MVP four times.

40. Coach Phil Jackson has won a "three-peat" three different times!

41. Jackson led the Chicago Bulls to three straight titles twice in the 1990s, and then took the Los Angeles Lakers to three straight beginning with the 2000 championship.

42. On February 7, 2023, LeBron James scored his 38,388th point,

breaking Kareem Abdul-Jabbar's all-time scoring record of 38,387 points.

43. The NBA career scoring record had stood for nearly 39 years.

16 Facts to Bowl You Over

1. Some historians trace bowling's roots back to 3200 B.C., while others place its origin in Europe in the third century A.D.

2. Legend has it that King Edward III banned bowling after his good-for-nothing soldiers kept skipping archery practice to roll.

3. Well into the 19th century, American towns were outlawing bowling, largely because of the gambling that went along with it.

4. Originally, it was a game of ninepin set up in a diamond formation.

5. The German immigrants (who popularized the game in the 1800s and saw it outlawed because of the gambling) added another pin, changed the formation to a triangle, and satisfied the courts that it was a different game. Modern bowling was born.

6. Bowling was originally outdoor fun, played in the sun.

7. In 1895, the American Bowling Congress (which is now known as the United States Bowling Congress) was formed, and local and regional bowling clubs began proliferating.

8. The first National Bowling Championship was held in Chicago, Illinois in 1901.

9. Frank Brill won the individual bowling championship with a score of 648.

10. In 1997, University of Nebraska sophomore Jeremy Sonnenfeld became the first bowler to "knock 900" by rolling three perfect 300 games in a row in an official tournament.

11. The world's largest bowling alley is found in Japan.

12. It takes pretty bad luck for someone to lose their bowling ball in their practice backswing during a professional event such as the Dayton Classic.

13. Doing it four times in one event will bring notice, as it did Fran Wolf in 1976. The last one got her an ovation!

14. In 1930, Wisconsinite Jennie Hoverson became the first woman to bowl a perfect game in the history of league bowling. Her recognition came much later, however.

15. African Americans were not allowed into sanctioned league bowling until 1951, four years after Jackie Robinson broke the color barrier in baseball.

16. A bowling pin needs to tilt at least 7.5 degrees to fall over.

34 Facts About Soccer

1. Evidence of games resembling soccer has been found in cultures that date to the third century B.C.

2. The Romans brought their version of the sport along when they colonized what is now England and Ireland.

3. On March 20, 1976, while playing for Britain's Aston Villa soccer team, footballer Chris Nicholl scored every goal in a 2–2 draw against Leicester City, including two "own goals," or goals for the opposing team.

4. Like golf, soccer (football) was outlawed in Scotland in 1457 because the sports were dangerous, time-wasting nuisances that detracted from more important pursuits—like archery.

5. During World War II, as Commonwealth forces prepared to make a stand at ancient Thermopylae in 1941, they held a scheduled soccer game and continued it even when strafed by Stukas.

6. Soccer is considered the most globally popular sport.

7. In 2003, Adidas erected a billboard in Tokyo's Shibuya district featuring two live human beings playing "vertical soccer."

8. The soccer players were suspended from the billboard with bungee cords, as was the soccer ball.

9. The first radio broadcast of a soccer game was in 1927.

10. Teddy Wakelam provided play-by-play and commentary of a match between Arsenal F.C. and Sheffield United on BBC radio. The match was a 1–1 tie.

11. In 2004, during an Olympic qualifying match between Peru and Argentina, frenzied Peruvian fans grew irate when referees disallowed a goal for the home team.

12. The resulting riot left 300 people dead and 500 injured.

13. In 1958, an airplane carrying the Manchester United soccer club home from a European Cup playoff match crashed on takeoff at Munich airport.

14. Eight team members and 15 other passengers died in the crash.

15. In 1908, the soccer club Inter Milan was founded.

16. The club has played in the Italian top tier of professional soccer since its inception, winning the league title more than 17 times.

17. Founded in 1904, the Fédération internationale de football association (FIFA) is the international governing body of association soccer.

18. FIFA oversees international competition and its membership now includes 211 national associations.

19. The first FIFA World Cup was held in Uruguay in 1930.

20. Uruguay won the title match over Argentina, 4–2, in front of a crowd of more than 68,000 spectators.

21. Sheffield F.C. was founded in 1857.

22. The association football—or soccer—club in Sheffield, England, is the oldest still in operation.

23. During an unofficial Christmas truce on the Western front during World War I, the sides exchanged prisoners and food, sang Christmas carols together, and played games of football, or soccer, with one another.

24. In 1969, Brazilian soccer star Pelé scored his 1,000th goal.

25. The striker played for Santos Football Club from 1956–74 and the New York Cosmos from 1975–77, as well as the Brazilian national team.

26. The first FIFA Women's World Championship was held in China in 1991.

27. The U.S. women won the first World Cup.

28. With five men's titles as of 2023, Brazil has won more World Cups than any other country.

29. Contrary to belief, soccer balls are actually oval-shaped, not round.

30. The checkered pattern creates an illusion of a perfect sphere.

31. Referees were not used in official soccer matches until 1881.

32. The fastest red card in history was given to player Lee Todd in 2000 for using an expletive two seconds into the game.

33. A soccer field is called a "pitch," because regulation fields are pitched, or sloped, five degrees upwards from one end to the other.

34. The teams switch sides at half so each has to play slightly uphill.

16 Rocky Marciano Facts

1. Rocky Marciano was the heavyweight champion of the world from 1952 to 1956.

2. Marciano was the only heavyweight champion to retire without a defeat or a draw.

3. Rocky Marciano was born Rocco Marchegiano in 1923.

4. Marciano survived a near-fatal bout of pneumonia in childhood.

5. As a young adult lacking a modern gym, Marciano practiced boxing by hanging an old mailbag to a tree in his backyard.

6. He quit high school and worked as a ditchdigger and at a coal company.

7. Marciano served during World War II in the Army.

8. He won the 1946 amateur armed forces boxing tournament.

9. He wasn't undefeated as an amateur, finishing the year with an 11–3 record.

10. Marciano played baseball as a child and even tried out for the Chicago Cubs in 1947.

11. He was cut in three weeks, though, and turned professional in boxing.

12. In 1951, Marciano competed against Joe Louis in the last match of Louis's career. Marciano won.

13. Marciano defended his title of heavyweight champion six times.

14. Marciano's last bout took place in 1955 against Archie Moore.

15. The retired boxer died in 1969 in a private plane crash.

16. Marciano was honored with a commemorative postage stamp in 1999.

36 Popular Toys & Games Facts

1. Bingo started sometime around 1929 as "Beano."

2. A smart entrepreneur named Edwin Lowe spotted a crowd playing Beano at a fair and ran with the idea.

3. The first boxed Stratego set in the U.S. came out in 1961.

4. An ancient Chinese game called Jungle or Animal Chess very much resembles Stratego, and a Frenchwoman patented its modern incarnation as *L'attaque* in 1910.

5. A barber from Ohio invented the popular all-ages card game Uno in 1969.

6. Creator Merle Robbins sold Uno to a game company in 1972 for $50,000 plus royalties.

7. Mike Marshall created the Hacky Sack, a version of the footbag, in 1972 to help his friend John Stalberger rehabilitate an injured leg.

8. Enrico Rubik introduced his puzzle cube in 1974.

9. It became popular in the 1980s, confounding millions of people worldwide with its 43 quintillion (that's 43 followed by 18 zeros) solutions.

10. In the early 1940s, a torsion spring fell off marine engineer Richard James's desk and tumbled end over end across the floor.

11. Since then, more than a quarter billion Slinkys have been sold worldwide.

12. The Milton Bradley Company released Twister in 1966.

13. It was the first game in history to use the human body as an actual playing piece.

14. The Super Soaker water gun was invented in 1988 by aerospace engineer Lonnie Johnson.

15. Silly Putty was developed in 1943 when James Wright, a General Electric researcher, was seeking a synthetic rubber substitute.

16. Silly Putty debuted as a toy in 1950.

17. Mr. Potato Head, with his interchangeable facial features, was patented in 1952.

18. Mr. Potato Head was the first toy to be advertised on television.

19. For the first eight years, parents had to supply children with a real potato until a plastic potato body was included in 1960.

20. Intending to create a wallpaper cleaner, Joseph and Noah McVicker invented Play-Doh in 1955.

21. Barbie came onto the toy scene in 1959, the creation of Ruth Handler and her husband Elliot.

22. Ruth and Elliot Handler, along with Harold Matson, founded the Mattel toy company.

23. Chatty Cathy, also released by the Mattel Corporation in 1959, was the era's second most popular doll.

24. Betsy Wetsy also made a splash with 1950s-era children.

25. Created by the Ideal Toy Company, Betsy's already-open mouth would accept a liquid-filled bottle.

26. Since 1963, when they were first introduced, more than 16 million Easy Bake Ovens have been sold.

27. A light bulb provided the heat source for baking mini-cakes in America's first working toy oven.

28. Toy lovers have to salute manufacturer Hasbro for its G.I. Joe action figure, which first marched out in 1964.

29. The 11-inch-tall doll for boys had 21 moving parts.

30. Hot Wheels screeched into the toy world in 1968, screaming out of Mattel's concept garage with 16 miniature autos.

31. Weebles were released by Hasbro in 1971.

32. At the height of their popularity, the Weeble family had its own tree house and cottage.

33. Sweet-smelling Strawberry Shortcake was created in 1977 by Muriel Fahrion for American Greetings.

34. The company expanded the toy line in the 1980s to include Strawberry's friends and their pets.

35. Xavier Roberts was a teenager when he launched his Babyland General Hospital during the 1970s in Cleveland, Georgia, allowing children to adopt a "baby."

36. In 1983, the Coleco toy company started mass-producing these dolls as Cabbage Patch Kids.

18 Facts About Pinball

1. Pinball was invented in the 1930s.

2. Pinball derived from the 19th-century game bagatelle.

3. Bagatelle involved a billiards cue and a playing field full of holes.

4. Some early pinball arcades "awarded" players for high scores.

5. In the mid-1930s, machines were introduced that provided direct monetary payouts.

6. This quickly earned pinball a reputation as a gambling device.

7. Starting in the 1940s, New York City Mayor Fiorello LaGuardia declared pinball parlors akin to casinos ("magnets for the wrong element"), ushering in an era of pinball prohibition.

8. Chicago, Los Angeles, and other cities followed, banning the game.

9. Despite or because of the ban, pinball became a favorite pastime among adolescents and teens in the 1950s.

10. Gottlieb's *Humpty Dumpty*, designed by Harry Mabs in 1947, was the first pinball game to feature flippers (three on each side) that allowed the player to use hand-eye coordination to influence gravity and chance.

11. In 1948, pinball designer Steven Kordek repositioned the flippers (just two) at the bottom of the playfield, and the adjustment became the industry standard.

12. New York's pinball embargo lasted until 1976.

13. City officials destroyed 11,000 machines before it was lifted.

14. The ban was lifted after council members voted 6–0 to legalize pinball in the Big Apple.

15. The turning point: Writer and pinball wizard Roger Sharpe called his shots during a demonstration in front of the New York City Council, proving that pinball was indeed a game of skill.

16. Though pinball popularity receded with the advent of the video game, it enjoyed the first of several revivals later in the 1970s, thanks to its association with such rock-and-roll luminaries as The Who, Elton John, and Kiss.

17. In 1991, Bally's introduced The Addams Family game to tie in with the release of the movie.

18. The Addams Family became one of the best-selling pinball games of all time, with 22,000 machines sold.

30 Popular Pinball Machines

1. Theatre of Magic
2. Monster Bash
3. Twilight Zone
4. Medieval Madness
5. Tales of the Arabian Nights
6. Cirqus Voltaire
7. Wizard of Oz
8. Funhouse
9. The Addams Family
10. The Simpsons Pinball Party
11. Cactus Canyon
12. Safe Cracker
13. Creature from the Black Lagoon
14. Star Wars
15. Lord of the Rings
16. Pirates of the Caribbean
17. South Park
18. Revenge from Mars
19. The Sopranos
20. NBA Fastbreak
21. Junk Yard
22. Scared Stiff
23. Guns N' Roses
24. Attack from Mars
25. Doctor Who
26. Elvis
27. Freddy: Nightmare on Elm Street
28. Harley-Davidson
29. Tales from the Crypt
30. F-14 Tomcat

31 Auto Racing Facts

1. In 1961, Phil Hill was the first American to win the Formula One World Championship.

2. Mario Andretti was the second, in 1978.

3. Al Unser Sr. and Jr. were the first father-son duo to find victory lane at the Indianapolis 500.

4. Formula One driver Sebastian Vettel of Germany was just 23 years old when he became the youngest world champion in history in 2010.

5. The International Motorsports Hall of Fame was founded

by NASCAR architect Bill France Sr. near Talladega Superspeedway in 1982.

6. The first running of the *24 Heures du Mans*, or 24 Hours of Le Mans, took place in 1923.

7. The most famous endurance race in the world will celebrate its 100th anniversary in 2023.

8. A.J. Foyt was the first driver to win the Daytona 500, Indy 500, 24 Hours of Daytona, and 24 Hours of Le Mans.

9. Foyt was also the first driver to win the Indy 500 in both a front-engine and rear-engine car.

10. In 2012, the Audi R18 E-Tron Quattro that won the 24 Hours of Le Mans was the first hybrid to win the event.

11. With 253 acres within its oval, the Indianapolis Motor Speedway, home of the Indianapolis 500, could house the Roman Colosseum, Churchill Downs, Yankee Stadium, Rose Bowl Stadium, and Vatican City.

12. Janet Guthrie, in 1977, was the first woman to qualify for and race in the Indy 500.

13. Mechanical problems led to an early exit from that race, but the following year she finished ninth.

14. The Indy 500 winner traditionally drinks milk in victory lane.

15. After three-time winner Louis Meyer drank buttermilk in victory lane in 1936, a dairy-industry executive made a pitch to keep the tradition going and it caught on.

16. Helio Castroneves was the first driver to take the checkered flag in his first two Indy 500 starts, in 2001 and 2002.

17. Honda provided the victorious engine for every Indy 500 win between 2004 and 2012.

18. Youngsters between the ages of 7 and 17 have been racing at the All-American Soap Box Derby, an annual summer festival based in Akron, Ohio, since 1935.

19. NASCAR's founding meeting was organized by Bill France Sr. on December 14, 1947, at the Streamline Hotel in Daytona Beach.

20. The circuit, called the "Strictly Stock" division, debuted two years later.

21. Richard Petty won ten consecutive NASCAR races in 1967, a NASCAR record that still holds as of 2023.

22. He won 27 of his 48 starts that year.

23. Jeff Gordon was the first NASCAR driver to host NBC's *Saturday Night Live*.

24. Danica Patrick became the first woman ever to win the pole for the Daytona 500 in 2013.

25. Driver Carl Edwards routinely performs a backflip to celebrate victories.

26. Jimmie Johnson won NASCAR's top series five years in a row from 2006 to 2010, the first (and so far only) NASCAR driver to win five consecutive Cup crowns.

27. Jeff Gordon was the fastest NASCAR driver in history to reach 50 career wins.

28. His win came in his 232rd race, the DieHard 500 at Talladega in 2000.

29. There was a tie in the NASCAR series points standings in 2011, requiring a tie-breaker to determine the Cup champion.

30. Tony Stewart and Carl Edwards were dead even in the final points. Stewart won his third Cup title based on total wins during the year.

31. The Netflix docuseries *Formula 1: Drive to Survive* has increased the popularity of international auto racing among Americans.

21 Early Video Games Facts

1. At MIT in 1962, Steve Russell programmed the world's first video game on a bulky computer known as the DEC PDP-1.

2. Spacewar featured spaceships fighting amid an astronomically correct screen full of stars.

3. Nolan Bushnell founded Atari in 1972, taking the company's name from the Japanese word for the chess term "check."

4. Atari released the coin-operated Pong later that year.

5. Its simple, addictive action of bouncing a pixel ball between two paddles became an instant arcade hit.

6. In 1975, the TV-console version of Pong was released.

7. It was received with great enthusiasm by people who could play hours of the tennis-like game in the comfort of their homes.

8. After runaway success in the Soviet Union in 1985 (and in spite of the Cold War), Tetris jumped the Bering Strait and took over the U.S. market the next year.

9. Released in 1978, Midway's Space Invaders was a hit that generated a lot of money and also presented the "high score" concept.

10. A year later, Atari released Asteroids and outdid Space Invaders by enabling the high scorer to enter his or her initials for posterity.

11. The 1980 Midway classic Pac-Man was the world's most successful arcade game, selling some 99,000 units.

12. In 1980, Nintendo's first game Donkey Kong marked the debut of Mario, soon to become one of the most recognizable fictional characters in the world.

13. Originally dubbed Jumpman, Mario was named for Mario Segali, the onetime owner of Nintendo's warehouse in Seattle.

14. Designers originally wanted the title character of Q*bert (released 1982) to shoot slime from his nose.

15. But it was deemed too gross.

16. From 1988 to 1990, Nintendo sold roughly 50 million home-entertainment systems.

17. In 1996, Nintendo sold its billionth video game cartridge for home systems.

18. In 1981, 15-year-old Steve Juraszek set a world record on Williams Electronics' Defender.

19. His score of 15,963,100 got his picture in *Time* magazine—and it also got him suspended from school.

20. He played part of his 16-hour game when he should have been attending class.

21. Atari opened the first pizzeria/arcade establishment known as Chuck E. Cheese in San Jose in 1977.

25 Facts About Hockey

1. The NHL's "Original Six" franchises included the New York Rangers, Toronto Maple Leafs, Detroit Red Wings, Boston Bruins, Chicago Blackhawks, and Montreal Canadiens.

2. On November 1, 1959, goaltender Jacques Plante was the first goalie to wear a full protective mask.

3. He did so after taking a puck to the face that split his lip.

4. Clint Benedict had worn a half-mask for a brief time in 1930, but said it blocked his vision and scrapped it after a few games.

5. A goal, an assist, and a fight in the same game are said to comprise a "Gordie Howe hat trick."

6. Wayne Gretzy totaled 894 goals and 1,963 assists for 2,857 points during his illustrious career, the all-time NHL leader in those categories by far.

7. Coach Scotty Bowman holds the record for coaching wins, with 1,244 career victories.

8. Bowman was also the first coach to win Stanley Cups with three different teams: five with the Montreal Canadiens, and two each with the Pittsburgh Penguins and the Detroit Red Wings.

9. Goalie Martin Brodeur of the New Jersey Devils holds the record for most career victories for a goaltender, with more than 600.

10. Brodeur also holds the league's records for shutouts.

11. In 2012–13, the Chicago Blackhawks scored at least one point in 24 straight games to open the season, a feat unmatched in NHL history.

12. Dave "Tiger" Williams was a terror on the ice during the 1970s and 1980s, racking up 3,966 penalty minutes.

13. The NHL Winter Classic takes place outdoors. Brrrr!

14. Father and son Bobby and Brett Hull scored more than 1,350 goals between them.

15. Two years after retiring, Gordie Howe returned to the ice in his mid-40s, suiting up for the World Hockey Assocciation's Houston Aeros with his son. He won the WHA scoring title, too.

16. In the 2006–07 season, 20-year-old Sidney Crosby became the youngest player to win a scoring title.

17. Crosby was also the youngest player to reach 200 career points.

18. The Montreal Canadiens hold the record for the franchise that has won the Stanley Cup the most times, with 24 wins as of 2022.

19. Lord Stanley of Preston, then the Governor General of Canada, paid $50 for the Stanley Cup in 1893.

20. In the eight seasons goaltender Ken Dryden played for the Montreal Canadiens, he won six Stanley Cups.

21. The Los Angeles Kings won their first ten road games of the 2012 Stanley Cup playoffs, a playoff record.

22. The first woman to have her name inscribed on the Stanley Cup was Detroit Red Wings president Marguerite Norris in 1955.

23. Chris Chelios played for the Stanley Cup in 24 different seasons with three different teams.

24. Henri Richard of the Montreal Canadiens has his name inscribed on the Cup eleven different times as a player, a record.

25. However, Jean Beliveau has his name inscribed seventeen times—ten times as a player and seven as an executive.

21 Simone Biles Facts

1. Gymnast Simone Biles was born on March 14, 1997 in Columbus, Ohio.

2. Her middle name is Arianne.

3. Biles stands 4'8".

4. She has been diagnosed with ADHD.

5. Biles has citizenship in Belize through her mother.

6. She began training with coach Aimee Boorman at age eight.

7. Biles was 14 when she began competing at an elite level.

8. She was 16 when she won her first World all-around title.

9. Biles has four skills named after her: one on vault, one on balance beam, and two on floor exercise.

10. Each named skill is the most difficult element on the apparatus in question.

11. She is so far the only gymnast to have performed them in international competition.

12. Biles won the gold medal for the All-Around competition at World Championships in 2013, 2014, 2015, 2018, and 2019, a record for female gymnasts.

13. She did not compete in 2017, and the competition was not held in 2020.

14. Biles has also won numerous gold medals in individual events, and holds a total of 32 Olympic and World Championship medals.

15. She has won a World medal in every individual event, the first American female gymnast to do so.

16. Biles was a flag bearer in the closing ceremonies at Rio de Janeiro.

17. Biles and fellow Olympian Katie Ledecky were sponsors of a U.S. Navy aircraft carrier in 2016, the USS *Enterprise*.

18. Biles competed in the 24th season of *Dancing with the Stars* in 2017, when she was taking a hiatus from gymnastics.

19. She came in fourth place.

20. During the 2020 Olympics (held in 2021), Biles withdrew from several events after experiencing "the twisties"—a phenomenon causing a gymnast to lose air awareness.

21. President Joe Biden awarded Biles the Presidential Medal of Freedom in July 2022.

26 Olympic Games Facts

1. The first modern Olympic marathon in 1896 was 24.8 miles instead of the current 26.2 miles, run from Marathon to Athens.

2. Greece's Spiridon Louis won in just under three hours.

3. Carl Lewis won four consecutive Olympic long jump gold medals from 1984 to 1996.

4. With 11 Olympic medals, Allyson Felix is the most decorated woman in U.S. track and field history.

5. The city of Los Angeles has hosted the Summer Games twice, in 1932 and 1984.

6. Los Angeles is slated to host again in 2028.

7. Swimmer Michael Phelps is the most decorated Olympian of all time, having won 23 gold medals, three silver, and two bronze for a total of 28 medals.

8. The second-most decorated Olympian is gymnast Larisa Latynina, who competed for the Soviet Union between 1956 and 1964.

9. Latynina won a total of 18 medals, including nine gold medals.

10. In 1972, wrestler Dan Gable cruised to a gold medal without surrendering a single point in any match.

11. The 1972 Munich Olympic Games were suspended for about 34 hours because of a hostage crisis.

12. Israeli athletes were killed by Palestinian terrorists.

13. After a 60-year drought for tennis fans, it was reinstated as a medal sport for the 1988 Games in Seoul.

14. Steffi Graf, competing for West Germany, won the women's single title in those games, defeating Gabriela Sabatini.

15. The Spinks brothers, Leon and Michael, brought home two Olympic boxing gold medals from the Montreal Games in 1976, Leon in light heavyweight and Michael in middleweight.

16. U.S. diver Greg Louganis was the first man in Olympic history to sweep both the 3-meter springboard and 10-meter platform gold medals in back-to-back games, in 1984 and 1988.

17. Though golf was played in the 1900 and 1904 Olympics, it then took a very long hiatus.

18. It was reinstated for the 2016 Summer Games in Rio.

19. The first Winter Games were played in 1924 in Chamonix, France.

20. At the time, they were referred to not as the Olympics but as "International Winter Sports Week."

21. At those first Winter Games, 16 nations competed in six sports.

22. Twenty years before the famed "Miracle on Ice," the U.S. won its first gold medal in Olympic ice hockey in 1960, on home ice in California.

23. Although Michelle Kwan entered the U.S. Figure Skating Hall of Fame as the most decorated figure skater in U.S. history, she never won a gold medal at the Olympic Games, settling for silver and bronze.

24. The first American to win two individual gold medals in men's figure skating was Dick Button in 1948 and 1952.

25. In 1980, American speed skater Eric Heiden became the first athlete in any Olympics to win five gold medals in one year.

26. He set Olympic records in all five events, including a world record in the 10,000 meters.

30 College Sports Team Names Facts

1. Many land-grant schools focused on teaching agriculture, so a number of schools have teams called the Farmers or Aggies.

2. Texas A&M, New Mexico State, and Utah State all field teams of Aggies.

3. A bright yellow banana slug represents the teams of University of California–Santa Cruz.

4. The slug lives amid the redwoods on campus, and the students chose it as their mascot in a protest against competition at the cost of all else.

5. While the school at one point had two mascots—the more dignified sea lion was the "official" mascot chosen by the school's chancellor—the students prevailed after several years and "Sammy the Slug" is now the official mascot.

6. Male players at the University of Arkansas–Monticello go by the Boll Weevils.

7. The women's teams have chosen the Cotton Blossom to represent them.

8. The St. Louis College of Pharmacy, Missouri, fields the Eutectics.

9. The word refers to a chemical process.

10. The mascot is a furry creature dressed up in a lab coat.

11. The Stanford Cardinals actually use a logo of a redwood.

12. The "cardinal" refers to the color red, not the bird.

13. Student athletes at Austin Peay State University in Tennessee are known as the Governors. (Austin Peay was a Tennessee governor.)

14. They were known at one point as the "Normalities."

15. At Washburn University in Kansas, the men's teams are the Ichabods, while the women are the Lady Blues.

16. The name makes sense when you know the school was renamed for a philanthropist who saved it during a time of financial struggle, Ichabod Washburn.

17. Men's teams go by the Mules at Central Missouri State.

18. Women's teams are known as the Jennies, female donkeys.

19. Before 1922, those students were known as "Normals" and "Teachers," so Mules and Jennies are a step up.

20. Student athletes at Furman University in South Carolina are known as the Paladins.

21. For a long time, players at Whittier College in California were known as the Poets.

22. The mascot, Johnny Poet, dressed up as poet John Greenleaf Whittier in colonial garb and carried an oversized pen.

23. Johnny Poet retired in fall 2017 and was replaced by Woolly Whittier, a mammoth.

24. Whittier College was President Richard M. Nixon's alma mater, so he might be the most Poetic president.

25. Sports teams at the University of Louisiana–Lafayette are known as the Ragin' Cajuns.

26. The University of Hawaii lets its teams name themselves by sports.

27. Names include the Warriors, the Rainbows, the Rainbow Warriors, and the Rainbow Wahine ("wahine" means women).

28. Oglethorpe University in Georgia uses the Stormy Petrel, a shorebird, as its mascot.

29. The University of Louisiana–Monroe, fields teams of Warhawks.

30. While the logo depicts a bird, the name actually refers to the World War II fighter planes used by Louisianan Claire Chennault and his "Flying Tigers."

10 Wacky Sports Injuries

1. Red Sox pitcher Clarence "Climax" Blethen wore false teeth, but he believed he looked more intimidating without them. During a 1923 game, he had the teeth in his back pocket when he slid into second base. The chompers bit his backside.

2. In 2004, baseball player Ryan Klesko was in the middle of pregame stretches when he jumped up for the singing of the national anthem and pulled an oblique/rib-cage muscle, which sidelined him for more than a week.

3. In 1927, New York Giants pitcher "Fat Freddie" Fitzsimmons was napping in a rocking chair when his pitching hand got caught under the chair and was crushed by his substantial girth. He missed three weeks.

4. After assisting on a goal in a 2004 match, newlywed soccer player Paulo Diogo celebrated by jumping up on a perimeter fence, accidentally catching his wedding ring on the wire. When he jumped down he tore off his finger.

5. During a 2006 playoff game, ice hockey player Jaromir Jagr threw a punch at an opposing player. Jagr missed, his fist slicing through the air so hard that he dislocated his shoulder.

6. During a publicity stunt for the Jacksonville Jaguars in 2003, a tree stump and ax were placed in the locker room to remind players to "keep chopping wood," or give it their all. Punter Chris Hanson took a swing and missed the stump, sinking the ax into his non-kicking foot.

7. Rookie shortstop Clint Barmes was sidelined from the Colorado Rockies lineup for nearly three months in 2005 after he broke his collarbone when he fell carrying a slab of deer meat.

8. In the late 1990s, professional British soccer player Darren Barnard was sidelined for five months with knee ligament damage after he slipped in a puddle of his puppy's pee on the kitchen floor.

9. Australian rugby player Jamie Ainscough's arm became infected in 2002, and doctors feared they might need to amputate. But after closer inspection, physicians found the source of the infection—the tooth of a rugby opponent had become lodged under his skin, unbeknownst to Ainscough.

10. In May 2004, Sammy Sosa sneezed so hard that he injured his back, sidelining the Chicago Cubs all-star outfielder and precipitating one of the worst hitting slumps of his career.

24 Paralympic Games Facts

1. The Paralympic Games had several predecessor events.

2. In 1948, in conjunction with the opening of the 1948 Summer Olympics in London, World War II veterans with spinal injuries competed in the 1948 International Wheelchair Games.

3. Several competitions of that sort are collectively called the Stoke Mandeville Games, named after the Stoke Mandeville Hospital rehab facility in England.

4. The IWAS World Games (formerly known as the World Wheelchair Games and the Stoke Mandeville Games) are still held in non-Olympic years.

5. The first official Paralympic Games were held in Rome in 1960.

6. Originally, only athletes in wheelchairs were eligible for competition.

7. The rules changed in the 1976 Summer Games.

8. The first Winter Paralympic Games were held in 1976 in Sweden.

9. More than 4,300 athletes from 159 countries competed in the Paralympic Games in Rio de Janeiro in 2016.

10. The Paralympic motto is "Spirit in Motion."

11. The Summer Paralympic Games include more than 20 sports and more than 500 medal events.

12. Sports include archery, badminton, cycling, judo, powerlifting, table tennis, volleyball, and wheelchair fencing.

13. A sport named boccia, similar to bocce, has no Olympic equivalent.

14. It was designed to be played by athletes with disabilities, especially cerebral palsy.

15. The Winter Paralympic Games include five sports: alpine skiing, para ice hockey, Nordic skiing, wheelchair curling, and para snowboarding.

16. A number of athletes have competed in both the Paralympic and Olympic Games.

17. Archer Neroli Fairhall of New Zealand, for example, competed in the 1980 Summer Paralympics and then in the 1984 Olympic Games in Los Angeles.

18. Fairhall was the first paraplegic competitor in the Olympic Games, although athletes with disabilities had competed previously.

19. Gymnast George Eyser competed with a prosthetic leg in the 1904 Olympic Games and won several gold medals.

20. Hungarian fencer Pal Szekeres won a bronze medal in the 1988 Summer Olympics.

21. A bus accident rendered him disabled, and he transitioned to wheelchair fencing at the Paralympic Games.

22. U.S. swimmer Trischa Zorn is the most decorated Paralympian, with 55 medals in various events, including 41 gold medals.

23. Norwegian Ragnhild Mykleburst holds the record for most medals earned at the Winter Paralympic Games: 22 medals, 17 of them gold.

24. Mykleburst competed in various skiing, biathlon, and ice sledge events.

29 Babe Didrikson Zaharias Facts

1. Olympian Babe Didrikson Zaharias was known as one of the best all-around athletes of her time.

2. She was born Mildred Ella Didriksen (then spelled with an e) in either 1911 or 1914 (accounts vary).

3. She was a native Texan, born in Port Arthur.

4. As a kid, she practiced a wide array of sports, including basketball, diving, swimming, tennis, bowling, and lacrosse.

5. Sports were not her only interest.

6. She entered sewing competitions at the Texas State Fair and enjoyed music, singing, and playing harmonica.

7. She earned her nickname for her prowess as baseball, a comparison to Babe Ruth.

8. She dropped out of high school before graduating, moving to Dallas to play basketball.

9. Didrikson played basketball for the Golden Cyclones, an industrial league team affiliated with an insurance company.

10. Between 1930 and 1932, she guided the Golden Cyclones to the Amateur Athletic Union national championship.

11. The insurance company sponsored her involvement in the 1932 AAU championships, which served as an Olympic-qualifying tournament.

12. Didrikson competed in eight of ten events, winning five gold medals while setting world records in the javelin, 80-meter hurdles, high jump, and baseball throw, thus earning her place in the 1932 Summer Games in Los Angeles.

13. At those Games, Didrikson won Olympic gold in both the javelin and the 80-meter hurdles, breaking her own world records in the process.

14. She also set a world record in the high jump, but her effort was downgraded from gold to silver-medal status because of her method.

15. She cleared the record height by diving headfirst over the bar—a method so revolutionary that Olympic officials refused to award her a gold medal.

16. In 1934, Didrikson barnstormed the country with the House of David amateur baseball team.

17. House of David played against several professional teams.

18. Babe famously faced off against Joe DiMaggio—and struck him out.

19. In 1934, she also pitched in two major-league spring training games in Florida.

20. She threw the first inning of a Philadelphia Athletics match against Brooklyn, walking one batter but not allowing a hit.

21. Two days later, Didrikson pitched an inning for the Cardinals against the Red Sox, yielding her first runs.

22. In her post-Olympic years, Didrikson also toured the country as a vaudeville act.

23. She began to play golf and won the second tournament she entered—the Texas Women's Amateur Championship in April 1935.

24. Didrikson went on to record 82 career amateur and professional golf victories—including 10 major titles.

25. Didrikson married wrestler George Zaharias in 1938.

26. She captured a trio of U.S. Open crowns in 1948, 1950, and 1954 and strung together a remarkable and unprecedented 17 consecutive wins from April 1946 to August 1947, a feat no other duffer of either gender has been able to equal.

27. Didrikson Zaharias was a founding member of the Ladies Professional Golf Association (LPGA) and continued to win

tournaments with uncanny ease until 1954, when she was diagnosed with colon cancer.

28. Fourteen weeks later, she returned to the links and won the U.S. Women's Open.

29. Though noticeably slowed by her illness, she captured another pair of titles before succumbing to the disease in September 1956.

48 Facts About Chess

1. Although the exact origins of chess are a bit murky, most historians agree that the oldest precursor to the game was a sixth century Indian strategy game called *chaturanga*.

2. In Sanskrit, the word means "four arms" or "four divisions."

3. Today, there is a yoga pose called *chaturanga* that describes the "four-limbed staff pose."

4. The "four divisions" were infantry, cavalry, elephantry, and chariotry.

5. Each were represented by a game piece.

6. From India, this pastime made its way to Persia, where the rules began to evolve, and the game more closely resembled what we know today.

7. *Shatranj*, as it was known in Persia, embraced the battle theme of its predecessor, adding a game piece to represent a king, and the objective became to capture the opponent's king.

8. Players exclaimed *Shāh Māt*—Persian for "the king is helpless"—when attack was imminent.

9. In Arabic, the exclamation used was *sheikh-mat*, which, over time, became the well-known *checkmate*.

10. By the year 1000, chess had spread from Middle Eastern lands into Russia and Europe.

11. The Moors carried the game with them to Northern Africa, Sicily, and the Iberian Peninsula.

12. Buddhist monks and Silk Road traders introduced the game to the Far East, where it evolved into a game called *xiangqi*, or "elephant chess," a variant of the game which is still played in China today.

13. The oldest known chess pieces were discovered at the ancient site of Afrasiyab in Uzbekistan.

14. Consisting of a king, a chariot, a vizier (a high-ranking political advisor in Muslim culture), a horse, an elephant, and two soldiers, the ivory pieces were dated to around A.D. 760.

15. Chess grew in popularity in Europe over the next several centuries, becoming a favorite among the upper classes and earning the nickname "the game of kings."

16. Ancient chess pieces often corollate with militaristic symbols of the time, including chariots, elephants, and soldiers.

17. But as the game gained popularity in Europe, the pieces took on more of the look of a noble court.

18. Prior to about 1850, there was no standard "chess set" for competitors to use, and chess boards and pieces came in a myriad of styles, depending on country and geographical region.

19. The first modern chess tournament was held in London in 1851.

20. The benchmark chess set today is called the "Staunton Chess Set."

21. It was named after Howard Staunton, who organized that 1851 chess tournament.

22. The set was designed by London architect and journalist Nathaniel Cooke.

23. The column-like appearance of the king, queen, and bishop in Cooke's set was reflected in much of Victorian London's neoclassical architecture.

24. The Italian-influenced balustrades popular on balconies and stairways at the time are seen in the design of his pawns.

25. Traditionally, chess boards are made of alternating dark and light wood.

26. Wooden boards, like that in the Staunton chess set, are also used for competitive chess championships sponsored by the World Chess Federation, also known by its French name, *Fédération Internationale des Échecs*, or FIDE.

27. The two bishops are topped with a "mitre," or the tall, folding cap worn by bishops in Western Christianity.

28. The rooks look like castles with crenellated battlements.

29. On the chess board, "ranks" are the rows of squares running horizontally across the board.

30. "Files" are the rows that run vertically, top to bottom.

31. The knight is the only piece that can "jump" over other pieces.

32. While in English-speaking countries the knight is sometimes nick-named the "horse," other languages often call this piece the "jumper."

33. The knight's unusual L-shaped jumping movement is one of the oldest defined movements in the game, dating all the way back to the invention of *chaturanga* in India.

34. Unlike the other pieces on its rank, the knight can immediately move from its starting position, thanks to its ability to jump over obstacles.

35. While today the queen is the most powerful piece on a chess board, this was not always the case.

36. Before the 15th century, this piece was one of the weakest in the game, able to advance only one diagonal square at a time.

37. As the game gained in popularity, the queen's move evolved into a combination of the rook's straight-line moves and the bishop's diagonal moves, giving her the ability to reach up to 27 squares from one single position.

38. Ruy López de Segura (~1530–~1580) was a Spanish priest and chess player who wrote an early book on chess.

39. A common opening sequence is called the "Ruy López" or the Spanish opening.

40. The Queen's Gambit opening was mentioned in one of the earliest known writings about chess, the Göttingen manuscript of 1490.

41. The Queen's Gambit is still commonly played today.

42. Actor Anya Taylor-Joy played Beth Harmon, the chess prodigy protagonist of Netflix series *The Queen's Gambit*, released in 2020.

43. The miniseries, based on a 1983 novel, won two Golden Globes.

44. Books about chess strategy for opening moves have been published as far back as 1497.

45. Experts estimate that theoretically, a chess game could encompass 5,949 total moves.

46. At a move a minute, that game would last four days!

47. By the end of World War II there were 24 annual international chess tournaments.

48. Today, there are more than a thousand.

23 Fencing Facts

1. Today, competitive fencing is comprised of three disciplines: foil, épée, and saber (or sabre).

2. Foil swords, light and maneuverable, weigh less than a pound.

3. For foil and the heavier épée, the point is scored with the tip of the blade.

4. Hits with the side of the blade do not count.

5. In terms of weight, the saber comes between the two other blades.

6. The saber uses the cutting edge of the blade, not just the point.

7. Most new fencers begin with foil.

8. Fencing has been a part of every single modern Olympic Summer Games.

9. The referee traditionally uses French terms to begin a match.

10. "En garde!" signals for the fencers to get in ready, or on-guard, position.

11. The second command is the French word for ready: "Prêts?"

12. When fencers are ready, the referee says "Allez!" the French word for "Go!"

13. Traditionally fencers salute each other and the referee.

14. Depending on the competition, they may salute the audience and the judges.

15. There is a fencing movement called a flunge that combines a lunge and another movement called a *flèche*.

16. Electronic scoring is used in major competitions.

17. Though it was used for épée events in the Olympics as far back as 1936, it was only introduced for saber events in 1988.

18. International events are governed by the International Fencing Federation, or *Fédération Internationale d'Escrime* (FIE). (Do disgruntled fencers ever say "Fie on FIE"?)

19. Fencing as a sport was popularized by Spanish fencers.

20. A Spanish manual teaching fencing dates back to the 1400s.

21. In foil, the fencers make valid hits on their opponent's torso, not the arms and legs.

22. In épée, the entire body can be targeted.

23. In saber, any touch above the waist is a valid point, with the exception of your opponent's weapon hand.

27 Larry Doby Facts

1. Lawrence "Larry" Doby was born in 1923 in South Carolina.

2. He went to high school in New Jersey, where he played baseball, basketball, and football, and also competed in track.

3. Between 1942 and 1947, Doby played baseball in the Negro Leagues, with an interruption when he served in the Navy during World War II.

4. In 1947, Doby became the first African American to play baseball in the American League when he joined the Cleveland Indians.

5. He was the second Black player in Major League baseball, joining 11 weeks after Jackie Robinson broke the color barrier in the National League.

6. Manager Bill Veeck, who had been searching for a player to integrate the American League, brought him aboard.

7. Unlike Robinson, Doby did not receive the benefit of playing minor-league ball, which would have allowed him to make a gradual transition to the majors.

8. Instead, Veeck brought Doby directly from the Negro Leagues to the Indians.

9. Doby's new team was not especially welcoming.

10. Doby related that some of his teammates refused to shake his hand when they were introduced.

11. Two even turned their backs.

12. Doby also faced racist opposition from opposing teams and fans.

13. Doby's debut season did not unfold as dramatically as Robinson's did.

14. While Robinson played well enough to win Rookie of the Year and helped the Dodgers advance to the World Series, Doby played sparingly and flailed at the plate, hitting only .156 in 32 at-bats.

15. During spring training, Doby was not allowed to stay at the hotel where his teammates were living.

16. Instead, he stayed with a local Black family.

17. Doby rebounded in 1948.

18. He became the Indians' regular center fielder, hit .301 with 14 home runs, and helped Cleveland clinch the AL pennant.

19. In 1948, famed pitcher Satchel Paige also joined the Cleveland Indians.

20. The Indians made it to the World Series in 1948, battling the Boston Braves. The Indians won.

21. Doby hit a home run in Game 4, becoming the first Black man to hit a home run in the World Series, and Paige became the first Black pitcher to take the mound in the World Series.

22. In 1949, Doby played in his first All-Star Game.

23. That game was the first to include Black players: Jackie Robinson, Roy Campanella, Don Newcombe, and Larry Doby all played.

24. By the time his career ended, Doby had qualified for seven All-Star teams, led the American League in home runs twice, and finished second in the MVP voting in 1954.

25. Doby later became the second Black person to manage a Major League team in 1975, when he held a manager's position with the Chicago White Sox. Once again, Bill Veeck hired him.

26. Doby was inducted in the Baseball Hall of Fame in 1998.

27. He passed away in 2003 at the age of 79.

23 Swimming Facts

1. The world swimming association, the International Swimming Federation, was founded in 1908 at the end of the 1908 Olympic Games held in London.

2. The federation is abbreviated as FINA after its French name, *Fédération internationale de natation*.

3. In Egypt, cave art depicting people making arm motions like they might be swimming was created as early as 10,000 years ago.

4. The cave is called the Cave of Swimmers.

5. Monenjo-Daro, a famous archaeological site in Pakistan that preserves an early city from about 2800 B.C., contains a site called the Great Bath that might have been a public pool.

6. Leonardo da Vinci's sketches contain one of a life preserver.

7. Cow bladders and reed bundles were used as early swimming aids to increase buoyancy.

8. The first indoor swimming pool in England, St. George's Baths, dates back to 1828.

9. The front crawl is perceived to be the fastest swimming style.

10. Gertrude Ederle swam across the English Channel using the front crawl in 1926.

11. Native American swimmers introduced the front crawl to the British in 1844.

12. Swimming events at the first Olympic Games of the modern era (1896, Athens) included the 100-meter, the 500-meter, the 1200-meter freestyle, and the 100-meter for sailors.

13. The second modern Olympic Games in Paris (1900) included additional swimming events, one of which was an obstacle swimming course in the Seine.

14. Obstacle swimming is also included as an event in the Military Pentathlon.

15. The four major styles used today in competition are the butterfly, backstroke, breaststroke, and freestyle.

16. World Championship pools are 160 feet long by 82 feet wide (50 by 25 m) and hold ten lanes.

17. Johnny Weissmuller, more famous for playing the title character in *Tarzan*, was a competitive swimmer who won five Olympic gold medals and more than 50 U.S. national championships.

18. Michael Phelps was the American flag bearer in Rio de Janeiro's opening ceremonies, his fifth Olympics.

19. Kathleen "Katie" Ledecky holds the medal record for female swimmers, with seven Olympic golds and 19 world championship gold medals.

20. Australian swimmer Ian Thorpe's nickname, given to him for speed, was "Thorpedo."

21. Some freedivers can hold their breath for more than 10 minutes.

22. The usual average is about 1½ to 2 minutes.

23. A few people can hold their breath for more than 20 minutes.

30 Facts About Monopoly

1. Monopoly was first published in 1935.

2. An earlier version, The Landlord's Game, dates back to 1903 and was created by Elizabeth "Lizzie" Magie.

3. Magie was a progressive feminist who followed the writings of economist Henry George, who believed in a land value tax.

4. The game was devised to illustrate the dangers of monopolies.

5. The Landlord's Game was patented in January 1904.

6. The patent expired in 1921; at the time, the game had only been self-published.

7. Magie patented a revised version of the game in 1924, and in 1932 this version was published by Adgame Company.

8. When Parker Brothers published a modified version of the game in 1935, it was credited to Charles Darrow, who received a patent for it.

9. The company eventually bought the rights to Magie's patent as well in order to establish clear ownership.

10. The game today is published by Hasbro.

11. During World War II, the British Secret Service sent information and maps to prisoners of war smuggled in Monopoly sets.

12. The rich character seen on the set is known as "Mr. Monopoly" and "Rich Uncle Pennybags."

13. The street names in the United States version derive from real locations in Atlantic City, New Jersey.

14. There is no Marvin Gardens in Atlantic City, but there is a Marven Gardens neighborhood.

15. The value of a property reflects the real estate prices in the Atlantic City neighborhoods when the game was developed.

16. Due to segregation, the cheaper places were streets in African American neighborhoods, while Park Place and Ventnor Avenue were situated in wealthy, all-White neighborhoods.

17. The British version of the game contains locations in London.

18. The four railroads are represented by King's Cross station, Marylebone station, Fenchurch Street station, and Liverpool Street station.

19. The dark blue squares representing the most expensive properties are Park Lane and Mayfair, the equivalent of the U.S. Park Lane and Boardwalk.

20. More money is now included in the Monopoly set.

21. Older standard sets included a total of $15,140, but newer sets provide $20,580.

22. The tokens have changed over time, in both shape and material.

23. During World War II, for example, wood tokens were used due to a shortage of metal.

24. In the late 1930s and early 1940s, tokens included a rocking horse, a purse, and a lantern.

25. For a long stretch between the end of World War II and 1998, the ten standard tokens were battleship, boot, cannon, horse and rider, iron, racecar, Scottie dog, thimble, top hat, and wheelbarrow.

26. In 1998, a contest was begun to introduce a new token, which ended up being a sack of money. The token was retired in 2007.

27. The horse and rider, wheelbarrow, thimble, boot, cannon, and iron have also been retired over time.

28. A cat, penguin, rubber duck, and T-rex were introduced as new tokens in 2013 and 2017.

29. Hasbro conducts a worldwide Monopoly tournament at intervals.

30. The tournament held in 2015 in Macau was won by Italian Nicolò Falcone.

21 Facts About Cycling

1. The bicycle we know today had a number of predecessors.

2. The dandy horse had two wheels but no pedals, so riders had to walk or run as they went.

3. Velocipedes had varying numbers of wheels in varying sizes.

4. Some early velocipedes had one very large wheel and one small one, for example.

5. In the Netherlands, a popular event for families is the *fietsvierdaagse*, which involves cycling for four days in the countryside to enjoy nature.

6. Cyclo-cross racing takes place in fall and winter and involves a variety of terrain. Cyclists have to dismount to get over obstacles.

7. Velodromes are facilities geared towards cyclists.

8. Velodromes have banked turns and are used for a form of riding called track cycling.

9. In the late 1800s and early 1900s, track racing in velodromes was incredibly popular.

10. In 1899, Marshall "Major" Taylor became the first African American to be a cycling world champion when he won the World Track Championship.

11. Some races last days, with six-day cycling being a popular event in Europe.

12. The Tour de France lasts 23 days.

13. It began in 1903.

14. Lance Armstrong won the race seven times from 1999 to 2005.

15. He was later stripped of his wins because of his use of performance-enhancing drugs.

16. In 1935, adventurer Fred Birchmore traveled around the globe on boat and bicycle, cycling some 25,000 miles.

17. Cycling has been an event at every Summer Olympic Games in the modern era, beginning in 1896.

18. Mountain biking became an Olympic sport in 1996 at the Atlanta Summer Games.

19. BMX racing—off-road biking—entered the Olympics in 2008.

20. Tandem bicycling, with two people on a bike, was an Olympic event for many years, but was dropped after 1972.

21. Bicycles where the two riders sit side by side are called sociables.

13 Duke Kahanamoku Facts

1. His full name was Duke Paoa Kahinu Mokoe Hulikohola Kahanamoku.

2. Born in 1890 in Honolulu, the Native Hawaiian was a swimmer, beach volleyball player, and surfer.

3. As a young man, Kahanamoku surfed at Waikiki Beach.

4. He competed in the 1912 Summer Olympics in Stockholm as a swimmer, winning one gold and one silver medal.

5. In the 1920 Olympic Games in Antwerp, he won two more gold medals in the 100 meters and the relay.

6. In the 1924 Olympic Games in Paris, he won the silver in the 100 meters. His brother won the bronze medal.

7. He left his mark on the sport of swimming by developing the flutter kick.

8. Kahanamoku popularized the sport of surfing through exhibitions.

9. In 1914, he competed in a surfing exhibition in Freshwater Beach in Australia that helped establish surfing as an international sport.

10. Between 1932 and 1961, Kahanamoku was the Sheriff of Honolulu.

11. He also acted, appearing in a variety of small roles in television and films.

12. He married Nadine Alexander in 1940.

13. When he died at age 77, his ashes were scattered in the ocean.

20 Facts About Volleyball

1. William Morgan invented volleyball (then called Mintonette) in 1895 in Holyoke, Massachusetts.

2. By Morgan's rules, there could be any number of players.

3. Currently, each team can only touch the ball three times (usually bump/pass, set, and spike), and the same player can't touch the ball consecutively.

4. That's a change from Morgan's original rules, where any number of contacts with the ball was allowed.

5. The "set" and "spike" were developed by players in the Philippines.

6. The International Volleyball Federation was founded in 1947 in Paris, France.

7. The U.S. demonstrated volleyball at the 1924 Summer Olympics in Paris, but it didn't become an official Olympic sport until 1964.

8. Beach volleyball was added to the Olympic Games in 1992 as a demonstration sport, and officially included in 1996.

9. The nation of Brazil has been dominant in Olympic beach volleyball, winning 13 medals between 1996 and 2016.

10. Only the U.S. has won more total beach volleyball medals.

11. The net is slightly lower in women's competition than men's competition, by about 7.5 inches (.19 m).

12. The volleyball used indoors traditionally has 18 panels.

13. The ball used for beach volleyball is generally larger than the ball used indoors.

14. Invented in Brazil, biribol is similar to volleyball but played in water.

15. A one-handed block is called a Kong, as in, "King Kong."

16. When the ball is at the top of the net and opposing players are both pushing against it to try to tip it over to the opposing side, it's known as a joust.

17. Vocabulary terms include moves known as pancakes, crepes, and waffles. Now we're hungry!

18. When a defensive player raises their arm a certain way, it's called a chicken wing.

19. Want to insult someone in beach volleyball? Call them a Lombardo.

20. It means they don't have any skill whatsoever.

19 Archery Facts

1. It sounds toxic, but an expert archer—or even just one who likes it very much—is a toxophilite.

2. There's a reason the words sound similar. The Greek used the word *toxon* for bow and arrow, and *toxikon* for a poison-tipped arrow. The Latin then picked up *toxicum* for any poison.

3. Bows and arrows have been used by human hunters since prehistory.

4. Evidence of arrowheads dating back at least 60,000 years has been found in South Africa.

5. Several instances of cave art in Spain that date back to the Mesolithic era (10,000 B.C. to 8,000 B.C.) show two groups of archers in combat.

6. Archery transitioned to a sport after firearms were developed, when they were no longer used in combat.

7. Papingo, or popinjay, is a shooting sport where competitors shoot fake birds off their perches.

8. One popinjay tournament in Scotland may date back to the 1480s.

9. The World Archery Federation was established in 1931 in Poland.

10. Archery was an event in the early days of the modern Olympics between 1900 and 1920, but was then dropped.

11. It returned as an Olympic sport in 1972.

12. Of all the Olympic medals awarded since 1972, South Korean athletes have won more than any other nation's archers.

13. All archery events at the Olympics use the recurve bow, in which the bow curves away from the archer.

14. Other types of bows include the longbow, compound bow, and crossbow.

15. Longbows require a lot of strength and are used today primarily by hunters and target shooters.

16. Compound bows use a system of pulleys and cams and help the archer hold the bow and aim.

17. Crossbows are short bows that are mounted sideways.

18. The word "fletcher," referring to the person who makes fletching on an arrow, comes from the French *flèche*, meaning arrow.

19. Traditionally bird feathers were used as fletching, but nowadays plastic fletching is common.

17 Facts About Scrabble

1. Scrabble was designed by Alfred Mosher Butts in 1938.

2. He first designed a predecessor game called Lexiko, which had tiles but no board.

3. To determine how many tiles of each letter would be included, he pored over *The New York Times* to study the frequency of each letter in the alphabet.

4. Currently Hasbro distributes the game in the United States and Canada, while Mattel does so throughout the rest of the world.

5. If you're in France and see Scrabble with a slightly different logo, that's why!

6. National Scrabble Day is celebrated on April 13 each year, Butts's birthday.

7. In competitive Scrabble, the wooden tiles are all completely smooth, so that no one can tell by feel what is printed on the tile.

8. In some countries, you'll see tiles of letters that aren't present in the English alphabet.

9. Because of this, the number of tiles varies from language to language.

10. English-language sets have 100 tiles.

11. Competitive player Benjamin Woo scored 1,782 points on the single word Oxyphenbutazone. (Oxyphenbutazone is a type of anti-inflammatory drug, no longer on the market due to negative side effects.)

12. Players can go the Scrabble page on Hasbro's web site and type in a word to determine whether it's in the Official Scrabble Players Dictionary published by Merriam-Webster.

13. Words are added to the dictionary periodically.

14. In 2018, it became "OK" to use the word "OK" in the U.S.

15. If you're a Scrabble purist who thought, "ew," to the thought of "OK" counting as a word, well, the word "ew" was also added in 2018.

16. A *Scrabble* game show hosted by Chuck Woolery ran from 1984 to 1990.

17. Playing all seven tiles is officially called a Bingo.

19 High-value Scrabble Words

1. **quixotry:** quixotic (extremely idealist) action or thought—77

2. **muzhiks:** a Russian peasant in czarist times—75

3. **bezique:** a card game—77

4. **caziques:** native chiefs of West Indian aborigines—77

5. **oxazepam:** a tranquilizing drug—77

6. **mezquit:** a spiny tree or shrub—77

7. **zombify:** to turn into a zombie—76

8. **quizzers:** those who quiz—75

9. **whizbang:** slang term for a highly explosive shell or firecracker—76

10. **quetzal:** a bird with brilliant plumage—75

11. **highjack:** same as hijack, to seize a vehicle while in transit—78

12. **musquash:** a British expression for the fur of the muskrat—72

13. **packwax:** a strong ligament that supports the back of the head—75

14. **squeezer:** one who squeezes or presses forcibly together—76

15. **quartzy:** resembling quartz or taking on the qualities of it—78

16. **quickly:** with speed, rapidly, very soon—76

17. **asphyxy:** the condition that results from interruption of respiration—75

18. **zinkify:** to zincify, which is to coat with zinc—76

19. jukebox: an automated phonograph that is usually coin-operated—77

23 Williams Sisters Facts

1. Venus Williams was born June 1980 in California.

2. Serena Williams was born September 1981 in Michigan.

3. Their parents are Richard Williams and Oracene Price.

4. Their parents acted in the capacity of coach, particularly early in their careers.

5. The family moved to West Palm Beach, Florida, when the sisters were nine and ten, so that they could train at Rick Macci's tennis academy.

6. As a younger tennis player, Venus held a 63–0 record on the United States Tennis Association junior tour.

7. Venus turned professional in 1984 at the age of 14.

8. Together, the pair has won 14 Grand Slam doubles titles.

9. Each has won four Olympic gold medals, one in women's singles and three in women's doubles, a shared record.

10. They won their medals in women's doubles together in the 2000, 2008, and 2012 Games.

11. Venus won her gold medal in singles in the 2000 Summer Games in Sydney, while Serena won in 2012 in London.

12. Venus also won a silver medal in mixed doubles in 2016 in the Summer Games in Rio de Janeiro.

13. Serena Williams has won 23 Grand Slam singles titles, 14 in women's doubles, and two in mixed doubles.

14. Venus won her first Grand Slam singles titles in 2000, winning the US Open and Wimbledon.

15. In 2000, the sisters also won their first doubles title at Wimbledon.

16. Venus successfully defended her singles titles in 2001.

17. Serena won her first Wimbledon Championship in singles in 2002, defeating her sister and earning the world's No. 1 ranking.

18. The sisters have played against each other in 15 Grand Slam tournaments, with Serena taking the prize in 10 of those.

19. In 2011, Venus was diagnosed with an autoimmune disease, Sjögrens Syndrome.

20. Venus holds an associate degree in fashion design and a Bachelor of Science in Business Administration.

21. Serena married Reddit co-founder Alexis Ohanian in 2017, and gave birth to daughter Alexis Olympia later that year.

22. The two sisters purchased a minority stake in the Miami Dolphins in 2009.

23. The 2021 film *King Richard* portrays the history of the Williams sisters and their father's coaching.

21 Figure Skating Facts

1. Figure skating was first included in the Olympic Games in 1908.

2. While most of those Games took place during the summer, figure skating took place in October.

3. It was the first "winter" sport included in the Games.

4. The events included men's singles, ladies' singles, pairs, and men's special figures.

5. The blade on a figure skate has two edges.

6. Skaters are expected to have "clean edges," that is, to skate and land from jumps on either the inside or outside edge, but not both.

7. In pairs skating, the move where the man spins the woman while she is almost parallel to the ice is called the "death spiral."

8. The one-handed death spiral was first performed at the 1948 Olympic Games by a Canadian team.

9. Unlike hockey skates, figure skates have "toe picks," a set of serrated edges, at the front of the blade.

10. The toe picks differ slightly in ice dancing blades.

11. Ice dancing became an Olympic sport in 1976.

12. In ice dancing, a "twizzle" is a type of one-foot turn.

13. Ice dancers are not allowed to lift their partner above the shoulder.

14. American figure skater Kristi Yamaguchi won two gold medals at the 1988 World Junior Championships, one for ladies' singles and one for pairs (with Rudy Galindo).

15. Both skaters went on to successful singles careers.

16. In every senior competition she entered, South Korean skater Yuna Kim ("Queen Yuna") earned a podium spot.

17. In 1998, French skater Surya Bonaly did a backflip in Olympic competition, landing on one blade.

18. The backflip is currently prohibited in figure skating in Olympic competition, though it is permitted in exhibitions.

19. When Canadians Tessa Virtue and Scott Moir won gold on their home ice in 2010, they were the first ice dancers from North America to win Olympic gold.

20. Japanese skater Hanyu Yuzuru was the first man to break the 100-point barrier in the men's short program and the 200-point barrier in the men's free skate.

21. American Nathan Chen scored 224.92 in the men's free skate during the 2019–20 Grand Prix Final.

24 Muhammed Ali Facts

1. One of the greatest boxers of all time, he was born Cassius Marcellus Clay Jr. in Louisville, Kentucky, in 1942.

2. He was only 18 when he won gold at the 1960 Summer Olympics, boxing in the light heavyweight division.

3. He was not expected to win in 1964 when he fought Sonny Liston for the World's Heavyweight Championship and won in seven rounds.

4. When they fought again a year later, Ali knocked out Liston in the first round.

5. He took the name Cassius X, and then Muhammed Ali, not long after that first fight, in 1964.

6. Ali's membership in Elijah Muhammed's Nation of Islam was a matter of press and public controversy both before and after the fight.

7. In March 1966, Ali refused to be drafted, saying it was contrary to his religious beliefs.

8. He was stripped of his passport and his boxing license.

9. Ali was dyslexic. For a time, this had kept him from the draft, as he was classified as someone to be called up "only in times of national emergency."

10. In 1966, the Army changed its standards to allow him to be drafted.

11. Ali lost several professional years in his late 20s to the boxing ban, which was rescinded in 1971.

12. On March 8, 1971, he took his first loss in a professional fight in a bout against Joe Frazier.

13. The two men fought at Madison Square Garden on January 28, 1974.

14. This time, Ali won.

15. They fought a third time in 1975 in Manila.

16. While Ali won, it was a hard-fought victory that exhausted him.

17. Ali popularized the "rope-a-dope" strategy, taking punches while cornered against the rope in a protective defensive stance that tires out one's opponent.

18. The technique was particularly in play during his 1974 bout against George Foreman, the "Rumble in the Jungle."

19. Ali left the Nation of Islam and transitioned to mainstream Sunni Islam, making the Hajj pilgrimage to Mecca.

20. By the time of his final, brutal fights, Ali was already showing signs of ill health.

21. He was diagnosed with Parkinson's disease in 1984.

22. Ali was chosen to light the flame at the 1996 Summer Olympics in Atlanta.

23. He was the flag bearer at the opening ceremonies of the 2012 Summer Olympics in London.

24. Ali died at the age of 74 in 2016. He was buried in Kentucky.

27 Horse Racing Facts

1. Former steeplechase jockey Richard "Dick" Francis wrote a series of best-selling mystery novels, often set in the horse-racing world.

2. In 1920, the famous thoroughbred Man o' War shared the honors of "athlete of the year," as proclaimed by *The New York Times*, with human Babe Ruth.

3. Man o' War was the grandsire of another famous racehorse, Seabiscuit.

4. Man o' War was also one of the horses shown in a 1925 silent film directed by John Ford, *Kentucky Pride*.

5. Seabiscuit was not a winner in his early races.

6. His statistics improved over time.

7. Britain's Jockey Club was founded as far back as 1750.

8. For a long time it governed British horse racing, but those responsibilities are now handled by the British Horseracing Authority.

9. The Triple Crown is awarded when a three-year-old horse wins the Kentucky Derby, the Preakness Stakes, and the Belmont Stakes.

10. Thirteen horses have won as of 2022.

11. When Secretariat won the Triple Crown in 1973, he was the first winner in 25 years.

12. There was an even longer draught in Triple Crown winners between Affirmed (1978) and American Pharoah (2015).

13. Horse racing in the United States predates the American Revolution.

14. A racing meet was recorded as far back as 1665.

15. The quarter horse is named for the breed's ability to run quickly at races of about a quarter mile.

16. Freehold Raceway in New Jersey was the site of racing in the 1830s, making it the oldest racetrack still operating in the U.S.

17. Churchill Downs in Louisville, Kentucky, where the Kentucky Derby is held, was opened in 1875.

18. The Kentucky Derby is generally held on the first Sunday in May.

19. The race is 1.25 miles long.

20. The first Kentucky Derby took place in 1875, at a length of 1.5 miles.

21. A colt named Aristides won.

22. The Kentucky Derby was first televised in 1949.

23. The longest horse race in the world takes place in Mongolia and is called the Mongol Derby.

24. It runs 621 miles (1,000 km), with the exact course changing each year.

25. The endurance race lasts ten days.

26. Both jockeys and horses face injuries and death in racing.

27. By some estimates, hundreds of horses die annually in the U.S.

Chapter 6

Health and the Human Body

18 Facts About Cells

1. The exact cell count in the human body is unknown, and figuring it out is tricky.

2. One popular estimate is 30 trillion cells; another says 37 trillion.

3. About 300 million cells die every minute.

4. Red blood cells are the most common type of cell.

5. More than 80 percent of the cells in the human body are erythrocytes, or red blood cells.

6. Cells of the same type group together to form tissue.

7. Fifteen million blood cells are produced and destroyed in the human body every second.

8. Mitochondria use oxygen to produce energy for the cell.

9. The longest cells in the human body are the motor neurons.

10. They can be up to four-and-a-half-feet long and run from the lower spinal cord to the big toe.

11. The longest-living cells in the body are brain cells, which can live a human's entire lifetime.

12. The cells that control smell and taste are the only sensory cells that are replaced during a person's life span.

13. There are probably about as many bacteria and microbe cells in our body as there are human cells.

14. For a long time people thought bacteria might outnumber human cells by 10 to 1, but later revisions brought the estimate down.

15. Fat cells are called adipocytes or lipocytes.

16. Epithelial cells line your skin, the outsides of your organs, and the insides of some organs.

17. They protect your body and organs.

18. Stem cells generate other types of cells.

21 Heart & Circulatory System Facts

1. The circulatory system moves blood throughout the body.

2. Its main components are the heart, blood vessels, and blood.

3. Laid end to end, there are about 60,000 miles of blood vessels in the human body.

4. The hard-working heart pumps about 2,000 gallons of blood through those vessels every day.

5. The human heart creates enough pressure while pumping to squirt blood 30 feet.

6. Blood consists of red blood cells, white blood cells, plasma, and platelets.

7. Newborn babies only have about 1 cup of blood.

8. The average adult has about 5 liters of blood.

9. The right side of your heart pumps blood into your lungs.

10. The left side of your heart pumps blood back through your body.

11. Arteries carry oxygenated blood from the heart into smaller sections called arterioles and then into even smaller sections called capillaries.

12. Nutrients carried in the blood diffuse into body tissues from the capillaries.

13. The word capillary comes from the Latin *capillus*, hair.

14. Veins carry the deoxygenated blood from body tissues back to the heart.

15. The human heart beats over 100,000 times per day.

16. The average woman's heart is 8 beats per minute faster than a man's heart.

17. The first open-heart surgery was performed in 1893 by Daniel Hale Williams, one of the few Black cardiologists in the U.S. at the time.

18. The human heart weighs less than 1 pound.

19. A man's heart, on average, is 2 ounces heavier than a woman's heart.

20. Approximately 20 percent of the oxygenated blood flowing from the heart is pumped to the brain.

21. Other than the cornea, every cell in the body gets blood from the heart.

25 Facts About Your Head

1. The human head is one-quarter of our total length at birth.

2. But the head is only one-eighth of our total length by the time we reach adulthood.

3. If your mouth was completely dry, you would not be able to distinguish the taste of anything.

4. Your nose is not as sensitive as a dog's, but it can remember 50,000 different scents.

5. The air from a human sneeze can travel at speeds of 100 miles per hour or more.

6. Each of your nostrils registers smell differently.

7. The right nostril detects the more pleasant smells, but the left one is more accurate.

8. Women are usually better than men at identifying specific smells.

9. Most people only have a 50 percent success rate at detecting a single drop of perfume in a three-room apartment.

10. Don't stick out your tongue if you want to hide your identity!

11. Similar to fingerprints, everyone also has a unique tongue print.

12. The human eye blinks an average of 3.7 million times per year.

13. The pupil of the eye expands as much as 45 percent when a person looks at something pleasing.

14. The retina of the eye is the only part of the central nervous system that can be seen from the outside of the body—but you have to look directly through the pupil to see it.

15. When we blink (about 20,000 times a day) our brain keeps things illuminated so the world doesn't go dark each time.

16. It takes six muscles to move each of your eyeballs.

17. There's a facial muscle called the levator labii superioris alaeque nasi muscle.

18. It's the longest muscle name.

19. Perhaps because of its long name, it's also known as the Elvis muscle, because it's used to raise your upper lip in the way Elvis famously did.

20. The body's smallest bones are found in the middle ear.

21. Changes to the cartilage in your ears means that they lengthen throughout the course of your lifetime.

22. You have rocks in your body!

23. Your inner ear has tiny crystals that help you maintain balance.

24. The average skull is about 0.26 to 0.27 inches (6.5 to 7 mm) thick.

25. Women's skulls are slighter thicker on average than men's.

22 Sweaty Facts

1. Sweat enables us to cool off when the exterior temperature rises (due to changes in the weather) or when our interior temperatures rise (due to exercise, anxiety, or illness).

2. Sweat is one of the mechanisms that our bodies use to keep us at a steady—and healthy—98.6 degrees Fahrenheit.

3. Humans have about 2.6 million sweat glands.

4. Not all of these glands produce the same kind of sweat.

5. Sweat has two distinct sources: eccrine and apocrine glands.

6. Eccrine glands exist all over the body and are active from birth.

7. They constantly release a salty, nearly odorless fluid onto the skin.

8. Apocrine glands are concentrated in the armpits, on the soles of the feet, in the palms of the hands, and in the groin.

9. They become active during puberty. Yes, puberty and perspiration go hand in hand.

10. Apocrine glands don't secrete liquid directly onto the skin.

11. Instead, each gland empties into a hair follicle.

12. When a person is under emotional or physical stress, the tiny muscle around the hair follicle contracts, pushing the liquid onto the skin, where it becomes sweat.

13. Apocrine glands carry lipids and proteins, as well as water and salt.

14. When these substances mix with the sebaceous oils in the hair follicles and then meet the bacteria on the skin, well, that's when you begin to hold your nose.

15. Apocrine sweat has been found to contain androsterone pheromones, those mysterious musky odors that are responsible for sexual arousal.

16. Deodorants are based on mildly acidic compounds that dry the skin before the odor starts.

17. Antiperspirants, another popular option, actually block sweat with aluminum salts.

18. Excessive sweating is officially known as hyperhidrosis.

19. Lack of sweat is called anhidrosis.

20. Both hyperhidrosis and anhidrosis are genuine medical conditions with serious complications.

21. Fortunately, both are treatable.

22. For most of us, dealing with sweat is fairly simple: Take a shower and wear loose, absorbent clothing.

8 Most Common Blood Types

1. O+: 37.4 percent

2. A+: 35.7 percent

3. B+: 8.5 percent

4. O-: 6.6 percent

5. A-: 6.3 percent

6. AB+: 3.4 percent

7. B-: 1.5 percent

8. AB-: 0.6 percent

37 Blood Type Facts

1. There are four major blood groups (A, B, AB, O) determined by the presence or absence of two antigens, A and B, on the surface of red blood cells.

2. Group A has only the A antigen on red blood cells (and B antibody in the plasma).

3. Group B has only the B antigen on red blood cells (and A antibody in the plasma).

4. Group AB has both A and B antigens on red blood cells (but neither A nor B antibody in the plasma).

5. Group O has neither A nor B antigens on red blood cells (but both A and B antibody are in the plasma).

6. Like eye color, blood type is inherited.

7. Whether your blood group is A, B, AB, or O is based on the blood types of your biological parents.

8. In addition to the A and B antigens, there is a protein called the Rh factor, which can be either present (+) or absent (−), creating the eight most common blood types: A+, A-, B+, B-, AB+, AB-, O+, and O-.

9. The most common blood type is O+, which occurs in about 37 percent of the U.S. population.

10. About 53 percent of Latinos have type O+ blood, as do 47 percent of African Americans, 39 percent of Asian Americans, and 37 percent of Caucasians.

11. O+ blood can be given to a person with A+, B+, AB+, or O+ blood.

12. A person with O+ blood can receive blood from O+ or O- donors.

13. A+ blood occurs in about 35 percent of the population.

14. Type A+ occurs in about 33 percent of Caucasians, 29 percent of Latinos, 27 percent of Asian Americans, and 24 percent of African Americans.

15. A person with A+ blood can receive A+, A-, O+, or O- blood.

16. However, A+ blood can be given only to a person with the A+ or AB+ blood types.

17. B+ blood occurs in about 8 percent of the U.S. population.

18. About 25 percent of Asians have B+ blood, as do 18 percent of African Americans and 9 percent of Latinos and Caucasians.

19. B+ blood can be given only to those with either AB+ or B+ blood.

20. This blood type can receive blood from B+, B-, O+, or O- donors.

21. O- is considered the universal donor because it can be given to anyone, regardless of blood type.

22. However, a person with O- blood can receive blood only from other O- donors.

23. About 8 percent of Caucasians have type O-, as do 4 percent of African Americans and Latinos, and 1 percent of Asians.

24. A- blood can be given to a person with AB-, A-, AB+, or A+.

25. A person with type A- can only receive blood from O- or A- donors.

26. Approximately 7 percent of Caucasians have type A-, as do 2 percent of Latinos and African Americans, and 0.5 percent of Asians.

27. AB+ is considered a universal receiver because people with this blood type can receive blood of any type.

28. But AB+ blood can only be given to a person who also has AB+.

29. Approximately 7 percent of Asian Americans have type AB+, as do 4 percent of African Americans, and 2 percent of Latinos and Caucasians.

30. B- blood can be given to those with B-, AB-, B+, or AB+ blood.

31. A person with B- blood can receive blood from O- or B- blood types.

32. About 2 percent of Caucasians, 1 percent of African Americans and Latinos, and 0.4 percent of Asians have type B- blood.

33. AB- is the rarest of the eight common blood types, occurring in about 0.6 percent of the U.S. population.

34. People with type AB- can give blood to AB+ or AB- blood types.

35. People with type AB- must receive blood from O-, A-, B-, and AB- blood types.

36. AB- blood occurs in about 1 percent of Caucasians, 0.3 percent of African Americans, 0.2 percent of Latinos, and 0.1 percent of Asian Americans.

37. The universal plasma donor has type AB blood.

26 History of the Lobotomy Facts

1. There's a reason why lobotomies have taken a place in the Health Care Hall of Shame.

2. A lobotomy is a surgical procedure that severs the paths of communication between the prefrontal lobe and the rest of the brain.

3. This prefrontal lobe—the part of the brain closest to the forehead—is a structure that appears to have great influence on personality and initiative.

4. So the obvious question is: Who the heck thought it would be a good idea to disconnect it?

5. It started in 1890, when German researcher Friederich Golz removed portions of his dog's brain.

6. He noticed afterward that the dog was slightly more mellow—and the lobotomy was born.

7. The first lobotomies performed on humans took place in Switzerland two years later.

8. The six patients who were chosen all suffered from schizophrenia.

9. While some did show post-op improvement, two others died.

10. Apparently this was a time when an experimental medical procedure that killed 33 percent of its subjects was considered a success.

11. Despite these grisly results, lobotomies became more commonplace.

12. One early proponent of the surgery even received a Nobel Prize.

13. The most notorious practitioner of the lobotomy was American physician Walter Freeman.

14. From the 1930s to the 1960s, Freeman performed the procedure on more than 3,000 patients.

15. Rosemary Kennedy, the sister of President John F. Kennedy, was one such patient.

16. Freeman pioneered a surgical method in which a metal rod (known colloquially as an "ice pick") was inserted into the eye socket, driven up into the brain, and hammered home.

17. This is known as a transorbital lobotomy.

18. Freeman and other doctors in the U.S. lobotomized an estimated 40,000 patients before an ethical outcry over the procedure prevailed in the 1950s.

19. Although the mortality rate had improved since the early trials, it turned out that the ratio of success to failure was not much higher.

20. A third of the patients got better, a third stayed the same, and a third became much worse.

21. The practice had generally ceased in the United States by the early 1970s.

22. It is now illegal in some states.

23. Lobotomies were performed only on patients with extreme psychological impairments, after no other treatment proved successful.

24. The frontal lobe of the brain is involved in reasoning, emotion, and personality, and disconnecting it can have a powerful effect on a person's behavior.

25. Unfortunately, the changes that a lobotomy causes are unpredictable and often negative.

26. Today, there are far more precise and far less destructive manners of affecting the brain through antipsychotic drugs and other pharmaceuticals.

21 Facts About Trepanation

1. Trepanation (also known as "trephination") is the practice of boring into the skull and removing a piece of bone, thereby leaving a hole.

2. It is derived from the Greek word *trypanon*, meaning "to bore."

3. This practice dates back 7,000 years and is still practiced today.

4. It was performed by the ancient Greeks, Romans, and Egyptians, among others.

5. Hippocrates, considered the father of medicine, indicated that the Greeks might have used trepanation to treat head injuries.

6. However, evidence of trepanning without accompanying head trauma has been found in other civilizations.

7. Speculation abounds as to its exact purpose.

8. Since the head was considered a barometer for a person's behavior, one theory is that trepanation was used as a way to treat headaches, depression, and other conditions that had no outward trauma signs.

9. In trepanning, the Greeks used an instrument called a *terebra*, an extremely sharp piece of wood with another piece of wood mounted crossways on it as a handle and attached by a thong.

10. The handle was twisted until the thong was extremely tight.

11. When released, the thong unwound, which spun the sharp piece of wood around and drove it into the skull like a drill.

12. Although it's possible that the *terebra* was used for a single hole, it is more likely that it was used to make a circular pattern of multiple small holes, thereby making it easier to remove a large piece of bone.

13. The Incas were also adept at trepanation.

14. The procedure was performed using a ceremonial tumi knife made of flint or copper.

15. The surgeon held the patient's head between his knees and rubbed the tumi blade back and forth along the surface of the skull to create four incisions in a crisscross pattern.

16. When the incisions were sufficiently deep, the square-shaped piece of bone in the center was pulled out.

17. Doctors still use this procedure, only now it's called a craniotomy.

18. It still involves removing a piece of skull to get to the underlying tissue.

19. The bone is replaced when the procedure is done.

20. If it is not replaced, the operation is called a craniectomy.

21. That procedure is used in many different circumstances, such as for treating a tumor or infection.

28 Medical Slang Terms Decoded

1. **Appy:** a person's appendix or a patient with appendicitis

2. **Baby Catcher:** an obstetrician

3. **Bagging:** manually helping a patient breathe using a squeeze bag attached to a mask that covers the face

4. **Banana:** a person with jaundice (yellowing of the skin and eyes)

5. **Blood Suckers/Leeches:** those who take blood samples, such as laboratory technicians

6. **Bounceback:** a patient who returns to the emergency department with the same complaints shortly after being released

7. **Bury the Hatchet:** accidentally leaving a surgical instrument inside a patient

8. **CBC:** complete blood count; an all-purpose blood test used to diagnose different illnesses and conditions

9. **Code Brown:** a patient who has lost control of his or her bowels

10. **Code Yellow:** a patient who has lost control of his or her bladder

11. **Crook-U:** similar to the ICU or PICU, but referring to a prison ward in the hospital

12. **DNR:** do not resuscitate; a written request made by terminally ill or elderly patients who do not want extraordinary efforts made if they go into cardiac arrest, a coma, etc.

13. **Doc in a Box:** a small health-care center, usually with high staff turnover

14. **FLK:** funny-looking kid

15. **Foley:** a catheter used to drain the bladder of urine

16. **Freud Squad:** the psychiatry department

17. **Gas Passer:** an anesthesiologist

18. **GSW:** gunshot wound

19. **MI:** myocardial infarction; a heart attack

20. **M & Ms:** mortality and morbidity conferences where doctors and other health-care professionals discuss mistakes and patient deaths

21. **MVA:** motor vehicle accident

22. **O Sign:** an unconscious patient whose mouth is open

23. **Q Sign:** an unconscious patient whose mouth is open and tongue is hanging out

24. **Rear Admiral:** a proctologist

25. **Shotgunning:** ordering a wide variety of tests in the hope that one will show what's wrong with a patient

26. **Stat:** from the Latin *statinum*, meaning immediately

27. **Tox Screen:** testing the blood for the level and type of drugs in a patient's system

28. **UBI:** unexplained beer injury; a patient who appears in the ER with an injury sustained while intoxicated that he or she can't explain

15 Facts About Taste Buds

1. Babies are born with taste buds on the insides of their cheeks.

2. Overall babies have more taste buds than adults, but they lose them as they grow older.

3. Adults have, on average, around 10,000 taste buds.

4. An elderly person might have only 5,000.

5. One in four people is a "supertaster" and has more taste buds than the average person.

6. Supertasters have more than 1,000 taste buds per square inch.

7. Twenty-five percent of humans are "nontasters" and have fewer taste buds than other people their age.

8. Nontasters have only about 10 taste buds per square inch.

9. A taste bud is 30 to 60 microns (slightly more than $1/1000$ inch) in diameter.

10. Taste buds are not just for tongues—they also cover the back of the throat and the roof of the mouth.

11. Attached to each taste bud are microscopic hairs called microvilli.

12. Taste buds are regrown every two weeks.

13. About 75 percent of what we think we taste is actually coming from our sense of smell.

14. Along with sweet, salty, sour, and bitter, there is a fifth taste, called umami, which describes the savory taste of foods such as meat, cheese, and soy sauce.

15. The absolute threshold of taste is one teaspoon of sugar mixed into two gallons of water—most people can only taste it 50 percent of the time.

64 Brain & Nervous System Facts

1. The weight of the human brain triples during the first year of life, going from about 10.5 ounces to about 32 ounces (300 grams to about 900 grams).

2. At birth, babies have about 100 billion brain cells, but most of their neurons aren't yet connected.

3. That process is complete by age 3.

4. Only 4 weeks after conception, a human embryo's brain is already developing at an astonishing pace.

5. During this stage of early fetal development, neurons are forming at a rate of 250,000 per minute!

6. The average adult human brain weighs about 3 pounds.

7. It makes up about 2 percent of the total body weight.

8. After you're 30 years old, the brain shrinks a quarter of a percent in mass each year.

9. From early childhood through puberty, synapses in the human neocortex are lost at a rate of 100,000 per second.

10. Fevers are controlled by the part of the brain called the hypothalamus.

11. Temperatures greater than 109 degrees can be fatal.

12. Located in the lower back portion of the brain, the cerebellum controls such things as posture, walking, and coordination.

13. Scientists also think the cerebellum plays a role in the way scents are processed.

14. The neurons in the human brain allow information to travel at speeds up to 268 miles per hour.

15. The brain itself does not feel pain, so neurosurgeons can perform brain operations while patients are awake.

16. The brain contains on average 86 billion neurons.

17. The common figure once cited was 100 billion neurons, but more specific tests lowered the estimate.

18. There are 1,000 to 10,000 synapses for each neuron in the brain.

19. An active brain produces new dendrites, which are the connections between nerve cells that allow them to communicate with one another.

20. Fat makes up about 10 percent of your brain.

21. A fatty substance called myelin wraps and insulates a number of your brain's neurons.

22. Without the insulator myelin, which increases brain efficiency, the brain would be ten times bigger and each person would have

to eat ten times as much food to provide enough energy for the brain to function.

23. The frontal lobes of your brain create feelings of self-awareness.

24. Evidence suggests that children develop self-awareness at around 18 months of age.

25. The outer part of your brain—the cortex—is split into right and left hemispheres.

26. They are connected by a bundle of 50 million neurons.

27. Brain scans of cab drivers in London showed that their hippocampi—the part of the brain that helps us navigate—was larger than those of other people.

28. A piece of human brain the size of a grain of sand contains 100,000 neurons and 1 billion synapses, all "talking" to one another.

29. The first cervical dorsal spinal nerve and dorsal root ganglia, which help bring sensory information into the brain and spinal cord, are missing in 50 percent of all people.

30. The part of your brain that keeps risky behavior in check isn't fully formed until 25 years of age.

31. Dates, statistics, and other factual memories (such as trivia) are stored in the front left side of the brain.

32. The human brain is about 75 percent water.

33. Your brain uses fatty acids from fats to create the specialized cells that allow you to think and feel.

34. Your amygdala is responsible for generating negative emotions such as anger, sadness, fear, and disgust.

35. Working on non-emotional mental tasks inhibits the amygdala, which is why keeping yourself busy can cheer you up when you're feeling down.

36. Alcohol weakens connections between neurons and makes new

cells grow less quickly, which interfere with brain activity and causes serious damage.

37. Scientists have found that it is impossible to learn something well enough to create a "permanent" memory.

38. All memories have a limited lifetime.

39. The first person to record electrical activity in the brain was Richard Caton.

40. He accomplished this feat in 1875.

41. An estimated 3 million Americans stutter, mostly children under age six.

42. By adulthood, fewer than 1 percent stutter.

43. Some scientists believe that stutterers have neurological differences from non-stutterers.

44. The human spinal cord houses about 1 billion neurons.

45. An image of a single item, such as a house or a face, activates at least 30 million neurons in the visual cortex of the brain.

46. The average person produces between 14 and 17 fluid ounces (400 and 500 milliliters) of cerebrospinal fluid every day.

47. Wearing a helmet when riding a bicycle can reduce the risk for brain injury by up to 88 percent.

48. The brain uses a whopping 17 percent of the body's energy.

49. The stress caused by frequent jetlag and changing work hours (like those experienced by airline employees and shift workers) can damage memory and the temporal lobe of the brain.

50. People who have damage in their brain's frontal lobe may not be able to tell when the punch line of a joke is funny.

51. Your brain can tell the difference between your own touch and someone else's—that's why you can't tickle yourself.

52. The amygdala allows you to read people's faces to determine how they are feeling.

53. Male and female brains have different reactions to bodily pain.

54. An ancient Greek doctor named Alcmaeon was the first to conclude—in 450 B.C.!—that the brain, not the heart, is the origin of thoughts and feelings.

55. Scientists have found that the planum temporale is the area of the brain responsible for giving musicians perfect pitch.

56. Over time, human brains have increased in size—at a rate of about 0.5 percent per decade.

57. Since an individual brain neuron measures only four microns in thickness, about 30,000 neurons would fit on the head of a pin.

58. The group of spinal nerves at the lower end of the spinal cord is called the cauda equina.

59. The name is perfectly descriptive: It's Latin for "tail of a horse."

60. People with the rare disorder agnosia (damage to areas of the occipital or parietal brain lobes) can't recognize and identify objects, and may not know whether a person's face is familiar to them.

61. Prolonged stress can kill cells in the hippocampus, the part of your brain that's critical for memory.

62. Thankfully, we're able to grow new neurons in this area again, even as adults.

63. Your brain is swaddled in several layers of membranes called meninges.

64. The fluid between these layers produces a water cushion that protects your brain if you bump your head.

24 Facts About Skin & Nails

1. Skin is your largest body organ, accounting for about 15 percent of your body weight.

2. Your skin covers an area of about 21 square feet.

3. Humans shed about 600,000 particles of skin every hour.

4. That works out to about 1.5 pounds each year, so the average person will lose around 105 pounds of skin by age 70.

5. New skin cells take about a month to reach the surface as the cells above them die off.

6. Every square inch of human skin has about 32 million bacteria on it, but fortunately, the vast majority of them are harmless.

7. A pair of feet have 500,000 sweat glands and can produce more than a pint of sweat a day.

8. If you're clipping your fingernails more often than your toenails, that's only natural.

9. The nails that get the most exposure and are used most frequently grow the fastest.

10. Fingernails grow fastest on the hand that you write with and on the longest fingers.

11. On average, nails grow about one-tenth of an inch each month.

12. Fingerprints are formed while a fetus is growing and are the result of DNA and environmental influences in the womb.

13. Factors such as contact with amniotic fluid and the pressure of bone growth affect the unique patterns.

14. By the second trimester of pregnancy, the ridges and loops in our digits are permanently etched into our skin.

15. Fingerprint identification is far from an exact "science."

16. Analysts look for points of similarity, but there are no universal standards, and no research dictates the number of points that establish a match with certainty.

17. Goosebumps happen when the muscles around the follicles of your body hair contract.

18. In animals with more fur, this reflex helps keep the body warm.

19. Your legs contain fewer oil glands than any other part of your body.

20. You have between 2 and 4 million eccrine sweat glands—those are the glands that are found all over your body.

21. Receptors—cells that communicate with sensory nerves—differ in the sizes of their receptive fields.

22. The receptors in your fingertips have smaller fields than the receptors on other parts of your body, making them better able to distinguish touches.

23. Your hands have about 17,000 touch receptors, concentrated in your fingertips.

24. Your eyelids have the thinnest skin of your body.

13 Medical Device Facts

1. A spirometer measures how much air your lungs take in and expel.

2. An autoclave uses heat to sterilize medical equipment.

3. A portable Automated External Defibrillator, or AED, delivers an electric shock to restore heart rhythm.

4. The device that doctors use to look inside your ear is called an otoscope.

5. Light and a magnifying lens make it easier for the doctor to see ear problems.

6. Built along the same line, a rhinoscope is used for examinations of the nose.

7. An ophthalmoscope is used to examine the eye.

8. The fancy name for a blood pressure monitor is a sphygmomanometer.

9. It's from Greek words for heartbeat and measurement.

10. The reflex hammer used to test reflexes comes in several models, including the Queen Square, the Babinski, the Buck, the Berliner, and the Stookey.

11. Electrocardiograph machines test the electrical function of the heart.

12. An ECG (or EKG) can detect heart rhythm abnormalities and blockages.

13. A tympanometer tests middle-ear function.

56 History of Anesthesia Facts

1. Today, it's nearly unimaginable to consider having surgery without the aid of anesthesia.

2. Yet for most of human history, people had no surefire way to guarantee pain-free surgery.

3. Humans have long tried to find ways to dull pain.

4. Alcohol was probably the earliest method, used in ancient Mesopotamia, where the Sumerians also cultivated opium poppy.

5. The opium latex obtained from the poppy is about 12 percent morphine, and was described by the Sumerians as *hul gil*, or "plant of joy."

6. The Babylonians soon became aware of opium's numbing effects, and the plant, along with their empire, next spread to Persia and Egypt.

7. The ancient Egyptians also made crude sedatives out of the mandrake fruit, which contains psychoactive agents that can cause hallucinations.

8. In India and China, cannabis incense was used to promote a feeling of relaxation, sometimes along with a cup of wine.

9. According to eyewitness reports, Bian Que, an early Chinese physician who practiced in the mid-300s B.C., was said to have used a "toxic drink" to induce a comatose-like state in two patients for several days.

10. Five centuries later, Chinese physician Hua Tuo is said to have used a combination of sedating herbs he called *mafeisan*, which would be dissolved in wine for a patient to drink before surgery.

11. According to surviving records, Hua Tuo performed many surgeries with the help of this mixture.

12. But its exact recipe was lost when the physician burned his own notes shortly before his death.

13. During the Middle Ages, Arabic and Persian physicians discovered that anesthetics could be inhaled as well as ingested.

14. Persian physician Ibn Sina, also known by the Latinized name Avicenna, described in the 1025 encyclopedia *The Canon of Medicine* a method of holding a sponge infused with narcotics beneath a patient's nose.

15. By the 13th century, the English were using a mixture of bile, opium, bryony, henbane, hemlock, lettuce, and vinegar called *dwale*, as a sedative.

16. This potion was not administered by physicians, who warned against its use, possibly because some of the ingredients can be poisonous.

17. Recipes to make the mixture were found in remedy books that ordinary housewives could tuck right next to their cookbooks.

18. By 1525, Swiss physician Paracelsus had discovered that the compound diethyl ether had analgesic properties.

19. But it would take more than two centuries before physicians would start to consider using it as an anesthetic.

20. These practices all had various levels of success and plenty of risks, but nothing was foolproof when it came to surgical anesthetic.

21. In fact, by the turn of the 19th century, surgery was performed only as a last resort—usually with very little, if any, anesthetic.

22. Not surprisingly, this caused immense fear and pain for patients, and extreme anxiety for the doctors who had to operate on them.

23. Thankfully, the discovery of gases like oxygen, ammonia, and nitrous oxide in the late 1700s gave scientists new ideas, and physicians and patients some new options.

24. English physician Thomas Beddoes was one of the first to take notice of the new discovery of gases.

25. Beddoes founded a medical research facility called the Pneumatic Institution in 1798.

26. His aim was to find therapeutic ways to use different gases.

27. Beddoes hoped to treat breathing issues like asthma and tuberculosis.

28. While working at the Pneumatic Institution, chemist Humphry Davy (who would later go on to invent the field of electrochemistry) discovered that nitrous oxide had analgesic effects.

29. He also noted that it produced a feeling of euphoria, which prompted him to call it "laughing gas."

30. Davy suggested that nitrous oxide gas should be used during surgical procedures.

31. This idea was not immediately acted upon, perhaps because Davy himself was not a physician.

32. In 1813, Davy was joined by a new assistant, Michael Faraday (now best known for his work with electromagnetism).

33. Faraday began studying the inhalation of ether and found it to also have analgesic effects, as well as having the ability to cause sedation.

34. Even after he published his findings in 1818, ether was mostly ignored by surgeons, too.

35. But strangely, in the United States, these gases found another use, when people realized they could be inhaled recreationally at "laughing gas parties" and "ether frolics."

36. By the mid-1800s, traveling lecturers and showmen would hold these unusual gatherings.

37. Members of the audience were encouraged to inhale nitrous oxide or ether while other audience members laughed at the mind-altering results.

38. One of the participants of these "ether frolics" was a young dentist named William Morton, who was intrigued by the analgesic potential of the gas.

39. Morton tested it on animals and then successfully used it for several patients.

40. After that, Morton was so confident in ether's ability to anesthetize that he offered to demonstrate his method to Dr. John Warren, the surgeon at Massachusetts General Hospital.

41. On October 16, 1846, before a large audience in what is now known as the "Ether Dome," Morton administered diethyl ether to a young patient named Edward Gilbert Abbott.

42. Warren then removed a tumor from Abbott's neck.

43. To the surprise of everyone in attendance, even the surgeon, Abbott appeared to remain comfortable during the entire procedure.

44. Afterwards, Abbott reported that he'd felt a scratching sensation, but no pain.

45. News of Abbott's surgery quickly spread around the world.

46. By December 1846, physicians in Great Britain were making use of inhalation anesthesia.

47. Scottish obstetrician James Young Simpson first used chloroform in 1847.

48. The use of ether and chloroform made surgery a less distressing venture for patient and surgeon alike.

49. The success of the compounds led to more anesthesia research.

50. The first intravenous anesthesia, sodium thiopental, debuted for human use in March 1934.

51. In the second half of the 20th century, Belgian doctor Paul Janssen synthesized more than 80 pharmaceutical compounds, including drugs used for anesthesia.

52. Ether and chloroform eventually fell out of favor as safer anesthesia alternatives became available.

53. Ether was highly flammable, so could not be used once electricity was used to monitor patients or when wounds were being cauterized.

54. Chloroform became associated with a high number of cardiac arrests.

55. Modern anesthesia is among the safest of all medical procedures.

56. Surprisingly, however, scientists are still uncertain exactly how anesthesia is able to affect our conscious minds.

18 Facts About Bones

1. An adult has fewer bones than a baby.

2. We start off life with 350 bones, but because bones fuse together during growth, we end up with only 206 as adults.

3. People stop growing taller when the growth plates at the ends of their bones fuse.

4. This tends to happen around 14 or 15 for girls, and is tied to the age at which they began menstruation.

5. Most boys reach their full height around age 16 to 18, but it can occur later.

6. While bones stop growing, bone cells regenerate.

7. More than a quarter of your bones are found in your hands and wrists: 27 bones in each.

8. Another quarter of your bones are located in your feet: 26 in each foot.

9. The largest bone of your foot, the heel bone, is called the calcaneus.

10. A child's bone heals more rapidly from a break than an adult's.

11. The hyoid bone in your throat isn't connected to a joint, unlike any other bone.

12. The femur is the longest, strongest, and largest bone in the body.

13. The tibia is named for its resemblance to a Latin pipe called a tibia.

14. The radius is named after its resemblance to the spoke of a wheel.

15. The fibula is named for its resemblance to a clasp in a brooch.

16. Your phalanges—or finger bones—are tied linguistically to the term phalanx, referring to an army formation.

17. The enamel on your teeth is stronger than your bones.

18. Your big toe is known as the hallux.

27 Digestive System Facts

1. In a lifetime, the average person produces about 25,000 quarts of saliva—enough to fill two swimming pools!

2. You get a new stomach lining every three to four days.

3. If you didn't, the strong acids your stomach uses to digest food would also digest your stomach.

4. Your stomach secretes about half a gallon of hydrochloric acid each day.

5. The small intestine is about four times as long as the average adult is tall.

6. Its loops would stretch out to 18 to 23 feet.

7. Your large intestine, though thicker than the small intestine, is actually shorter: only about 5 feet.

8. Your appendix is found right off your large intestine and is technically part of your gastrointestinal (GI) tract.

9. Scientists aren't entirely sure of the function of the appendix, and humans can easily live without it, but one hypothesis is that it stores "good" bacteria.

10. Food travels from the mouth, through the esophagus, and into the stomach in seven seconds.

11. Your esophagus is about 10 inches long.

12. The fancy name for the sound your digestive system makes when your stomach growls is borborygmus.

13. That noise is made by gas moving through your small intestine, not just your stomach, but "my small intestine is growling" doesn't have the same ring.

14. When you eat a meal, it takes anywhere from two to five days for it to work its way through your system.

15. The bulk of that time is spent in the large intestine.

16. Doctors can track how long it takes for food to make its way through your system—and where it might be getting stuck—with a bowel transit time test in which you swallow a pill with a wireless transmitter.

17. Different foods are digested at different rates.

18. Your body processes carbohydrates more quickly than proteins or fats, so fattier foods like meat takes longer to digest than fruits and vegetables.

19. That means that if you want to feel full for longer, a snack with a little bit of fat (cheese and crackers rather than just crackers) will help tide you over until dinner.

20. Some people naturally have more of the enzyme that helps you digest beans and other "gassy" vegetables.

22. Because the digestive system uses muscles, not gravity, to do its work, you could technically eat upside down.

23. In the 1820s, a fur trapper named Alexis St. Martin was accidentally shot and left with a hole in his side that meant his doctor William Beaumont could perform experiments to learn about stomachs and the digestive system.

24. Beaumont conducted experiments for a decade, treating

St. Martin as a servant during that time, before St. Martin left the area and refused to return.

25. Each day, a healthy individual releases a minimum of 17 ounces of gas due to flatulence.

26. Most gas is composed of odorless hydrogen, nitrogen, and carbon dioxide.

27. In some humans—about 30 percent of the adult population—the digestive process also produces methane.

25 Facts About Hair

1. The average human head has 100,000 hair follicles.

2. Each of those follicles is capable of producing 20 individual hairs during a person's lifetime.

3. Hair color helps determine how dense the hair on your head is.

4. Blondes (only natural ones, of course) have the densest hair, averaging 146,000 follicles.

5. People with black hair tend to have about 110,000 follicles.

6. Those with brown hair are about average with 100,000 follicles.

7. Redheads have the least dense hair, averaging about 86,000 follicles.

8. Each hair grows approximately 5 inches per year.

9. Nearly 80 percent of people in the United States say they spend more money on hair products than any other grooming goods.

10. Despite all the effort we put into coifing and pampering our locks, the hair we see is biologically dead.

11. Hair is alive only in the roots, which are fed by small blood vessels beneath the skin's surface.

12. Hair cells travel up the shaft and are eventually cut off from the blood supply that is their nourishment.

13. The cells die before being pushed out of the follicle onto the head—or back, arm, or anywhere else.

14. Hair is incredibly strong.

15. The average head of hair can support roughly 12 tons of weight—much more than the scalp it's attached to.

16. Hair grows in cycles.

17. During the first phase, hair is actively growing.

18. In the second phase, it rests in the follicle until it is pushed out of the root.

19. Healthy hair grows about 0.39 inch in a month.

20. Each hair is completely independent from the others.

21. When a hair falls out, another one may not grow directly in its place.

22. The average person loses between 50 and 100 resting hairs each day.

23. We are born with every hair follicle we'll ever have, though the composition, color, and pattern of the hair changes over time.

24. Hair is made of protein.

25. Eating a healthy diet that contains sufficient protein, vitamins, minerals, and water is the best way to ensure healthy hair.

55 Phobias and Their Definitions

1. **Ablutophobia:** fear of washing or bathing

2. **Achluophobia:** fear of darkness

3. **Acrophobia:** fear of heights

4. **Aerophobia (or aviophobia):** fear of flying

5. **Agoraphobia:** fear of open spaces, crowds, or leaving a safe place

6. **Aichmophobia:** fear of pointed objects

7. **Ailurophobia:** fear of cats

8. **Alektorophobia:** fear of chickens

9. **Algophobia:** fear of pain

10. **Amaxophobia:** fear of riding in a car

11. **Anthropophobia:** fear of people or society

12. **Anuptaphobia:** fear of staying single

13. **Arachibutyrophobia:** fear of peanut butter sticking to the roof of your mouth

14. **Arachnophobia:** fear of spiders

15. **Astraphobia:** fear of thunder and lightning

16. **Atychiphobia:** fear of failure

17. **Autophobia:** fear of being alone

18. **Barophobia:** fear of gravity

19. **Bathmophobia:** fear of stairs or steep slopes

20. **Bibliophobia:** fear of books

21. **Catoptrophobia:** fear of mirrors

22. **Chronomentrophobia:** fear of clocks

23. **Chronophobia:** fear of time or the passage of time

24. **Claustrophobia:** fear of closed spaces

25. **Coulrophobia:** fear of clowns

26. **Cynophobia:** fear of dogs

27. **Dystychiphobia:** fear of accidents

28. **Emetophobia:** fear of vomiting

29. **Gamophobia:** fear of marriage

30. **Genuphobia:** fear of knees

31. **Glossophobia:** fear of public speaking

32. **Hemophobia:** fear of blood

33. **Hippopotomonstrosesquipedaliophobia:** fear of long words...go figure!

34. **Iatrophobia:** fear of doctors

35. **Ichthyophobia:** fear of fish

36. **Koumpounophobia:** fear of clothing buttons

37. **Lockiophobia:** fear of childbirth

38. **Melanophobia:** fear of the color black

39. **Mageirocophobia:** fear of cooking

40. **Mysophobia:** fear of germs or dirt

41. **Necrophobia:** fear of death or dead things

42. **Nomophobia:** fear of being without your mobile phone

43. **Nosocomephobia:** fear of hospitals

44. **Nyctophobia:** fear of the dark or of night

45. **Ophidiophobia:** fear of snakes

46. **Ornithophobia:** fear of birds

47. **Philemaphobia:** fear of kissing

48. **Philophobia:** fear of being in love

49. **Pupaphobia:** fear of puppets

50. **Pyrophobia:** fear of fire

51. **Scoptophobia:** fear of being stared at

52. **Trichophobia:** fear of hair

53. **Trypophobia:** fear of small holes or bumps

54. **Xanthophobia:** fear of the color yellow

55. **Zuigerphobia:** fear of vacuum cleaners

46 Facts About Sleep

1. By 60 years of age, 60 percent of men and 40 percent of women will snore.

2. While snores average around 60 decibels, the noise level of normal speech, they can reach more than 80 decibels.

3. Eighty decibels is as loud as the sound of a pneumatic drill breaking up concrete, and noise levels over 85 decibels are considered hazardous to the human ear.

4. Snoring can be a sign of sleep apnea, a life-threatening sleep disorder.

5. As many as 10 percent of people who snore have sleep apnea.

6. Sleep apnea can cause people to stop breathing as many as 300 times every night and can lead to a stroke or heart attack.

7. The longest time a human being has gone without sleep is 11 days and 25 minutes.

8. The world record was set by American 17-year-old Randy Gardner in 1963.

9. On average, humans sleep three hours less per night than other primates.

10. Chimps, rhesus monkeys, and baboons sleep ten hours per night.

11. When we sleep, we drift between rapid-eye-movement (REM) sleep and non-REM sleep in alternating 90-minute cycles.

12. Non-REM sleep starts with drowsiness and proceeds to deeper sleep, during which it's harder to be awakened.

13. During REM sleep, our heart rates increase, our breathing becomes irregular, our muscles relax, and our eyes move rapidly beneath our eyelids.

14. REM sleep was initially discovered years before the first studies that monitored brain waves overnight were conducted in 1953, though scientists didn't understand its significance at first.

15. Studies have shown that our bodies experience diminished capacity after we've been awake for just 17 hours.

16. We behave as if we were legally drunk.

17. After five nights with too little sleep, we actually get intoxicated twice as fast.

18. During their first year, babies cause between 400 and 750 hours of lost sleep for parents.

19. A newborn (0–3 months) needs 16.5 hours of sleep a day.

20. Infants (4–11 months) need about 12 to 15 hours of sleep per day.

21. Toddlers (1–2 years) should get 11 to 14 hours of sleep a day.

22. Preschoolers (3–5 years) should average 10 to 13 hours a day.

23. The average adult needs 7 to 9 hours of sleep per day.

24. Some adults called "short sleepers" naturally require less than 6 hours of sleep at night.

25. At the other extreme are "long sleepers" who need 9 or more hours of sleep per night.

26. There is no evidence that seniors need less sleep than younger adults.

27. If you average 8 hours of sleep a night, you'll have slept for more than 100,000 hours by the time you're 35.

28. During sleep, the body manufactures a hormone that prevents movement, which is why you don't actually act out your dreams.

29. At least 10 percent of all people sleepwalk at least once in their lives.

30. Men are more likely to sleepwalk than women.

31. Sleepwalking occurs most commonly in middle childhood and preadolescence (11 to 12 years of age), and it often lasts into adulthood.

32. When we sleep, our bodies cool down.

33. Body temperature and sleep are closely related.

34. Most people sleep best in moderate temperatures.

35. Caffeine can overcome drowsiness, but it actually takes about 30 minutes before its effects kick in, and they are only temporary.

36. Our eventual need for sleep is due in part to two substances the body produces—adenosine and melatonin.

37. While we're awake, the level of adenosine in the body continues to rise, signaling a shift toward sleep. While we sleep, the body breaks down adenosine.

38. When it gets dark, the body releases the hormone melatonin, which prepares the brain and body for sleep and helps us feel drowsy.

39. Most people average about four dreams per night, each lasting roughly 20 minutes.

40. People reaching 80 years old will have had approximately 131,400 dreams in their lifetimes.

41. Most people take about 6 seconds to yawn.

42. If you see someone yawn, there's a 55 percent chance you'll also yawn within 5 minutes.

43. There's a 65 percent chance you'll start yawning soon, just because you've been reading about yawning!

44. A study from 2003 found that 18 percent of people surveyed never or rarely dreamed in color.

45. In 1942 as many as 71 percent never or rarely dreamed in color.

46. Brain waves are actually more active during dreams than they are when a person is awake.

20 Elizabeth Blackwell Facts

1. Elizabeth Blackwell was the first woman to receive a medical degree in the United States.

2. She was born in 1821 in England.

3. When Blackwell was a teenager, her family moved to New York and later to Ohio.

4. Blackwell was a schoolteacher and an active social reformer who believed in the abolition of slavery and women's rights.

5. Blackwell applied to and was rejected by several medical schools.

6. She was finally accepted in 1847 to Geneva Medical College in New York after the current crop of medical students voted to accept her.

7. The vote had to be unanimous: If even one male student had voted against her, Blackwell would have been denied.

8. Her graduate thesis discussed typhus fever.

9. She received her medical degree in 1849.

10. Blackwell did post-graduate work in Paris at a maternity ward.

11. While working with an infant patient with an eye infection, she herself became infected and lost her sight in one eye.

12. Blackwell's sister Emily also pursued a medical degree.

13. The two sisters, along with another female doctor, set up the New York Infirmary for Indigent Women and Children in 1853.

14. Today, that facility is the New York-Presbyterian Lower Manhattan Hospital.

15. Rebecca Cole, the second Black woman in the United States to become a physician, interned there.

16. Blackwell mentored Elizabeth Garrett Anderson, the first woman to become a physician in Britain.

17. Blackwell was one of the founders (along with Florence Nightingale and several others) to create the London School of Medicine for Women, a medical school for female doctors.

18. She thought inoculation was overrated.

19. Blackwell was a proponent of eugenics, which had a strong current of support in the United States.

20. Blackwell never married but did adopt an orphan, Kitty.

40 Facts About Obscure Illnesses & Disorders

1. Progeria, which speeds up the aging process, causes people to grow old and die within just a few years.

2. It affects about 1 in 20 million people.

3. People afflicted with foreign accent syndrome (FAS) wake up one day suddenly speaking with a completely different accent—often from countries they've never even been to.

4. Doctors think FAS is caused by brain injury, and it can happen after strokes.

5. Children with the extremely rare disorder harlequin ichthyosis are born with thick, scaly patches of skin covering their face, like a suit of armor.

6. Unfortunately, the armor harms more than protects.

7. Most afflicted with harlequin ichthyosis die in childhood.

8. Kuru is a rare neuromuscular disease, but you probably don't need to worry about catching it—unless you're a cannibal.

9. That's because the disease is only transmitted by eating infected human brain tissue.

10. Pantothenate kinase-associated neurodegeneration is a rare degenerative brain disease that causes spasms, tremors, loss of speech, and blindness.

11. It commonly strikes children before the age of 10, making it both terrifying and heartbreaking.

12. Luckily, doctors estimate that only one in a million individuals are affected.

13. Sleeping Beauty syndrome—more officially known as Kleine-Levin syndrome—is no fairy tale.

14. Sufferers of this rare hypersomniac condition go through long stretches of their life sleeping.

15. Worse, when they're awake they're spaced out and nonfunctional.

16. Officially known as sirenomelia, "mermaid syndrome" is a birth defect in which an infant is born with its legs fused together.

17. The syndrome only strikes about 1 in 100,000 births.

18. People suffering from cold urticaria are allergic to cold weather: They develop rashes and hives when exposed to it.

19. People with hyperthymesia never forget what day their anniversary falls on—or anything else, for that matter.

20. Those with this extremely rare disorder (less than 100 confirmed cases as of 2022) remember every detail of every day for most of their lives.

21. People with the unusual and rare mental disorder of reduplicative paramnesia believe that they are in a place different from where they actually are.

22. Capgras syndrome is a rare psychological disorder that makes sufferers suspicious of their loved ones or even their own reflections.

23. People with Fregoli delusion believe they're being followed by someone and that everybody they see is that person dressed up in disguise.

24. The neuromuscular condition Fields is so rare that there are only two known cases in the history of recorded medical science.

25. The disease is named after British identical twins Catherine and Kirstie Fields, the only two people known to have the affliction.

26. People suffering from the little-seen mental disorder known as the Cotard delusion take low self-esteem to its limits.

27. At the extreme, patients believe they do not exist.

28. Others believe that organs are putrefying, limbs have vanished, or blood is disappearing from the body.

29. In the rare childhood neurological disorder Landau Kleffner syndrome, children suddenly lose the ability to comprehend and express language.

30. Even more strangely, sufferers of this condition sometimes completely regain speech within a few years.

31. One of the rarest of all conditions, *craniopagus parasiticus* describes a birth defect in which a "parasitic" twin head is attached to a newborn's head.

32. Fewer than 15 cases of this condition have been reported in the history of medical literature.

33. Don't tell a person suffering from subjective-double syndrome that you saw somebody who looked like him or her on the street.

34. He or she already believes that they have one or more doppelgängers.

35. Alien Hand syndrome is pretty much what it sounds like— the sensation that a force completely beyond your control is manipulating your hands.

36. Dancing eyes-dancing feet syndrome isn't nearly as fun as it sounds.

37. Symptoms of this obscure condition include irregular, rapidly twitching eyes and random muscle spasms that make sitting and standing nearly impossible.

38. A rare metabolic disorder called Fish Odor Syndrome (also known as trimethylaminuria, or TMAU) results in the afflicted releasing an enzyme called trimethylamine through their sweat, urine, and breath.

39. This enzyme also happens to give off a strong "fishy" smell.

40. Dietary changes can help control the disorder.

13 Health Cures in Tudor England

1. Urine wasn't one of the humors, but Tudor doctors did take urine samples—to taste.

2. Tasting urine goes back before Tudor England, though.

3. If urine tasted sweet, that was a sign of what is now known as diabetes.

4. If you had the gout in Tudor England, the preferred remedy was goat grease mixed with saffron.

5. Everyone in Tudor England—men, women, and children— drank beer.

6. The water was entirely too filthy to drink.

7. During plague times, doctors wore long face masks, called "beaks," and filled them with herbs to ward off the epidemic (or at least its olfactory cues).

8. One odd Tudor concept was "weapon salve."

9. If some antisocial Tudor Englishwoman slashed your arm with a sickle, you'd find her sickle and anoint the weapon with a healing potion, supposedly healing your arm.

10. Got a headache? In Tudor England, you'd use willow bark tea (naturally containing aspirin).

11. If your head hurt too much to scrounge around for willow bark, you could resort to hair and urine.

12. You'd take a lock of your hair, and boil it in your urine. Then you'd toss the hair into the fire. Headache cured!

13. To remedy lung problems, Tudor English took licorice and comfrey root.

5 Ways to Get Poisoned

1. Ingestion (through the mouth)

2. Inhalation (breathed in through the nose or mouth)

3. Ocular (in the eyes)

4. Dermal (on the skin)

5. Parenteral (from bites or stings)

30 Facts About Poison

1. More than half of poison exposures occur in children under age six, and most poisonings involve medications and vitamins, household and chemical personal-care products, and plants.

2. Close to 90 percent of all poisonings occur at home.

3. If you or someone in your house ingests something poisonous, stay calm and call 911 (if the person has collapsed or is not breathing) or your local poison control center.

4. Three-quarters of exposures can be treated over the phone with guidance from an expert.

5. Mystery novels are filled with stories of characters choosing to off their enemies with arsenic.

6. Colorless and odorless, this close relative of phosphorous exists in a variety of compounds, not all of which are poisonous.

7. Women in Victorian times used to rub a diluted arsenic compound into their skin to improve their complexions.

8. Some modern medications used to treat cancer actually contain arsenic.

9. When certain arsenic compounds are concentrated, they're deadly.

10. Arsenic has been blamed for widespread death through groundwater contamination.

11. Many historians believe that Napoleon died of arsenic poisoning while imprisoned, because significant traces of arsenic were found in his body by forensics experts 200 years after his death.

12. At the time Napoleon lived, wallpaper and paint often contained arsenic-laced pigments, and so he may have been exposed to the poison in his everyday surroundings.

13. Emerald green, a color of paint used by Impressionist painters, contained an arsenic-based pigment.

14. Some historians suggest that Van Gogh's neurological problems

had a great deal to do with his use of emerald green paint in large quantities.

15. Every part of the perennial herb belladonna is poisonous, but the berries are especially dangerous.

16. The poison attacks the nervous system instantly, causing a rapid pulse, hallucinations, convulsions, ataxia, and coma.

17. Wolfsbane was used as an arrow poison by the ancient Chinese, and its name comes from the Greek word meaning "dart."

18. Wolfsbane takes a while to work, but when it does, it causes extreme anxiety, chest pain, and death from respiratory arrest.

19. Meadow saffron can be boiled and dried, and it still retains all of its poisonous power.

20. As little as seven milligrams of this stuff could cause colic, paralysis, and heart failure.

21. Hemlock, used to execute Socrates, is poisonous down to the last leaf and will often send you into a coma before it finishes you for good.

22. Ricin is a waste product left behind when castor beans are processed for their oil.

23. It is extremely lethal, and in its concentrated form just a few grains can kill.

24. Even eating castor beans can cause damage.

25. Ricin has been used experimentally in cancer treatment, to kill targeted cells.

26. A Bulgarian dissident living in London was killed by ricin carried on the tip of an umbrella by an assassin.

27. The skin of two frogs in Colombia yields batrachotoxin that has been used in poison darts. Even tiny amounts cause heart failure.

28. Anthrax is caused by bacteria, *Bacillus anthracis*.

29. Anthrax can be found in soil and sicken grazing animals.

30. Tetrodotoxin, found in the pufferfish, is one of the most toxic substances known.

18 Medical Specialties Facts

1. Nephrologists study diseases of the kidneys.

2. Oncologists study cancer.

3. Neurologists deal with the nervous system, including the brain.

4. Hematologists study diseases of the blood.

5. Gastroenterologists study the digestive system.

6. Endocrinologists study hormones and the endocrine system.

7. Ophthalmologists are medical doctors who specialize in the eye and vision.

8. They went to a standard medical school.

9. Optometrists also specialize in eye care.

10. They get a post-college degree in an optometry school, earning a doctor of optometry degree.

11. Pathologists study diseases and their effects on the human body.

12. Subspecialties of pathology include fields like hematopathology (focused on blood), dermatopathology (focused on skin), and forensic pathology (focused on cause of death).

13. Radiologists specialize in medical imaging such as X-rays.

14. Rheumatologists deal with rheumatic diseases such as rheumatoid arthritis.

15. Pulmonologists specialize in the respiratory system.

16. Dermatologists deal with skin disorders.

17. While the term ENT (ear, nose, and throat specialist) is more common, another word for doctors who specialize in treatment for those areas is an otolaryngologist. (Whew! That's a mouthful!)

18. Podiatrists work with the lower limbs and feet.

30 Black Death Facts

1. The Black Death killed 25 million Europeans from 1347 to 1353, or one-third of the continent's population.

2. Europe would not regain its previous population level until more than a century and a half later.

3. The plague is caused by the bacteria *Yersinia pestis*, which mainly spreads when a flea bites an infected rat (or other rodent) for breakfast and then bites a human for lunch or dinner, thus passing on the bacteria.

4. The plague comes in three flavors: bubonic, pneumonic, and septicemic.

5. Bubonic is the cover girl of the bunch—the one most people associate with the term "plague."

6. The telltale symptom of bubonic plague is buboes, infected lymph nodes on the neck, armpit, and groin. They turn black and ooze blood and puss.

7. Pneumonic plague occurs when a person inhales the bacteria from someone who is infected. "Cover your mouth when you sneeze," has never made more sense.

8. Septicemic plague is when *Yersinia pestis* gets into your bloodstream.

9. Septicemic plague can cause gangrene due to tissue death in extremities like fingers and toes, turning them black.

10. The gangrene and the black buboes and lesions all contributed to the term "Black Death."

11. The Black Death infected the lymphatic system, resulting in high fever, vomiting, enlarged glands, and—in the case of pneumonic plague—coughing up bloody phlegm.

12. Bubonic plague was fatal in 30 to 75 percent of cases.

13. Pneumonic plague was fatal in 75 percent of cases.

14. Septicemic plague was always fatal.

15. Between 1347 and 1350, the plague spread across Europe, Asia, and North Africa.

16. The disease may have killed as many as 200 million people overall.

17. Improvements in sanitation helped bring the Black Death to an end, but the plague still pops up now and then in isolated outbreaks.

18. In India, between August and October 1994, 693 people contracted the plague, and 56 of them died.

19. Contemporary English-language chronicles called it the plague or the great pestilence; the phrase "Black Death" was applied later.

20. The plague was reported first in 1347 in Crimea.

21. Genoese traders then spread it to Italy, where it spread to France, Spain, Portugal, and England, and from there to Scandinavia.

22. Ships from Constantinople brought the disease to Egypt and from there to the Middle East and other parts of North Africa.

23. In Cairo, somewhere between 30 and 40 percent of the population died.

24. People were buried in mass graves.

25. The death rate among clergy was higher than that in the general population since they often cared for others.

26. Rural and more isolated areas did not have as many victims.

27. Between the 16th and 19th centuries, a "Little Ice Age" took place.

28. Some scientists attribute the global cooling period at least in part to the Black Death; as humans left areas with sparse populations, forests returned.

29. In many areas, Christians blamed and massacred Jewish communities.

30. Hundreds of Jewish communities were destroyed.

17 Facts About Body Terms

1. Another word for a burping or belching is eructation.

2. The act of sneezing is known as sternutation.

3. A hiccup is medically known as singultus.

4. Lacrimation is the act of crying.

5. When you stretch and yawn after waking up, it's called pandiculation.

6. The indentation above your upper lip is called a philtrum, after the Greek word for "love potion."

7. The clinical term for your belly button is umbilicus.

8. Excess perspiration is known medically as diaphoresis.

9. Peristalsis refers to the muscle movements that your body takes as it digests food.

10. Polydactylism refers to cases where people have extra fingers or toes.

11. Bollywood actor and heartthrob Hrithik Roshan, for example, has an extra thumb on one hand.

12. Deglutition is the act of swallowing.

13. If you're undergoing reverse peristalsis, watch out. It occurs when the contractions push the food back up instead of down— when you're about to vomit.

14. Cerumen is the medical term for earwax.

15. The front of your elbow is called the antecubitis.

16. The back, the bony end to your elbow, is called the olecranon.

17. Sphenopalatine ganglioneuralgia is the scientific name for "brain freeze," the headache you sometimes get when you eat something cold.

24 Fascinating Freud Facts

1. Sigmund Freud was born on May 6, 1856, in the Moravian town of Freiberg in the Austrian Empire (now Příbor, Czech Republic).

2. He qualified as a doctor of medicine in 1881 at the University of Vienna.

3. A neurologist and psychiatrist, Freud's research on human behavior left a lasting impact on the field of psychology.

4. He developed the concepts of the id, the ego, and the superego.

5. Freud is most famous for being the "Father of Psychoanalysis."

6. Psychoanalysis is a set of theories and therapeutic techniques used to study the unconscious mind.

7. Put together, they form a method of treatment for mental disorders.

8. Freud himself was not without issues.

9. He was a heavy smoker—smoking as many as 20 cigars a day for most of his life.

10. As a result, Freud endured more than 30 operations for mouth cancer.

11. In the 1880s, he conducted extensive research on cocaine, advocating use of the drug as a cure for a number of ills, including depression.

12. Reports indicate that Freud was probably addicted to cocaine for several years during this time period.

13. A friend for whom he prescribed cocaine was later diagnosed with "cocaine psychosis" and subsequently died.

14. Biographers referred to this as the "cocaine incident."

15. Freud suffered psychosomatic disorders and phobias, including agoraphobia (a fear of crowded spaces) and a fear of dying.

16. *The Interpretation of Dreams*, the book Freud considered his most significant work, was initially a commercial failure.

17. Only 351 copies were sold in the first six years.

18. Freud's famous couch was a gift from a patient.

19. When he began to employ the "talking cure," Freud had patients recline on the couch.

20. Though his *Theory of Sexuality* was being widely denounced as a threat to morality, Freud decided that sexual activity was incompatible with accomplishing great work and stopped having sexual relations with his wife.

21. Yet Freud is thought to have had a long affair with his wife's sister, Minna Bernays, who lived with the couple.

22. Freud denied these persistent rumors, but in 2006, a German researcher uncovered a century-old guest book at a Swiss hotel in which Freud registered himself and Minna as "Dr. Freud and wife."

23. Freud fled his native Austria after the Nazi Anschluss in 1938 and spent his last year of life in London.

24. Dying from mouth cancer, in September 1939, Freud convinced his doctor to help him commit suicide with injections of morphine.

16 Facts About Smallpox

1. Smallpox was one of the most devastating illnesses in human history, killing more than 300 million people worldwide during the 20th century alone.

2. Smallpox marks were found on Egyptian mummies, so we know the disease has been around for a very long time.

3. Before vaccination, people practiced variolation, inhaling or introducing into their system a small amount of fluid from a smallpox sore.

4. People infected through variolation often had a milder course to their disease.

5. You couldn't catch smallpox twice, so survivors were often tasked with helping other sufferers.

6. Smallpox has been wiped off the face of the earth, except for samples of the virus held in labs for research purposes.

7. Scientists declared the world free of smallpox in 1979.

8. Symptoms of smallpox included a high fever, head and body aches, malaise, vomiting, and a rash of small red bumps that progressed into sores that could break open and spread the virus.

9. The virus could also be spread via contact with shared items, clothing, and bedding.

10. The mortality rate could be as high as 30 percent.

11. A British commander during the French and Indian War used smallpox as a bioweapon.

12. Smallpox was an entirely human disease; it did not infect any other animal or insect on the planet.

13. Thus, once vaccination eliminated the chances of the virus spreading among the human population, the disease disappeared.

14. People who contracted cowpox did not contract smallpox.

15. While English doctor Edward Jenner was not the first to realize this, he put the knowledge to use and his work led to vaccination replacing variolation as a prevention method.

16. The United States has not vaccinated for smallpox since 1972.

18 Cures for the Hiccups

1. **Drinking Cure:** Swallowing water interrupts the hiccupping cycle, which can quiet nerves. Gargling with water may also have the same effect, but swallowing is probably the fastest way to cure hiccups.

2. **Pineapple Juice Cure:** Some say the acid in pineapple juice obliterates hiccups, but it's probably just the swallowing action that comes from drinking.

3. **Gulp Cure:** Just like drinking water, swallowing any food or drink is a good way to dispel the dreaded hiccups. If water or juice bores you, why not have a snack?

4. **Little Brother Cure:** If you stick out your tongue, you'll stimulate your glottis, the opening of the airway to your lungs. Since a closed glottis is what causes hiccups in the first place, this usually works pretty well.

5. **Drink Upside Down Cure:** This one is a bit unusual, but it's not totally illogical. In addition to swallowing the water, it's pretty hard to figure out how to drink upside down. The concentration needed might equalize the breathing and cure the hiccups.

6. **Cotton Swab Cure:** This cure works just like the Little Brother Cure. Take a cotton swab and tickle the roof of your mouth. People will wonder what you're doing, but it's better than drinking upside down.

7. **Sugar Cure:** Especially popular among the six-year-old set, a lump of sugar not only tickles the glottis, it gets the hiccupping person swallowing—a double threat to the hiccups.

8. **Scaredy-Cat Cure:** The effectiveness of this is dubious at best, since once you ask someone to scare you, you're not going to be truly surprised. However, if you have a friend with ESP, he or she might be able to help. Losing your breath or gasping might just reset your glottis automatically.

9. **Squeeze Cure:** Can't stop hiccupping? Squeeze those suckers outta there! Sit in a chair and compress your chest by pulling your knees up to your chin. Lean forward and feel those hiccups magically disappear.

10. **Sternum Cure:** Some hiccup experts claim that by massaging the sternum, hiccups will melt away. There's not a lot of science to substantiate this claim, but we've never met a massage we didn't like.

11. **Hear No Evil Cure:** This cure was reported in the medical journal *Lancet*, so it has to work, right? The article claims that if you plug your ears, you will, in effect, short-circuit your vagus nerve, which controls hiccups.

12. **Brown Bag Cure:** It might be that breathing into a brown paper bag cures hiccups because the hiccupping person is taking in more carbon dioxide when inhaling. Or, it might be that the person is concentrating more on breathing, slowing it down and smoothing it out.

13. **Hold Your Breath Cure:** This is one of the oldest hiccup remedies, and it usually works pretty well. What is the science behind it? It probably works the same way a paper bag does—it forces a little more control over your breathing.

14. **Earlobe Cure:** Earlobes aren't just good for wearing earrings. If you rub them, you can cure your hiccups! This silly cure has no basis in logic or fact, but try it, what do you have to lose?

15. **Headstand Cure:** Not everyone can stand on their head, but if you can, you might have a good hiccup cure. By standing on your head, you're probably using a fair amount of concentration and messing with your breathing. This should lead to a cessation of the hiccups.

16. **Sound of Music Cure:** If you sing or yell as loudly as you can for at least two minutes or longer, you might notice your hiccups leave the building. But your friends might leave, too.

17. **Sleeper Cure:** Give your glottis, throat, and diaphragm a break—lie down on your back. This is a gentler way to get rid of those obnoxious hiccups.

18. **Run for It Cure:** Run. Fast. For ten minutes. See?

20 Facts About Left-Handedness

1. About 10 percent of the population identifies as left-handed.

2. About 1 percent are ambidextrous.

3. More men than women are left-handed.

4. Some psychologists have noted that lefties are more often found in creative fields.

5. There is no standard for what constitutes left-handedness, making research into handedness difficult.

6. Left-handed pitchers are called southpaws because in the early days of the game, baseball diamonds were designed so that the batters were facing east to prevent the late-afternoon sun from shining in their eyes.

7. This meant that left-handed pitchers were throwing from the south.

8. Historical lefties include Alexander the Great, Isaac Newton, Pablo Picasso, Leonardo da Vinci, Michelangelo, Mark Twain, Oprah Winfrey, and Jimi Hendrix.

9. Left-handers are more likely to stutter, have dyslexia, and suffer from allergies.

10. Left-handed and ambidextrous presidents include James Garfield, Herbert Hoover, Harry Truman, Gerald Ford, Ronald Reagan, George H.W. Bush, Bill Clinton, and Barack Obama.

11. At the 1976 All-Star Game, the athletic and ambidextrous Gerald Ford thrilled fans by throwing one pitch right-handed to Johnny Bench of the Cincinnati Reds and a second pitch left-handed to Carlton Fisk of the Boston Red Sox.

12. Ford could write with either hand and did some athletic moves with one versus the other.

13. President James Garfield could write Latin with one hand and Greek with the other—at the same time.

14. At one time, teachers would force students to write with their right hands, even when left-handed.

15. Researchers at Oxford University have found a gene for left-handedness.

16. Scientists have found that handedness develops *in utero*.

17. The hair of right-handed people, in the great majority of cases, swirls clockwise on the top of their head, but the hair of left-handed people can swirl in any direction.

18. International Left-Handers Day is August 13, and was first celebrated in 1976.

19. Southpaws and people who write with both hands have a corpus collosum—that's the bridge between the two hemispheres of the brain—about 11 percent bigger than right-handed people.

20. Left-handed adults find many workplaces inefficient or dangerous because they're designed for right-handed people.

15 Blood Transfusion Facts

1. Blood transfusions are performed about every two seconds somewhere in the world.

2. One of the earliest recorded blood transfusions took place in the 1490s. Pope Innocent VIII suffered a massive stroke and his physician advised a blood transfusion.

3. Unfortunately, the methods used were crude and unsuccessful; the pope died within the year and so did the three young boys whose blood was used.

4. Richard Lower, an Oxford physician in the mid-1600s, performed blood transfusions between dogs and eventually between a dog and a human.

5. The dog was kept alive via the transfusion, and the experiment was considered a success.

6. Several years later, cross-species transfusions were deemed unsafe.

7. In 1818, British obstetrician James Blundell performed the first successful transfusion of human blood.

8. Blundell's patient, a woman suffering postpartum hemorrhaging, was given a syringe-full of her husband's blood.

9. The woman lived and Blundell became a pioneer in the study of blood transfusions.

10. One unit of blood can be separated into several components: red blood cells, plasma, platelets, and cryoprecipitate, which is a substance helpful in the clotting process.

11. In 1916, scientists introduced a citrate-glucose solution that allowed blood to be stored for several days.

12. Due to this discovery, the first "blood depot" was established during World War I in Britain.

13. Nearly 16 million blood components are transfused each year in the U.S.

14. If you need a blood transfusion, try not to need one in the summer or around the holidays. Shortages of all blood types happen during these times.

15. Donating blood seems to appeal to folks with a sense of civic duty: 94 percent of blood donors are registered voters.

13 Medical Leeches & Maggots Facts

1. Maggots are nothing more than fly larvae—one of the most basic forms of life.

2. But they can be used to fight infection in people who do not respond to antibiotics.

3. Applied to a dressing that is made in the form of a small "cage," maggots are applied to almost any area that does not respond well to conventional treatment.

4. The maggot thrives on consuming dead tissue, a process called "debridement."

5. After several days, the maggots are removed—but only after they have consumed up to ten times their own weight in dead tissue, cleaned the wound, and left an ammonialike antimicrobial enzyme behind.

6. Leeches are small animal organisms that have been used by physicians and barbers (who, in the olden days, were also considered surgeons) for over 2,500 years for treating everything from headaches and mental illnesses to hemorrhoids.

7. Used extensively until the 19th century, the "Golden Age of Leeches" was usurped by the adoption of modern concepts of pathology and microbiology.

8. Hirudotherapy, or the medicinal use of leeches, has enjoyed a recent resurgence after their demonstrated ability to heal patients when other means have failed.

9. Leeches are raised commercially around the world with the majority coming from France, Hungary, Ukraine, Romania, Egypt, Algeria, Turkey, and the United States.

10. Leeches feed on the blood of humans and other animals by piercing the skin with a long proboscis.

11. Oftentimes this is the most effective way to drain a postsurgical area of blood, and it can actually facilitate the healing process.

12. At the same time leeches attach to their host, they inject a blood-thinning anticoagulant; they continue until they have consumed up to five times their body weight in blood.

13. The host rarely feels the bite because the leech also injects a local anesthetic before it pierces the skin.

65 Body Part Name Facts

1. The artery of Adamkiewicz is named after Albert Wojceich Adamkiewicz, a Polish pathologist.

2. Adamkiewicz studied the central nervous system and created a test for detecting tryptophan, the amino acid popularized in the *Seinfeld* episode as making you sleepy.

3. The myenteric plexus in the gastrointestinal tract is also known as Auerbach's plexus after the German anatomist credited with its discovery, Leopold Auerbach.

4. Auerbach's son Felix became a doctor as well, and another son, Friedrich, became a chemist.

5. Friedrich's daughter and Leopold's granddaughter Charlotte became a renowned geneticist who fled Germany and Nazism in 1933.

6. A band of cardiac muscle called Bachmann's Bundle is named after Jean George Bachmann.

7. Bachmann studied cardiac rhythms in dogs in the 1910s.

8. A set of ducts in the kidneys are known as the ducts of Bellini after Lorenzo Bellini, an Italian anatomist.

9. Bellini served as physician to Pope Clement XI in the 1700s.

10. The largest neurons in the central nervous system are called Betz cells after the Ukrainian scientist, Vladimir Betz, who discovered them.

11. Young Vladimir attended a university then known as Saint Vladimir University.

12. Broca's area is a part of the brain that's linked to speech production. French physician Pierre Paul Broca worked with patients who had speech issues after suffering injuries that affected that area.

13. Broca had an interest in the pseudoscience of craniology (phrenology) and invented measuring tools for the skull; the pursuit is now widely understood as racist.

14. Cajal-Retzius cells are seen in the developing brain.

15. Two different scientists working independently described them: Santiago Ramón y Cajal and Gustaf Retzius.

16. Spanish neuroscientist Santiago Ramón y Cajal received the Nobel Prize in Physiology or Medicine in 1906.

17. Swedish anatomist Gustaf Retzius was the son of an anatomist and the grandson of a chemist.

18. The Calyx of Held and the endbulb of Held, parts of the auditory system, are named after German anatomist Hans Held.

19. The structure resembles a flower in shape, hence the name calyx.

20. A canal in your eye called the hyaloid canal is also called Cloquet's canal after French anatomist Jules Cloquet or Stilling's canal after German anatomist Benedict Stilling.

21. Cloquet was also an artist known for the scientific illustrations in his work on anatomy.

22. Benedict Stilling had a private medical practice in Germany. His son Jakob was an ophthalmologist who described an eye disorder named after him.

23. The organ of Corti, located in the cochlea, was named after its discoverer, Italian anatomist Alfonso Corti.

24. Corti made an extensive study of cochleas in both humans and animals.

25. Darwin's tubercule, which describes an ear feature that some people and primates have, is named after Charles Darwin.

26. The perisinusoidal space in your liver is also called the space of Disse after German anatomist Joseph Disse.

27. Disse was a histologist—an anatomist who studies microscopic anatomy of cells and tissues.

28. A set of salivary glands, Von Ebner's glands are named after Victor von Ebner, an Austrian anatomist.

29. He also gave his name to Ebner's lines (lines in the dentin of the tooth) and a network of cells called Ebner's reticulum.

30. The Eustachian tubes in your ears are named after early Italian anatomist Bartolomeo Eustachi.

31. Eustachi, who was born early in the 1500s, was one of the founders of anatomical study.

32. The Fallopian tubes that connect the ovaries to the uterus are named after Gabriele Falloppio, an Italian anatomist.

33. In addition to his work as an anatomist studying the eye and the ear, and how condoms could prevent syphilis, Falloppio was also a Catholic priest and a professor of botany.

34. Italian anatomist Carlo Giacomini has several structures named after him: a vein in the thigh, a vertebrae, and a structure in the hippocampus called the band of Giacomini.

35. Giacomini's skeleton and brain were preserved due to his wishes, and exhibited at a Museum of Human Anatomy.

36. An organelle found in cells called the Golgi apparatus was discovered by Italian scientist Camillo Golgi.

37. A tendon reflex is also named after Golgi.

38. A large cerebral vein is traditionally called the vein of Galen after the famed Greek physician.

39. The Haversian canals found in bones, through which blood vessels and nerves pass, are named after English physician Clopton Havers.

40. Havers was the first to describe those structures.

41. Though he also described a set of connective tissues called Sharpey's fibres in his work in the 1600s, those are named after a Scottish anatomist of the 1800s who described them later.

42. A structure in the kidney, the loop of Henle is named after German anatomist Friedrich Gustav Jakob Henle.

43. Structures in the eyeball called the Crypts of Henle help secrete a substance used in tears.

44. Heschl's gyri, or the transverse temporal gyri, are folds on the brain that help process auditory information.

45. They're named after Austrian anatomist Richard Heschl.

46. Heschl was a professor at the University of Vienna and other Austrian schools.

47. The antrum of Highmore (also called the maxillary sinus), the largest sinus in the body, is named after British surgeon Nathaniel Highmore.

48. Highmore described the sinus in 1651 as part of a larger work on human anatomy.

49. The bundle of His (pronounce it like "hiss"), a set of cells in the heart, is named after Swiss cardiologist William His Jr.

50. His father (or: His's father), William His Sr., was also an anatomist who studied nerve development.

51. Folds in the small intestine called the valves of Kerckring are named after Dutch anatomist Theodor Kerckring.

52. Kerckring lived in the 1600s and had his painting done by a pupil of Rembrandt.

53. The islets of Langerhans, found in the pancreas, secrete insulin, and were discovered by German scientist Paul Langerhans.

54. Langerhans described "islands of clear cells" in the pancreas but did not know their function.

55. German anatomist Herbert von Luschka has a number of anatomical structures named after him: foramina, crypts, joints, and ducts.

56. The "von" in his name was added when he became a noble.

57. Macewen's triangle is an area of the brain named after Scottish surgeon William Macewen.

58. Macewen (1848–1924) was instrumental in the field of brain surgery.

59. If you have tenderness at a specific location of your torso called McBurney's point, it likely indicates appendicitis.

60. The spot is named after American surgeon Charles McBurney.

61. Though his work on appendicitis was valuable, McBurney was one of the doctors who treated President McKinley after he was shot and presented an overly-optimistic view of the president's condition to the public.

62. Wernicke's area is an area in the brain linked to speech (along with Broca's area).

63. German physician and psychiatrist Karl Wernicke studied forms of aphasia caused by damage to that area.

64. The circle of Willis is the place where several arteries join together at the bottom of the brain, and is named after English physician Thomas Willis.

65. Born in 1621, Willis produced a work of anatomy containing illustrations by Christopher Wren, the architect of St. Paul's Cathedral.

20 Facts About Twins

1. Identical twins result when a zygote (a fertilized egg) divides in half, forming two embryos.

2. The embryos develop in tandem and, at birth, are identical twin siblings.

3. Scientists have found that there are variations in identical twins' individual gene segments.

4. Researchers believe these disparities occur in the womb, when dividing cells cause small genetic differences in each twin.

5. This explains why one identical twin can develop a genetic disorder while the other twin remains healthy.

6. One identical twin can be right-handed and the other left-handed.

7. Even identical twins don't have the same fingerprints.

8. Although no two fingerprints are alike, identical twins often have similar patterns.

9. Identical twins are found in about 3 or 4 out of every one thousand pregnancies.

10. Twins are becoming more common, primarily due to fertility treatments.

11. Fraternal twins are the result of two eggs being fertilized by two different sperm.

12. Scientifically, fraternal twins are called dizygotic twins; a zygote is a fertilized egg.

13. Dizygotic twins are more common than identical twins; about two in three sets of twins are dizygotic.

14. Fraternal twins do not share more genetic information than any pair of siblings.

15. If you have fraternal twins, having a boy and a girl is more common than having two children of the same gender.

16. About half of twins are born prematurely.

17. Twins are more common in pregnancies in women over 30, pregnancies that were helped along by fertility treatments, and women with a higher body weight and/or height.

18. Diet can also affect whether a woman carries twins. Eating dairy foods can increase a woman's likelihood of carrying twins.

19. Fraternal twins run in families, due to the gene that causes a woman to produce two eggs, increasing her likelihood of having twins.

20. Twins react to each other in the womb, reaching out to touch each other.

20 *Gray's Anatomy* Facts

1. Henry Gray, the writer, was born in 1827.

2. He died at age 34 of smallpox in 1861.

3. He learned anatomy and surgery at St. George's Hospital, London, and later was a lecturer there.

4. He won a prize in medical school for an essay on eyeballs.

5. The first edition of *Gray's Anatomy* was published in 1858.

6. The first edition was dedicated to Sir Benjamin Collins Brodie, a surgeon who did extensive research into joint disease.

7. A revised edition was published in 1860, the year before Gray died.

8. The illustrator, Henry Vandyke Carter, was born 1831.

9. He was studying to become a doctor and worked at St. George's School of Medicine.

10. His knack for medical illustrations—both for Gray and for others—helped him pay for his schooling.

11. Carter at one point referred to Gray as a "snob" in his journal.

12. Before they worked together on the longer project, Carter provided illustrations to Gray for an essay about spleens.

13. A few years after he married, Carter discovered that his wife's name was actually an alias and that she had been previously married.

14. He married again later in life.

15. Carter lived in India for decades.

16. He died of tuberculosis at the age of 65.

17. As of January 2023, *Gray's Anatomy* is in its 42nd edition in the United States. The book is updated frequently to keep current.

18. Editions in the United States and Britain are numbered slightly differently.

19. The first edition totaled 750 pages.

20. The 42nd edition totals 1,606 pages and touts more than "150 new radiology images."

81 Global Disease & Vaccine Facts

1. According to the U.S. Census Bureau, the average life expectancy at the beginning of the 20th century was 47.3 years.

2. A century later, that number had increased to 77.85 years, due largely to the development of vaccinations and other treatments for deadly diseases.

3. The Centers for Disease Control and Prevention (CDC) states that childhood vaccination prevents four million deaths every year.

4. In 2021, global immunization coverage dropped to 81 percent, the lowest rate in over a decade.

5. Twenty-five million children under age one did not receive basic vaccines through routine immunization.

6. This is six million more than before the start of the COVID-19 pandemic in 2019.

Chicken Pox

7. According to the CDC, before the chicken pox vaccine was approved for use in the U.S. in 1995, there were 11,000 hospitalizations and 100 deaths from the disease every year.

8. Chicken pox, caused by the varicella-zoster virus, creates an itchy rash of small red bumps on the skin.

9. The virus spreads when someone who has the disease coughs or sneezes, and a nonimmune person inhales the viral particles.

10. The virus can also be passed through contact with the fluid of chicken pox blisters.

11. Most cases are minor but in more serious instances, chicken pox can trigger bacterial infections, viral pneumonia, and encephalitis (inflammation of the brain).

12. The varicella-zoster virus can lay dormant for many years and resurface as shingles, which can be extremely painful. A vaccine for shingles also now exists.

Diphtheria

13. Diphtheria is an infection of the bacteria *Corynebacterium diphtheriae* and mainly affects the nose and throat.

14. The bacteria spreads through airborne droplets and shared personal items.

15. *C. diphtheriae* creates a toxin in the body that produces a thick, gray or black coating in the nose, throat, or airway, which can also affect the heart and nervous system.

16. Even with proper antibiotic treatment, diphtheria kills about 10 percent of the people who contract it.

17. The first diphtheria vaccine was unveiled in 1913.

18. Although vaccination has made a major dent in mortality rates, diphtheria still exists in developing countries and other areas where people are not regularly vaccinated.

Hib Disease

19. Invasive H. flu, or Hib disease, is an infection caused by the *Haemophilus influenzae* type b (Hib) bacteria, which spreads when an infected person coughs, sneezes, or speaks.

20. Invasive H. flu is a bit of a misnomer because it is not related to any form of the influenza virus.

21. It can lead to bacterial meningitis (a potentially fatal brain infection), pneumonia, epiglottitis (severe swelling above the voice box that makes breathing difficult), and infections of the blood, joints, bones, and pericardium (the covering of the heart).

22. Children younger than five years old are particularly susceptible to the Hib bacteria because they have not had the chance to develop immunity to it.

23. The first Hib vaccine was licensed in 1985.

24. Despite its success in the developed world, the disease is still prevalent in the developing world.

Malaria

25. Malaria is a parasitic infection of the liver and red blood cells.

26. In its mildest forms it can produce flu-like symptoms and nausea.

27. In its severest forms, malaria can cause seizures, coma, fluid buildup in the lungs, kidney failure, and death.

28. Malaria is transmitted by female mosquitoes of the genus *Anopheles*.

29. When the mosquito bites, the parasites enter a person's body, invading red blood cells and causing the cells to rupture.

30. As the cells burst, they release chemicals that cause malaria's symptoms.

31. In 2021, there were about 247 million cases of malaria worldwide, including 619,000 deaths.

32. Children in sub-Saharan Africa account for most of the deaths.

33. Other high-risk areas include Central and South America, India, and the Middle East.

34. While there is a new vaccine, malaria is still mostly treated with a variety of drugs, some of which kill the parasites once they're in the blood and others that prevent infection in the first place.

35. If you can avoid the parasite-carrying mosquitoes, you can avoid malaria, so the disease is often controlled using mosquito repellent and bed netting, especially in developing countries.

Measles

36. Measles is a highly contagious viral illness of the respiratory system that spreads through airborne droplets when an infected person coughs or sneezes.

37. Although the first symptoms of measles mimic a simple cold, with a cough, runny nose, and red watery eyes, this disease is more serious.

38. As measles progresses, the infected person develops a fever and a red or brownish-red skin rash.

39. Complications can include diarrhea, pneumonia, brain infection, and even death, although these are seen more commonly in malnourished or immunodeficient people.

40. Measles has historically been a devastating disease, but WHO reported that between 2000 and 2018, the mortality rate dropped 73 percent due to a global immunization drive.

41. In 2021, there were around 128,000 measles deaths globally. In 1999, that figure had stood at 871,000.

42. Until 1963, when the first measles vaccine was used in the United States, almost everyone got the measles by age 20.

43. While measles was declared eliminated from the U.S. in 2000, it has recently seen a resurgence, partly due to anti-vaccine sentiment.

44. Most children today receive the measles vaccine as part of the MMR vaccination, which protects against measles, mumps, and rubella (German measles).

Pertussis

45. Pertussis, or whooping cough, is a highly contagious respiratory infection caused by the *Bordetella pertussis* bacteria.

46. The descriptive nickname comes from the "whooping" sounds that infected children make after one of the disease's coughing spells.

47. The coughing fits spread the bacteria and can last a minute or longer, causing a child to turn purple or red and sometimes vomit.

48. Severe episodes can cause a lack of oxygen to the brain.

49. Adults who contract pertussis usually have a hacking cough rather than a whooping one.

50. Although the disease can strike anyone, it is most prevalent in infants under age one because they haven't received the entire course of pertussis vaccinations.

51. The pertussis vaccine was first used in 1933.

52. Adolescents and adults become susceptible when the immunity from childhood vaccinations wanes and they don't get booster shots.

53. More than 85 percent are now covered by the vaccines globally.

Pneumococcal Disease

54. Pneumococcal disease is the collective name for the infections caused by *Streptococcus pneumoniae* bacteria, also known as pneumococcus.

55. The most common types of infections caused by *S. pneumoniae* are middle ear infections, pneumonia, bacteremia (blood stream infections), sinus infections, and bacterial meningitis.

56. There are more than 90 types of pneumococcus.

57. The ten most common types are responsible for 62 percent of the world's invasive diseases.

58. Those infected carry the bacteria in their throats and expel it when they cough or sneeze.

59. Like any other germ, *S. pneumoniae* can infect anyone.

60. But certain population groups are more at risk, such as the elderly, people with cancer or AIDS, and people with a chronic illness such as diabetes.

61. There are two types of vaccines available to prevent pneumococcal disease, which the CDC recommends that children and adults older than age 65 receive.

Polio

62. Polio is caused by a virus that enters the body through the mouth, usually from hands contaminated with the stool of an infected person.

63. In about 95 percent of cases, polio produces no symptoms at all (asymptomatic polio), but in the remaining cases of polio, the disease can take three forms.

64. Abortive polio creates flu-like symptoms, such as upper respiratory infection, fever, sore throat, and general malaise.

65. Nonparalytic polio is more severe and produces symptoms similar to mild meningitis, including sensitivity to light and neck stiffness.

66. Paralytic polio produces the symptoms with which most people associate the disease, even though paralytic polio accounts for less than 1 percent of all cases.

67. Paralytic polio causes loss of control and paralysis of limbs, reflexes, and the muscles that control breathing.

68. Dr. Jonas Salk's inactivated polio vaccine (IPV) first appeared in 1955, and Dr. Albert Sabin's oral polio vaccine (OPV) first appeared in 1961.

Tetanus

69. Tetanus is contracted when reproductive cells (spores) of *Clostridium tetani* are found in the soil and enter the body through a skin wound.

70. Once the spores develop into mature bacteria, the bacteria produce tetanospasmin, a neurotoxin (a protein that poisons the body's nervous system) that causes muscle spasms.

71. In fact, tetanus gets its nickname—lockjaw—because the toxin often attacks the muscles that control the jaw.

72. Lockjaw is accompanied by difficulty swallowing and painful stiffness in the neck, shoulders, and back.

73. The spasms can then spread to the muscles of the abdomen, upper arms, and thighs.

74. According to the CDC, tetanus is fatal in about 10 to 20 percent of cases, but fortunately, it can't be spread from person to person—you need direct contact with C. *tetani* to contract the disease.

75. Today, tetanus immunization is standard in the U.S., but if you are injured in a way that increases tetanus risk (i.e. stepping on a rusty nail, cutting your hand with a knife, or getting bitten by a dog), a booster shot may be necessary if it's been several years since your last tetanus shot.

Yellow Fever

76. Yellow fever is spread by mosquitoes infected with the yellow fever virus.

77. Jaundice, or yellowing of the skin and eyes, is the hallmark of the infection and gives it its name.

78. Most cases of yellow fever are mild and require only three or four days to recover, but severe cases can cause bleeding, heart problems, liver or kidney failure, brain dysfunction, or death.

79. People with the disease can ease their symptoms, but there is no specific treatment, so prevention via the yellow fever vaccine is key.

80. The vaccine provides immunity from the disease for ten years or more and is generally safe for everyone older than nine months.

81. Yellow fever occurs only in Africa, South America, and some areas of the Caribbean, so only residents or travelers who are destined for these regions need to be concerned about it.

27 Facts About Parasites

1. There are more than 130 parasites that can inhabit the human body.

2. Researchers suspect that instances of Crohn's disease, a once-rare inflammatory intestinal disorder, may be on the rise because of the lack of intestinal parasites in much of the developed world.

3. The parasitic worms that live in your intestines are called helminths.

4. Demodex mites are also called "face mites," because they live on human hair follicles, eyelashes, and nose hairs.

5. As they're only 0.0118 inch long, as many as 25 Demodex mites can live on a single hair follicle.

6. There are more than 3,000 species of lice in the world.

7. Head lice are parasites that live on human hair, gripping the shafts with their claws and drinking blood from the scalp.

8. Although head lice only live for a month in your hair, each one can lay up to 100 eggs during that time.

9. Leeches can suck ten times their body weight in blood.

10. Tapeworms can grow more than 60 feet long in human intestines.

11. The tapeworm's segmented tail contains eggs.

12. This is so that when the segments break off and are expelled from the host's body, the eggs can move on to another animal.

13. Instead of a head, the tapeworm has a hooked knob that it uses to cling to the intestinal walls as it sucks nutrients off the surface.

14. Tapeworms can only reproduce in humans.

15. When the eggs are eaten by another animal, they reside in the animal's muscle tissue until that flesh is consumed by a human as under-cooked meat.

16. Upon attaching to a human host, a chigger (also called red bugs or harvest mites) uses an enzyme to dissolve the flesh at the bite, and then it consumes the liquefied tissue.

17. Mosquitoes help transmit botfly eggs to humans, where they hatch and burrow into the skin.

18. To remove them, lay a slice of raw meat on the skin. The maggots will leave the body and enter the meat instead.

19. Roundworms grow to 15 inches long and lay as many as 200,000 eggs daily.

20. Roundworms are the most common form of intestinal parasite, with an estimated one billion hosts worldwide.

21. Rather than feeding on the material found in human intestines, hookworms attach themselves to feed on the blood and intestinal tissue.

22. Occasionally, a whipworm infection is discovered when a worm crawls out of the anus or up through the throat and out through the nose or mouth.

23. Whipworms can cause the loss of approximately one teaspoon of blood per day.

24. After entering the human body in larval form, the two-foot-long adult Guinea worm exits by creating a hole in the flesh of the leg.

25. Giardiasis is caused by a one-celled parasite found in dirty water; noticeable symptoms include sulfurous belches and flatulence.

26. One out of every three people worldwide is hosting an intestinal parasite.

27. Mosquitoes that harbor the malarial parasite plasmodia bite more people per night and live longer than uninfected mosquitoes, but they lay fewer eggs.

28 Bias, Racism, & Sexism in Medicine Facts

1. The pulse oximeter, a device that is clipped around the patient's finger to determine oxygen levels, is less accurate for people with darker skin.

2. It can give slightly higher values than are accurate for Black patients, leading doctors to treat low oxygen inadequately.

3. One 2020 evidence-review suggests that while one pulse oximeter on the market had solved the problem, others had not, suggesting that those devices had been calibrated by testing lighter-skinned people and variations in skin color not accounted for.

4. A statue of J. Marion Sims, a doctor called the "father of modern gynecology," stands at the South Carolina capitol.

5. Sims developed several gynecological procedures through experimentation on enslaved Black women with those conditions.

6. He often operated on the enslaved women without anesthesia.

7. A common lie during the time of slavery in the United States to justify brutality was that Black people did not feel pain in the same way due to "thicker" or "less sensitive skin."

8. A 2016 study of medical students showed that those beliefs had seeped into the culture and persisted.

9. Forty percent of trainees in the study believed that Black patients were not as sensitive to pain as White patients.

10. Women who complain of pain are more likely than men to be prescribed sedatives rather than pain medication.

11. Men and women often have different symptoms when they have a heart attack.

12. Women are more likely to be misdiagnosed as fine and discharged from the ER.

13. Men and women process pain differently. Studies that used only male mice to test efficacy of pain medication mean that pain medication might not be effective for women.

14. The infant mortality rate for Black babies is significantly higher than that of White children.

15. A study released in 2020 showed that when Black doctors cared for Black babies, the gap in mortality rates shrank greatly.

16. The term "Mississippi Appendectomy" refers to the widespread practice by which women, primarily Black women who were seeking treatment for other issues, were sterilized without their consent during the 20th century.

17. In one state, North Carolina, 7,600 women were sterilized between the 1930s and the 1970s, 5,000 of them Black women.

18. During the same time span, Puerto Rican women were also undergoing forced sterilization at a high rate.

19. As many as 1 in 3 women were targeted.

20. Because rules against oral contraception were stricter in the mainland United States, women in Puerto Rico were used in trials; they were generally not fully informed of risks.

21. In California, some 20,000 people were sterilized without consent, disproportionately Latina women.

22. Between 1973 and 1976 alone, more than 3,400 Native American women were sterilized without consent.

23. Compulsory sterilization of the disabled while they were in prison was legal for decades in the United States.

24. The Tuskegee experiments ran for decades and only ended in 1972.

25. Black men who sought treatment for syphilis were told they were getting free care. They were not. Researchers were charting the course of untreated syphilis.

26. In 2020, the American Heart Association released a statement that noted that more than half of LGBTQ patients reported discrimination from health care professionals, leading to adverse outcomes.

27. In 1995, EMTs were treating a woman named Tyra Hunter after a car accident. They abruptly stopped treating her, and when she was brought to the hospital she was given substandard care and died. The EMTs had stopped treating her when they realized she was trans.

28. In 2010, a study found that 19 percent of trans people had been refused medical care at least one time.

30 Facts About Peter Latham

1. English physician Peter Mere Latham revolutionized the field of medical education.

2. In the 19th century, he championed the idea that medical practitioners should also be teachers of clinical medicine.

3. Latham was born in 1761 to a medical family.

4. His grandfather, John Latham, was the President of the Royal College of Physicians and the founder of the Medical Benevolent Society.

5. Latham followed in his grandfather's footsteps, attending Oxford University, where he received a Doctor of Medicine and joined the College of Physicians.

6. He began working at a St. Bartholomew's Hospital as a resident physician in 1815.

7. Latham periodically returned to Oxford to deliver lectures.

8. His reputation grew, and he eventually became the personal physician to Queen Victoria.

9. In 1836, Latham published *Lectures on Subjects Connected with Clinical Medicine*.

10. It primarily discussed diseases of the heart and pulmonary system.

11. Latham's intent in publishing the book was to provide a medical manual for students, which was a fairly new idea at the time.

12. Until then, medical students primarily attended lectures given by physicians on medical topics.

13. Latham's *Lectures* could therefore be considered the forerunner of medical college textbooks of today.

14. At that time the concept of a full-time professor and medical researcher also did not yet exist.

15. Medical students attended college and listened to lectures, but did not go through any kind of residency program before beginning their practice.

16. There was no uniform means of collecting or analyzing clinical data.

17. The bulk of medical knowledge that was taught was based in existing and largely untested doctrine.

18. During his tenure at St. Bartholomew's, Latham taught undergraduate medical courses.

19. He pioneered a number of educational techniques that are taken for granted today.

20. Latham stressed the importance of what he called "self-learning" in clinical instruction, or observational learning.

21. He presented his students with medical cases that they could then see with their own eyes, thereby building a foundation of experience that would serve them when they entered practice.

22. He emphasized the value of lifelong learning, telling his students that "pathology is a study for your whole life."

23. Latham believed that the essential purpose of learning was to gain a base of wisdom that could equip the clinician with keen problem-solving ability.

24. Latham was also a pioneer of the use of teaching aids.

25. He encouraged his students to interact with patients, and said there were "a multitude of things" that could be learned at the patient's bedside.

26. Latham had students attend postmortem autopsies and view pathological specimens.

27. He thought that books and lectures should be used in medical education only sparingly and to supplement hands-on learning.

28. In spite of his extraordinary contributions, Latham is not widely known, even to students of medicine or clinicians.

29. Latham retired from practicing medicine in 1865, and moved to Torquay, England.

30. He died there in 1875.

6 Peter Latham Quotes

1. "Poisons and medicine are oftentimes the same substance given with different intents."

2. "The practice of physic is jostled by quacks on the one side, and by science on the other."

3. "The diagnosis of disease is often easy, often difficult, and sometimes impossible."

4. "We should always presume the disease to be curable, until its own nature prove it otherwise."

5. "It takes as much time and trouble to pull down a falsehood as to build up the truth."

6. "People in general have no notion of the sort and amount of evidence often needed to prove the simplest of fact."

19 Miscellaneous Facts

1. The human body has enough fat to produce seven bars of soap.

2. The average adult body consists of approximately 71 pounds of potentially edible meat, not including organ tissue.

3. The gluteus maximus, the muscle that makes up the buttocks, is the biggest muscle in the body.

4. The sartorius muscle in your thigh is the longest.

5. The liver is the largest internal organ and weights more than 3 pounds on average.

6. There aren't only five senses that deliver information to the brain. There are also somatic senses, including pain, temperature, and pressure.

7. Most adults breathe about 12 to 20 times a minute, faster during exercise.

8. Kids have a higher respiratory rate, with a 10-year-old breathing up to 30 times a minute, a toddler breathing up to 40 times a minute, and a newborn breathing up to 60 times a minute!

9. Temperature fluctuates throughout the course of the day. It's normal for your temperature at its highest point to be about 0.9 degrees Fahrenheit (0.5 degrees Celsius) higher than at its lowest point.

10. As you grow older, your body temperature drops.

11. Studies show that a normal body temperature a century ago may have been higher on average than it is now.

12. The left lung is smaller than the right lung in order to provide room for the heart.

13. The average human body contains enough iron to form a three-inch-long nail.

14. Synesthesia is a rare occurrence where an individual's senses are linked.

15. It is estimated that the phenomena affects about one person out of 20,000—this person can taste colors and hear shapes.

16. If the shoes you bought fit fine in the store but pinch your toes later, it might be because you bought them in the morning and are wearing them in the afternoon.

17. Your feet can swell during the day, up to half a size.

18. The Chinese emperor Shen-Nung was the first to use acupuncture as a medical treatment in 2700 B.C.

19. President Herbert Hoover's personal physician invented a game in an attempt to keep the president in shape. Hooverball uses a heavy medicine ball that is thrown over a volleyball-style net.

Chapter 7

Space

39 Solar System Facts

1. Humans have known about Mercury, Venus, Mars, Jupiter, and Saturn since prehistoric times because these planets are visible to the naked eye.

2. The word and original definition of "planet" are derived from the Greek *asteres planetai*, which means "wandering stars."

3. Planets are known as wanderers because they appear to move against the relatively fixed background of the stars, which are much more distant.

4. Stars and planets can be differentiated by two characteristics: what they're made of and whether they produce their own light.

5. Unlike stars, planets are built around solid cores.

6. They're cooler in temperature, and some are even home to water and ice.

7. The *Voyager 1* probe left Earth in 1977.

8. It (probably) reached interstellar space in 2013.

9. By NASA estimates, the heliosphere ends about 100 AU away from the sun, where an AU (astronomical unit) is the distance from Earth to the sun.

10. It's around 11 billion miles.

11. Scientists believe our solar system formed 4.6 billion years ago.

12. Neptune is about 30 AU away from the sun.

13. The Kuiper Belt, a ring containing icy objects, extends about 50 AU away.

14. The Kuiper Belt has been seen by the *New Horizons* probe.

15. Most of the mass of our solar system—more than 99 percent of it—is in the sun.

16. In fact, the mass of our solar system is "1.0014" solar masses, where one solar mass is the mass of the sun.

17. The nearest known star is Proxima Centauri, 4.25 light-years away.

18. There are objects in the solar system called centaurs.

19. Similar to asteroids and comets, centaurs have unstable orbits.

20. The Oort Cloud at the very edge of the solar system is named after Dutch astronomer Jan Oort.

21. Long-range period comets with orbits of longer than 200 years come from the Oort Cloud, whereas short-period comets come from the closer Kuiper Belt.

22. The term solar system appeared around 1700 in English.

23. Directives from the universe? "Try molasses" and "Slam oysters" are both anagrams of "solar system."

24. An asteroid belt lies between the rocky planets and the gaseous ones, located between Mars and Jupiter.

25. The four largest asteroids in this Main Asteroid Belt are Ceres (a dwarf planet), Vesta, Pallas, and Hygiea.

26. Scientists estimate that there are at least 1.1 million asteroids that measure a mile in diameter in the Main Asteroid Belt, and even more asteroids that are smaller than that.

27. The plane of Earth's orbit, and most of the other planets, is called the ecliptic.

28. Comets, however, orbit the sun at different angles.

29. All the planets in the solar system orbit the sun in the same direction.

30. Most other solar objects do as well, but some comets do not.

31. Halley's comet is one example of a comet with a retrograde orbit.

32. Halley's comet, a short-period comet, is visible from Earth every 75 years.

33. It's the only comet that we can see without a telescope that appears twice in a human life span.

34. Halley's comet will next appear in 2061.

35. During the formation of the solar system, there may have been hundreds of protoplanets.

36. Some joined together to become larger masses. Some were destroyed.

37. The objects we're familiar with—planets, asteroids, and so forth—are all found in the interplanetary medium.

38. In most other solar systems, the star is orbited by medium-sized planets that are larger than our rocky planets and smaller than our gas giants.

39. According to scientists, gold exists on Mars, Mercury, and Venus.

15 Animals in Space Facts

1. The Soviet Union launched the first living creature into orbit on November 3, 1957.

2. That creature was a three-year-old stray dog named Laika.

3. Weighing just 13 pounds, Laika's calm disposition and slight stature made her a perfect fit for the cramped capsule of Sputnik II.

4. In the weeks leading up to the launch, Laika was confined to increasingly smaller cages and fed a diet of a special nutritional gel to prepare her for the journey.

5. Sputnik II was a 250-pound satellite with a simple cabin, a crude life-support system, and instruments to measure Laika's vital signs.

6. When the Soviets announced that Laika would not survive her

historic journey, the mission ignited a debate in the West regarding the treatment of animals.

7. Initial reports suggested that Laika survived a week in orbit.

8. It was revealed many years later that Laika only survived for roughly the first five hours.

9. The craft's life-support system failed, and Laika perished from excess heat.

10. Despite her tragic end, the heroic little dog paved the way for occupied spaceflight.

11. Between 1957 and 1966, the Soviets successfully sent 13 more dogs into space—and recovered most of them unharmed.

12. Dogs were initially favored for spaceflight over other animals because scientists believed they could best handle confinement in small spaces.

13. To train for her spaceflight, Laika was confined to smaller and smaller boxes for 15 to 20 days at a time.

14. In 1959, the United States successfully launched two monkeys into space.

15. Named Able and Baker, the monkeys were the first of their species to survive spaceflight.

32 Facts About Space & the Cosmos

1. The cosmos contains approximately 50 billion galaxies.

2. If you could fly across our galaxy from one side to the other at light speed, it would take 100,000 years to make the trip.

3. If you attempted to count the stars in a galaxy at a rate of one every second, it would take about 3,000 years to complete the task.

4. Space smells like diesel fuel, barbecue, and hot metal.

5. Astronauts can't smell space while they're on a spacewalk, of

course, because they're in their suits. However, the scent clings to their gear.

6. The Milky Way is a spiral galaxy that formed approximately 14 billion years ago.

7. Our solar system is just a small part of it.

8. Most of the stars we can see are in our galaxy.

9. Our galaxy is about 100,000 light-years in diameter and 1,000 light-years thick.

10. The sun, along with Earth, is around 26,000 light-years from the center of our galaxy, halfway to the edge of the galaxy along the Orion spiral arm.

11. The Magellanic Clouds, Earth's nearest galaxies, are visible only in the Southern Hemisphere.

12. Both are irregular blobs that appear to the naked eye as fuzzy patches.

13. Visible to the naked eye and easily enjoyed with binoculars, Andromeda is the closest spiral galaxy (like our own Milky Way).

14. You can spot the Andromeda galaxy in a place with little light pollution in the Northern Hemisphere, ideally in November.

15. We watched the Crab Nebula blow itself apart (or rather, we watched the light reach us).

16. In A.D. 1054, Arab and Chinese astronomers noted a star visible during daytime—now it has puffed out a big gas cloud.

17. The Crab Nebula is found in Taurus (winter, mainly in the Northern Hemisphere), just above the tip of the lower "horn."

18. You can get a reasonable view of the cloud surrounding the star's wreckage with a four-inch telescope, but binoculars will reveal something, too.

19. Not visible to the naked eye but fascinating in pictures or through a big telescope, the Ring Nebula looks like a smoke ring surrounding a small star.

20. Just south of Vega in the constellation Lyra (summer, mainly in the Northern Hemisphere) is a little parallelogram of stars, and the Ring Nebula is in the middle of its bottom short side.

21. Take a good look at Vega (impossible to miss), because in 14,000 years it'll be our North Star again as Earth's axis cycles around.

22. The Coalsack Nebula is smack in the middle of the Milky Way and big enough to block out most of the "milk."

23. You have to get below the equator to see it, but its position just left of and below the Southern Cross makes it easy to spot all year round.

24. It's not at all hard to find the Orion Nebula (winter, Northern Hemisphere; summer, Southern Hemisphere) in the dagger of Orion's three-star belt, and you can easily see the haze with binoculars—great through a small telescope.

25. A star cluster, the Pleiades are one of the highlights of binocular astronomy, easy to spot in Taurus (winter, Northern Hemisphere; summer, Southern Hemisphere).

26. Look for a little coffee-cup shape of blue-white stars that show up sapphire in light magnification.

27. Algol is an eclipsing binary star (a bright star eclipsed at intervals by a dimmer nearby star).

28. For folks on Earth, Algol seems to vary in brightness, and you can tell with the naked eye; visibility will go from easy to difficult.

29. The eclipses last several hours and occur every three days.

30. A planet bigger than Jupiter? TrES-4 is 1,400 light-years from Earth and nearly twice the size of Jupiter.

31. TrES-4 is a planetary oddball, though, because of its low density—which is about the same as balsa wood.

32. The asteroid Psyche 16 is made of metal, mostly iron and nickel—so much metal that some estimate it is worth more than the entire global economy.

27 Facts About the Sun

1. Every year the sun loses 360 million tons.

2. In about five billion years, the sun will become a red giant.

3. The diameter of the sun is about 109 times the diameter of Earth.

4. In miles, the sun's diameter is 864,000 miles.

5. About three-quarters of the sun's mass is made of hydrogen.

6. Approximately one-quarter is made of helium and small amounts of oxygen, carbon, neon, and iron.

7. At its core, the sun converts hydrogen into helium.

8. The energy from that process creates the sun's light and heat.

9. The sun is brighter than about 85 percent of stars in the Milky Way.

10. The light from the sun takes an average of 8 minutes and 19 seconds to reach Earth.

11. The sun has a surface temperature of about 5,800 kelvin.

12. That's more than 9,900 degrees Fahrenheit!

13. At its core, the sun is far hotter: 15.7 million kelvin, or 27 million degrees Fahrenheit.

14. Sunspots appear and disappear, and levels of solar radiation rise and fall, in an 11-year cycle.

15. Sunspots are areas on the surface that are a bit cooler than the surrounding areas.

16. They're formed by fluctuations in the magnetic field.

17. When we say a bit cooler, we don't mean humans could live there—sunspots average about 3,800 kelvin instead of 5,800 kelvin.

18. The sun rotates faster at its center than at its poles.

19. The sun releases charged particles—the solar wind—from its upper atmosphere.

20. Auroras are caused by fluctuation and flares in the solar wind, as the particles meet the gases in Earth's atmosphere.

21. In auroras, the gas in our atmosphere determines the color.

22. Red and green are caused by oxygen, while blue and purple are associated with nitrogen.

23. Astronomers on Earth first spotted solar flares in 1859.

24. Richard Carrington identified a geomagnetic storm and tied it to solar activity.

25. Geomagnetic storms caused by solar flares that release electro-magnetic radiation can cause power outages on Earth.

26. The sun is almost a perfect sphere.

27. Earth's surface gravity is 9.8 meters per second squared. By comparison, the sun's is 274 meters per second squared.

5 Brightest Natural Objects

1. Sun (brightest star)

2. Moon (brightest natural satellite)

3. Venus (brightest planet)

4. Jupiter

5. Sirius (brightest night star)

30 Mercury Facts

1. Light travels from the sun to Mercury in 3.2 minutes.

2. Mercury travels around the sun every 88 days.

3. If you were on Mercury, the sun would appear three times larger than it does on Earth.

4. Mercury's temperature during the day can rise to 800°F.

5. But it doesn't stay warm at night, when temps drop to -290°F.

6. At the poles, the temperature never rises above the freezing point.

7. It never even gets near it, with a high temp of -135°F.

8. Mercury's orbit around the sun isn't a perfect sphere—it's more like an egg.

9. At its closest point it gets to 29 million miles away from the sun, but at its farthest point it moves to 43 million miles away.

10. You could fit 24,462 planets the size of Mercury, the smallest planet, into Jupiter, the largest one.

11. Mercury can be seen without a telescope, but not always easily.

12. It's easiest to spot near sunrise or dusk.

13. Mercury was first recorded by Assyrian and then Babylonian astronomers.

14. The Greeks were the ones to realize that when they sighted Mercury in the morning, it was the same "star" they spotted in the evening.

15. *Mariner 10* was the first spacecraft to visit Mercury back in the 1970s.

16. *MESSENGER* visited Mercury later, leaving Earth in 2004 and doing a flyby in 2008.

17. *MESSENGER* did a final pass in 2015 before crashing on Mercury.

18. *MESSENGER* was an acronym, standing for MErcury Surface, Space ENvironment, GEochemistry, and Ranging.

19. *MESSENGER* detected water ice at Mercury's north pole.

20. Mercury's core has a high iron content, higher than Earth's.

21. Mercury's surface has heavy cratering.

22. There are more than 400 named craters as of 2023.

23. Writers and artists from around the world and many different historical periods are represented.

24. Named craters include Ailey (named after choreographer Alvin

Ailey), Angelou (named after author Maya Angelou), and Brooks (named after poet Gwendolyn Brooks).

25. Eons ago, Mercury experienced a meteor strike that blasted a crater 800 miles wide called Caloris Basin.

26. The strike actually raised hills on the other side of the planet.

27. Mercury has wrinkles!

28. Well, it has compression folds from when the interior of the planet began to contract as it cooled and the surface formed "wrinkle ridges" and other folds.

29. Mercury, like Earth, has a magnetosphere.

30. *MESSENGER* detected "magnetic tornadoes."

31 Facts About Galileo

1. Galileo was born in Florence, Italy, in 1564.

2. Galileo pioneered the use of the telescope to observe celestial bodies, advanced our understanding of physics, and famously demonstrated that objects of unequal weight fall at the same speed.

3. In 1542, Nicolas Copernicus had proposed a heliocentric model, in which the planets orbit the sun.

4. Without physical evidence, however, it was widely rejected by authorities and leading scientists.

5. In 1609, Galileo first heard about the invention of a telescope in Holland.

6. Intrigued by this new invention, he set out to build his own.

7. Galileo's designs eventually improved the telescopes' magnification by as much as thirty times, fundamentally changing their use from terrestrial observation to astronomical observations.

8. He observed and drew the mountains and craters of the moon in great detail.

9. Galileo discovered several of the moons of Jupiter. Not everything in the solar system revolved around Earth.

10. He observed the phases of Venus.

11. In 1610, Galileo published a catalogue of his discoveries, the *Sidereus Nuncius*, which made him famous throughout Europe.

12. Armed with his discovery of the phases of Venus, which proved that Venus was orbiting the sun, Galileo began to publicly advocate for a heliocentric system.

13. This was seen as a direct challenge to the Church's authority.

14. In 1615 his writings were submitted to the Roman Inquisition.

15. In spite of the fact that Galileo travelled to Rome to defend the theory, in 1616 the Inquisition declared heliocentrism not only "foolish and absurd in philosophy," but heretical to boot.

16. The pope ordered Galileo to abandon heliocentrism and stop defending the idea altogether.

17. For ten years, Galileo complied, avoiding the controversy.

18. When a more liberal pope came to power in 1632, he encouraged Galileo to publish an unbiased work about the arguments for and against heliocentrism.

19. In the resulting work, *Dialogue Concerning the Two Chief World Systems*, Galileo instead openly advocated for heliocentrism.

20. He was summoned to Rome and tried for heresy.

21. Galileo maintained his innocence throughout the six-month trial, until he was threatened with torture if he did not admit his guilt and recant his beliefs.

22. He did recant, but reportedly whispered "and yet it moves" in a final act of defiance.

23. Convicted of being "vehemently suspect of heresy," Galileo was sentenced to house arrest, under which he would remain for the rest of his life.

24. While under house arrest, Galileo wrote one of his finest works, *Discourses and Mathematical Demonstrations Relating to Two New Sciences.*

25. In it, he laid out his findings over the past 30 years of scientific study.

26. The two new sciences were the motion of objects and strength of materials.

27. They laid the foundation for kinematics and material engineering.

28. In the same work, Galileo first proposed the idea that bodies of different mass fall at the same rate (this would be demonstrated by Apollo 15 astronauts on the moon), as well as the behavior of bodies in motion, the principle relativity of motion, and the nature of infinity.

29. Among his numerous contributions to astronomy, Galileo was the first to observe that the Milky Way was composed of stars.

30. The Father of Modern Science died in 1642 at the age of 77.

31. In 2016 the *Juno* spacecraft arrived at Jupiter, carrying a plaque in his honor.

23 Venus Facts

1. Aside from Earth's moon, Venus is the brightest object in Earth's night sky.

2. It is depicted in van Gogh's famous painting *The Starry Night.*

3. The Greeks thought Venus was two objects: a morning star and an evening one, named Phosphoros and Hesperos.

4. Its buttery cloud cover is mostly carbon dioxide mixed with sulfur dioxide.

5. Atmospheric pressure at the surface is equal to being half a mile below Earth's seas.

6. The surface temperature can reach 900°F.

7. In 1962, *Mariner 2* flew by Venus and scanned it.

8. It was the first exploration of another planet.

9. The Soviet Union sent a number of probes beginning in the 1960s.

10. *Venera 3*, which crash-landed in 1966, was the first spacecraft to land on the surface of another planet.

11. Several later Soviet probes did land on the planet and transmit information back.

12. Venus spins in the opposite direction as Earth.

13. A complete rotation—a day—on Venus takes 243 Earth-days.

14. Venus's orbit around the sun is a nearly perfect circle.

15. Venus is highly volcanic, and volcanism shaped its surface.

16. Many of the more than 1,600 volcanoes are now dormant or extinct.

17. Evidence exists that some features may still be volcanically active.

18. Venus has a weaker magnetic field than Earth.

19. Venus might have once had oceans on its surface before the sun's radiation and Venus's atmosphere caused them to evaporate.

20. Venus's clouds are made of sulfuric acid.

21. If you were somehow able to stand on the planet's surface, you wouldn't see a lot of sunlight because of the planet's cloud cover.

22. Wind speeds reach 185 miles per hour there.

23. Venus has no moons now, but it may have had them in the past.

32 Facts About Earth

1. Earth is the only planet not named after a god.

2. The word Earth comes from German and means "the ground."

3. Objects weigh slightly less at the equator than at the poles.

4. The distance between the Earth and the sun is about 93 million miles.

5. Earth's orbit isn't a perfect circle around the sun, though, so the distance can vary from 91 million miles (perihelion) to 94.5 million miles (aphelion).

6. Astronomers were able to calculate the distance from the sun to Earth as far back as the 1600s.

7. Earth is the largest of the "rocky" planets.

8. Our magnetic field can change directions.

9. This happens about once every 400,000 years, but not on a regular schedule.

10. When the magnetic field changes directions, the process lasts a few centuries (during which a compass would be completely unreliable) before things stabilize.

11. Earth is the densest planet in our solar system.

12. While Earth only has one moon, it has a lot of other objects orbiting it.

13. There are at least 22,000 "near-Earth asteroids."

14. Of those, about 2,000 pose danger to Earth because of their size and relative closeness.

15. One of those near-Earth asteroids, 3122 Florence (named after Florence Nightingale), even has two small moons of its own!

16. In 2017, Florence passed Earth from a "close" distance of 4,391,000 miles.

17. NASA estimates that about once every 100 years, a meteorite substantial enough in size to cause tidal waves wallops Earth's surface.

18. About once every few hundred thousand years, an object strikes that is large enough to cause a global catastrophe. So future hits—both larger and smaller—are inevitable.

19. The Ensisheim Meteorite, the oldest recorded meteorite, struck Earth on November 7, 1492, in the small town of Ensisheim, France.

20. The Tunguska Meteorite, which exploded near Russia's Tunguska River in 1908, is still the subject of debate more than one hundred years later.

21. It didn't leave an impact crater, which has led to speculation about its true nature.

22. The Hoba Meteorite, found on a farm in Namibia in 1920, is the heaviest meteorite ever found.

23. Weighing in at about 66 tons, the rock is thought to have landed more than 80,000 years ago.

24. Over the years, erosion, vandalism, and scientific sampling have shrunk the rock to about 60 tons.

25. Santa had to compete for airspace on Christmas Eve 1965, when Britain's largest meteorite sent thousands of fragments showering down on Barwell, Leicestershire.

26. Arizona would be short one giant hole in the ground if it wasn't for a 160-foot meteorite landing in the northern desert about 50,000 years ago.

27. It left an impact crater about a mile wide and 570 feet deep.

28. At 186 miles wide, Vredefort Dome in South Africa is the site of the biggest impact crater on Earth.

29. The Sudbury Basin in Sudbury, Ontario, is a 40-mile-long, 16-mile-wide, 9-mile-deep crater caused by a giant meteorite that struck Earth about 1.85 billion years ago.

30. It's also one of the most profitable—the metals from the 6- to 12-mile-wide asteroid that caused the crater brought a motherlode of nickel, copper, and platinum, making Sudbury a metal haven.

31. The 105-mile-wide Chicxulub crater in Yucatán Peninsula, Mexico—an impact crater identified by geologists in 1992—is

thought to be associated, or at least partially associated, with the extinction of the dinosaurs.

32. More than 150 impact craters have been identified on Earth, some on the surface and some hidden below the surface.

30 Constellation Facts

1. The word constellation comes from the Latin *cōnstellātiō*, meaning "set with stars."

2. Smaller groupings of stars are known as asterisms.

3. One example of an asterism is Venus' Mirror in the constellation Orion.

4. According to the International Astronomical Union (IAU), there are 88 officially recognized constellations.

5. Forty-eight of these constellations are known as ancient or original, meaning they were created and used by ancient civilizations.

6. The earliest evidence of constellations is found on Mesopotamian clay writing tablets dating back to 3000 B.C.

7. This is likely the origin of ancient Greek constellations.

8. Ancient constellations were recorded in Greek poet Aratus's *Phenomena* and Greek astronomer Ptolemy's *Almagest*.

9. Modern constellations, like the Peacock, Phoenix, and Sextant, were identified by astronomers of the 14th, 15th, and 16th centuries.

10. Your location on Earth determines what constellations you'll be able to see in the night sky, as well as how high they appear above the horizon.

11. Circumpolar constellations are constellations that never set below the horizon as viewed from a particular latitude.

12. Instead, they make a full circle around a celestial pole, like the North Star in the Northern Hemisphere.

13. The time of year also determines what constellations are visible.

14. Stars appear to move west across the sky during the year as the Earth revolves around the sun.

15. The brightest stars in a constellation are not necessarily the largest.

16. A star's brightness is also dependent on its temperature and distance from Earth.

17. Because each star in a constellation moves independently of one another, all constellations slowly change shape over time.

18. Constellations assist navigators in locating certain stars.

19. This is called celestial navigation.

20. It is especially useful when traveling across the ocean where other landmarks are not visible.

21. NASA astronauts are also trained in celestial navigation as a backup in the event their modern instruments fail.

22. Astronomers drew a boundary around each of the 88 official constellations, dividing the celestial sphere into 88 pieces.

23. Any star inside a constellation boundary is considered part of the constellation even if it is not actually part of the image.

24. Constellations are used by astronomers when naming stars and meteor showers.

25. This is helpful when conveying approximate locations because any given point in the sky lies in one of the modern constellations.

26. Astrology is the belief that the location of celestial objects like stars in constellations can describe a person's character or predict the future.

27. There are 12 astrological constellations of the zodiac, but 13 astronomical zodiac constellations: Capricornus, Aquarius, Pisces, Aries, Taurus, Gemini, Cancer, Leo, Virgo, Libra, Scorpius, Sagittarius, and Ophiuchus.

28. The cycle of the zodiac was used by ancient cultures to determine the time of year.

29. Due to a phenomenon called precession, Earth wobbles on its axis as it rotates.

30. This affects seasonal timing, causing them to shift with respect to zodiac constellations.

23 Facts to Know About Tycho

A golden nose, a dwarf, a pet elk, drunken revelry, and astronomy? Read about the wild life of this groundbreaking astronomer.

✳ ✳ ✳ ✳

1. Tycho Brahe was a Dutch nobleman who is best remembered for blazing a trail in astronomy in an era before the invention of the telescope.

2. Through tireless observation and study, Brahe became one of the first astronomers to fully understand the exact motions of the planets.

3. In 1560, Brahe, then a 13-year-old law student, witnessed a partial eclipse of the sun.

4. He was so moved by the event that he bought a set of astronomical tools and a copy of Ptolemy's legendary astronomical treatise, *Almagest*, and began a life-long career studying the stars.

5. Brahe differed from his forebears in his belief that new discoveries in astronomy could be made, not by guesswork and conjecture, but by rigorous and repetitious studies.

6. His work includes many publications and even the discovery of a supernova now known as SN 1572.

7. Brahe became one of the most widely acclaimed astronomers in all of Europe.

8. When King Frederick II of Denmark heard of Brahe's plans to move to the Swiss city of Basle, the king offered him his own island, Hven, located in the Danish Sound.

9. Once there, Brahe built his own observatory known as Uraniborg.

10. Brahe ruled Hven as if it were his own personal kingdom, forcing tenants to supply him with goods and services or be locked up in the island's prison.

11. At one point Brahe imprisoned an entire family—contrary to Danish law.

12. While famous for astronomy, Brahe is more infamous for his colorful lifestyle.

13. At age 20, he lost part of his nose in an alcohol-fueled duel (reportedly using rapiers while in the dark) after a Christmas party.

14. Portraits of Brahe show him wearing a replacement nose possibly made of gold and silver and held in place by an adhesive.

15. When his body was exhumed in 1901, green rings discovered around Brahe's nasal cavity led some scholars to speculate that the nose may actually have been made of copper.

16. While a considerable amount of groundbreaking astronomical research was done on Hven, Brahe also spent his time hosting legendarily drunken parties.

17. Such parties often featured a little person named Jepp who dwelled under Brahe's dining table and functioned as something of a court jester.

18. Brahe may have believed that Jepp was clairvoyant.

19. Brahe kept a tame pet elk, which stumbled to its death after falling down a flight of stairs—the animal had gotten drunk on beer at the home of a nobleman.

20. He was ostracized for marrying a woman from the lower classes, considered shameful for a nobleman such as Brahe.

21. Due to his shameful union, all of his eight children were considered illegitimate.

22. According to legend, Brahe died from a bladder complication

caused by not urinating, out of politeness, at a friend's dinner party where prodigious amounts of wine were consumed.

23. The tale lives on, but recent research suggests this version of Brahe's demise could be apocryphal: He may have died of mercury poisoning from his own fake nose.

15 Exoplanet Facts

1. An exoplanet is any planet outside our solar system.

2. There are billions of them, maybe as many as there are stars.

3. As of January 2023, there are more than 5,295 confirmed exoplanets.

4. Many of those planets have been discovered by TESS, the Transiting Exoplanet Survey Satellite.

5. Most planets, as in our solar system, orbit a star.

6. Those that do not are known as rogue, nomad, or sunless planets.

7. The first exoplanets were discovered in the 1990s.

8. 51 Pegasi B, a gas giant 50 light-years from Earth, was discovered in 1995.

9. It was the first confirmed exoplanet.

10. Only a few exoplanets can be seen by telescopes.

11. Most are determined when people spot stars getting dimmer because the planet passes in front of it, or when a star's light changes in certain ways.

12. Exoplanets are divided into gas giants (like Jupiter); Neptune-like; Super-Earth (rocky planets larger than Earth); and Terrestrial (rocky planets Earth's size or smaller).

13. Some gas giant exoplanets, including 51 Pegasi B, are larger than Jupiter.

14. In 2017, NASA announced the discovery of seven Earth-sized planets orbiting around a star.

15. The planets of this star system, named TRAPPIST-1, possibly contain liquid water and water vapor.

23 Facts About the Moon

1. Our moon is the fifth largest moon in the solar system, after three of Jupiter's and one of Saturn's.

2. The moon orbits Earth from a distance of about 240,000 miles.

3. The moon's equatorial circumference is about 6,784 miles, compared to Earth's 24,901 miles.

4. The moon helps stabilize Earth's rotation on its axis.

5. When a Soviet spacecraft flew by the moon in 1959, it was the first time that humans knew what the far side of the moon looked like.

6. The moon does have an atmosphere—it's just very, very thin.

7. The moon's magnetosphere is very weak, compared to Earth's.

8. Apollo missions have brought back more than 840 pounds of lunar material.

9. NASA's Artemis program plans to send people back to the moon by 2024, including the first woman.

10. The moon has more than 9,100 craters.

11. Asteroids and meteoroids continue to form craters.

12. Some of the craters are more than two billion years old.

13. The largest crater on the moon is the South Pole-Aitken Basin, which measures 1,600 miles in diameter.

14. The Oceanus Procellarum on the moon, a vast lunar plain, might have been formed by an impact crater.

15. If so, it would be even larger, with a diameter of 1,611 miles.

16. The "lunar maria" are plains that were formed by volcanoes.

17. The word "mare" was the Latin word for seas.

18. Lunar eclipses occur when Earth comes between the sun and the moon.

19. A "supermoon" happens when a full moon or a new moon happens at the same time that the moon and Earth are closest to each other.

20. An Indian mission in 2008 discovered evidence of water on the moon.

21. Further exploration found ice water at the lunar poles, trapped within craters.

22. The moon reaches high temperatures of 260°F, and lower temperatures of -280°F.

23. A point called the Selenean Summit, found on the rim of an impact crater, is the moon's highest point, standing about 35,387 feet above the lunar mean.

10 Biggest Moons

1. Ganymede (Jupiter)
2. Titan (Saturn)
3. Callisto (Jupiter)
4. Io (Jupiter)
5. Moon (Earth)
6. Europa (Jupiter)
7. Triton (Neptune)
8. Titania (Uranus)
9. Rhea (Saturn)
10. Oberon (Uranus)

43 Moon-visiting Men Facts

1. Twenty-four men have visited the moon.

2. Twelve walked on the surface and twelve others reached lunar orbit.

3. The Apollo 8 mission launched in 1968 with astronauts Frank Borman, Jim Lovell, and Bill Anders.

4. Frank Borman was the commander of Apollo 8, a fighter pilot and test pilot.

5. Jim Lovell went to the moon twice, with Apollo 8 and Apollo 13.

6. While he was scheduled to walk on the moon with Apollo 13, the mission's technical difficulties meant he did not.

7. Bill Anders took a famous color image of the earth called "Earthrise."

8. Apollo 10 launched in May 1969, as a test flight for the moon landing.

9. Astronauts John Young, Thomas Stafford, and Eugene Cernan were aboard.

10. John Young and Gene Cernan would later land on the moon as part of the Apollo 16 and Apollo 17 missions, respectively.

11. Stafford later served as Chief of the Astronaut Office and the commander of the Apollo-Soyuz Test Project flight.

12. The crew of Apollo 11 that made the first moon landing in July 1969 consisted of Neil Armstrong, Edwin "Buzz" Aldrin, and Michael Collins.

13. Armstrong and Aldrin walked on the moon, while Collins remained in orbit.

14. When Armstrong made his first spaceflight as pilot of *Gemini 8*, he was the first civilian astronaut to fly in space.

15. Aldrin, a Presbyterian, held a private religious ceremony on the moon.

16. Michael Collins was born in Italy, the son of an Army officer.

17. Charles "Pete" Conrad, Alan Bean, and Richard F. Gordon Jr. manned the Apollo 12 mission in November 1969, with Gordon orbiting the moon while the other men landed.

18. Gordon was slated to go to the surface of the moon himself with Apollo 18, but the mission was scrubbed due to budget cuts.

19. Conrad later commanded the first crewed Skylab mission in 1973.

20. Alan Bean painted as a hobby, and later painted pictures of the moon, using real dust gathered from his keepsake patches.

21. Jim Lovell, Jack Swigert, and Fred Haise were the Apollo 13 crew.

22. They spent four days in lunar orbit in a very cold lunar module, *Aquarius*.

23. Fred Haise was slated to return to the moon with Apollo 19, but the mission was cancelled for budgetary reasons.

24. Jack Swigert later ran for office and won his congressional election, but died of cancer before serving.

25. The crew of Apollo 14, who traveled to the moon in 1971, consisted of Stuart Roosa, Alan Shepard Jr., and Edgar Mitchell.

26. Roosa stayed in orbit while the other men went to the surface of the moon.

27. Roosa had been a firefighter with the U.S. Forest Service before joining the Air Force.

28. He carried seeds provided by the U.S. Forest Service with him on the mission, and "Moon Trees" were later grown from those seeds.

29. At age 47, Alan Shepard was the oldest man to walk on the moon, as well as the first to strike golf balls on the surface.

30. Mitchell was later one of the co-founders of the Institute of Noetic Sciences, which studied paranormal phenomena.

31. The 1971 Apollo 15 mission was manned by David Scott, James Irwin, and Alfred Worden; Worden remained in orbit.

32. Irwin experienced irregularities in his heart rhythm during his time on the moon, although it had settled into a regular rhythm by his return to Earth.

33. He later had several heart attacks, including one just a few years later.

34. While Worden was in the command module *Endeavour*, he was 2,235 miles away from his crewmates, earning the distinction of most isolated human being in history.

35. Scott had been the command module pilot of Apollo 9 mission in March 1969.

36. In 1972, John Young, Charles Duke, and Ken Mattingly comprised the crew of the Apollo 16 mission, with Mattingly staying in orbit while the other men landed.

37. At 36 years old, Charles Duke was the youngest man to walk on the moon.

38. Mattingly had been scheduled for the Apollo 13 mission, but was removed from that mission due to exposure to rubella.

39. Young had served as a naval aviator in the Korean War.

40. The final mission to the moon, Apollo 17, took place in December 1972, and was crewed by Eugene "Gene" Cernan, Harrison Schmitt, and Ron Evans.

41. As he re-entered the Apollo Lunar Module after Harrison Schmitt on their third and final lunar excursion, Cernan became the last person to step foot on the moon.

42. Evans, who stayed in orbit while the other men landed, performed an EVA during the return to Earth.

43. It was the final spacewalk of the Apollo program.

25 Facts About Jupiter

1. It takes 43.2 minutes for the light from the sun to reach Jupiter, across an average distance of 484 million miles.

2. It takes about 12 Earth-years for Jupiter to orbit the sun.

3. The surface gravity of Jupiter is more than two-and-a-half times greater than that of Earth.

4. Ammonia and water in the planet's atmosphere appear as swirls and markings through a telescope.

5. The main components of the planet's atmosphere are hydrogen and helium.

6. As of 2023, Jupiter has 92 known moons, though only 57 are named.

7. All of Jupiter's moons, together with smaller moonlets, form a satellite system called the Jovian system.

8. Jupiter's four largest moons (Io, Europa, Ganymede, and Callisto) are some of the largest in the solar system.

9. Ganymede is the biggest. It's even larger than Mercury!

10. In fact, Ganymede is the ninth largest object in our solar system.

11. It is the only moon we've discovered with a magnetic field.

13. The moon Io is considered to be the most geologically active object in our solar system.

14. Astronomers have watched some of Io's 400-plus active volcanoes erupt, and a few of its mountains rise higher than Mount Everest.

15. On March 2, 1998, the spacecraft *Galileo* discovered a liquid ocean on Jupiter's moon Europa.

16. The surface of the moon Callisto is shaped by impact craters.

17. Jupiter has an ocean—but not of water. It is made of liquid hydrogen.

18. Jupiter does have rings, but they're made of dust and pale in comparison to Saturn's.

19. They were spotted by *Voyager 1*.

20. Jupiter's diameter is more than 88,840 miles across (the diameter of Earth is around 7,918 miles).

21. *Pioneer 10* was the first spacecraft to perform a flyby.

22. Jupiter rotates in about 10 hours, giving it the shortest day of any planet.

23. Winds on Jupiter can reach speeds of more than 330 miles per hour.

24. The Great Red Spot, a storm, has been observed for at least three centuries.

25. Jupiter has a powerful magnetic field, significantly more powerful than Earth's, that causes beautiful aurora.

57 Moons of Jupiter

1. Metis
2. Adrastea
3. Amalthea
4. Thebe
5. Io
6. Europa
7. Ganymede
8. Callisto
9. Themisto
10. Leda
11. Himalia
12. Ersa
13. Pandia
14. Elara
15. Lysithea
16. Dia
17. Carpo
18. Valetudo
19. Euporie
20. Harpalyke
21. Hermippe
22. Euanthe
23. Thyone
24. Mneme
25. Iocaste
26. Praxidike
27. Ananke
28. Thelxinoe
29. Orthosie
30. Helike
31. Eupheme
32. Kale
33. Chaldene
34. Taygete
35. Herse
36. Kallichore
37. Kalyke
38. Pasithee
39. Philophrosyne
40. Cyllene
41. Autonoe
42. Megaclite
43. Eurydome
44. Pasiphae
45. Callirrhoe
46. Isonoe

47. Aitne

48. Hegemone

49. Sponde

50. Eukelade

51. Erinome

52. Arche

53. Eirene

54. Carme

55. Aoede

56. Kore

57. Sinope

28 Martian Facts

1. It takes 12.6 minutes for light from the sun to reach Mars.

2. Mars's red appearance comes from the iron oxide on its surface.

3. Mars, like Earth, has seasons.

4. Mars orbits the sun in 687 Earth-days.

5. Its average distance from the sun is 142 million miles.

6. At its further point, Mars can be 154 million miles from the sun.

7. At its closest point, Mars can be 128 million miles from the sun.

8. Olympus Mons, the largest volcano in the solar system, rises 16 miles above the Mars's surface—three times higher than Mount Everest—and is the size of Arizona.

9. The caldera (central crater) alone is large enough to swallow a medium-sized city.

10. It's a shield volcano (like the ones in Hawaii), the result of lava flowing out over the ages.

11. The Valles Marineris canyon stretches more than 3,000 miles, with its greatest depth sinking to more than 4 miles.

12. Mars is, of course, named after the Roman god of war.

13. Mars's two moons, Phobos and Deimos, are named after the twin sons of the Greek god of war, Ares, and Aphrodite.

14. The two named craters on Deimos are named Swift and Voltaire.

15. Voltaire wrote a short story in which he suggested that Mars had two moons.

16. Jonathan Swift suggested the same in *Gulliver's Travels*.

17. Areology is the study of the geology of Mars, from the Greek god of war, Ares.

18. The Mandarin name for Mars translates to "fire star."

19. *Mariner 4*, launched in 1964, was the first spacecraft to get close to Mars.

20. The first unsuccessful landing was by the Soviet *Mars 3* mission in 1971.

21. The U.S.'s *Viking 1* mission achieved a successful landing.

22. February of 2021 was a busy month for spacecraft arriving at Mars: the United Arab Emirates sent an orbiter called *Hope*; China sent an orbiter, a lander, and a rover; and NASA sent the *Perseverance* rover.

23. *Perseverance* began collecting the first samples of the Martian surface.

24. Mars is a colder planet than Earth, with a low temperature of -225°F.

25. Even at its warmest, Mars doesn't often exceed 70°F.

26. Mars probably used to have a magnetic field, but doesn't anymore.

27. The planet has ice caps at its poles.

28. Mars has large dust storms that affect the entire planet.

16 Facts About Stars

1. The brightest star in our sky (after the sun, of course) is Sirius.

2. In Greek, the word means "glowing."

3. Sirius is actually a binary star—a star system of two stars, in this case Sirius A and Sirius B.

4. Sirius will only become brighter over time, as it's moving closer to our solar system.

5. Sirius is found in the constellation Canis Major.

6. After Sirius, Canopus is the brightest star in the night sky.

7. Canopus is part of the constellation Argo.

8. Polaris, the Pole Star, is about 433 light-years away.

9. Polaris is a triple star, made up of a yellow supergiant and two smaller stars.

10. Red dwarf stars—the smallest stars—are the most numerous stars in the galaxy.

11. "Hypergiants" are massive stars that are larger, hotter, brighter, and emit more energy than our sun.

12. They are rare and relatively short-lived, with a life span of "only" a few million years.

13. Betelgeuse, seen in the constellation Orion, is a red supergiant.

14. The oldest stars are believed to be about 13.7 billion years old.

15. The color of a star reflects its temperature.

16. We often think of blue as a "cool" color and red as "hot," but actually blue stars are the hottest and red the coolest.

22 Saturn Facts

1. You can easily view Saturn's rings with a basic telescope.

2. Every so often, the planet of Saturn appears to lose its rings.

3. That has to do with their position at the time, when we're only seeing the edge directly. (Think of how thin a record might look if you were looking only at its edge.)

4. Seven hundred planets the size of Earth would fit in Saturn, while about 1,300 would fit in Jupiter.

5. Saturn is about 886 million miles from the sun.

6. It takes about 29 Earth-years for Saturn to orbit the sun.

7. Like Jupiter's atmosphere, Saturn's is made primarily of hydrogen and helium.

8. Saturn has seven rings made of ice and rock.

9. Some of the particles in Saturn's rings aren't very small—think the size of a building rather than a boulder.

10. Saturn is not a very dense planet. It's less dense than water!

11. It takes sunlight 79 to 80 minutes to travel from the sun to Saturn.

12. Some of Saturn's winds can reach speeds of 1,600 feet per second.

13. The planet has a very strong magnetic field.

14. Saturn has a big moon collection: There are 63 named moons as of January 2023, but it likely has 83.

15. Its largest moon, Titan, has an atmosphere (mostly nitrogen with some methane) and standing liquid on its surface.

16. Titan is much bigger than Earth's moon and is the second largest known moon in the solar system, after Jupiter's Ganymede.

17. Saturn's moon Enceladus is covered with ice.

18. Astronomer William Herschel discovered Enceladus.

19. Enceladus is home to cryovolcanoes—volcanoes that shoot water and other materials instead of lava.

20. *Pioneer 11* first visited Saturn in a flyby in September 1979.

21. *Cassini* studied the planet from orbit for 13 years and carried the Huygens probe that landed on Titan.

22. *Cassini* ended its mission by plunging through Saturn's rings and falling into Saturn's atmosphere, achieving purposeful vaporization.

26 Facts About Uranus

1. Uranus is the farthest planet from Earth visible to the naked eye, but you have to know exactly where to look.

2. William Herschel discovered the planet in 1781 using a telescope.

3. At first, he thought it might be a comet.

4. Herschel called his discovery Georgium Sidus, or George's Star, after his monarch King George III.

5. It was later named Uranus to fit in with the other planets.

6. The name was suggested by Johann Bode, another astronomer who had confirmed Herschel's observations.

7. The element uranium was discovered in 1789, and was named after the planet.

8. It takes Uranus 84 Earth-years to orbit the sun.

9. Uranus has a small rocky core surrounded by water, methane, and ammonia.

10. Its atmosphere consists primarily of hydrogen and helium.

11. Small amounts of methane in the atmosphere give the planet a blue-green color.

12. Despite it being an "icy" planet, scientists estimate that the temperature at Uranus's core is around 9,000°F.

13. The odd thing about Uranus is that it's oriented on its side.

14. Earth is tilted 23.5 degrees out of the plane of the solar system (our local "up" reference point in space).

15. Uranus is tilted 98 degrees, so its poles are in the middle and its equator runs from top to bottom.

16. Uranus has at least 13 rings.

17. It may have more that are yet to be discovered.

18. The only spacecraft to visit Uranus has been *Voyager 2*, and it was only a flyby that occurred in 1986.

19. Uranus has 27 known moons.

20. Many are named after Shakespearean characters, including Miranda and Ariel.

21. The largest moon is Titania, which may contain water.

22. It takes sunlight nearly 160 minutes to reach Uranus.

23. Wind speeds on Uranus can reach 560 miles per hour.

24. While most of Uranus's rings are gray, one is reddish and another is blue.

25. The rings are likely younger than the planet itself, forming later.

26. Uranus has a magnetosphere.

27 Edwin Hubble Facts

We've all heard of the Hubble Space Telescope, which has sent us so many amazing visuals of space. But who, exactly, was Hubble?

✳ ✳ ✳ ✳

1. Edwin Hubble was born in Marshfield, Missouri, in 1889.

2. He didn't set out to be a scientist, even though astronomy was a hobby when he was a boy.

3. Instead, he studied law at the University of Chicago and The Queen's College, Oxford, fulfilling a promise to his father, who preferred he study law.

4. But after his father died in 1913, Hubble went back to school, studying astronomy at the University of Chicago's Yerkes Observatory, where he earned a PhD in 1917.

5. In a foretelling of things to come, his dissertation was entitled, "Photographic Investigations of Faint Nebulae," for which he used of one of the world's most powerful telescopes at the time, housed at the observatory.

6. After serving in the U.S. Army during World War I, Hubble began working at the Carnegie Institution for Science's Mount Wilson Observatory, where he worked for the rest of his life.

7. Hubble's colleagues espoused theories about the cloudy patches of nebulae they observed through the institution's 100-inch Hooker Telescope, then the world's largest.

8. The prevailing theory at the time was that the Milky Way galaxy comprised the entire universe.

9. So these nebulae, Hubble's fellow astronomers believed, were all located within the galaxy.

10. But when Hubble began studying the nebulae his colleagues had been observing, he wasn't so sure that they were all confined to the Milky Way.

11. One of these nebulae, named the Andromeda nebula, contained a very bright star called a "Cepheid variable."

12. These stars pulsate in a predictable way that can help to determine their distance from an observer.

13. Over several months in 1923, Hubble took pictures of the Andromeda nebula and measured the brightness of the Cepheid variable star, determining that its brightness varied over a period of 31.45 days.

14. He also calculated that it was 7,000 times brighter than our own sun, and determined that it was 900,000 light-years away.

15. Scientist Harlow Shapley, who headed the Harvard College Observatory from 1921 to 1952, had just a few years prior calculated the distance across the Milky Way to be about 100,000 light-years.

16. Although Shapley at first criticized Hubble's findings, these measurements clearly showed that the Andromeda nebula was located far outside of the Milky Way galaxy, proving that the universe was much vaster than previously assumed.

17. Today, scientists know that there are actually two types of

Cepheid variable stars, and Hubble's calculations were a bit off: the Andromeda galaxy, as it's now called, is approximately two million light-years away.

18. As telescopes advanced, Hubble began to observe more galaxies, measuring their brightness and noticing a shift in the light toward the red end of the spectrum.

19. This "redshift" occurs due to the Doppler effect, which is a change in the frequency or wavelength of a wave relative to movement.

20. Hubble measured as many galaxies as he could, and in 1929 published a paper revealing his findings.

21. The universe, Hubble realized, was not a static cosmos as had been assumed, but rather it was moving outward.

22. All of the galaxies we can observe are flying apart from each other at great speeds, causing our universe to expand.

23. Hubble calculated that each galaxy is moving at a speed in direct proportion to its distance, a concept now known as Hubble's Law.

24. Galaxies that are farther away are moving more quickly than galaxies that are closer to us.

25. By measuring the rate of expansion, scientists determined the universe's approximate age, which is about 13.8 billion years old.

26. Scientists also realized that if the universe is continually expanding, that must also mean that at one time, it was much smaller and more compact.

27. This realization eventually gave rise to the Big Bang theory.

27 Facts About Neptune

1. Johanne Galle discovered Neptune in 1846, using sophisticated mathematical predictions.

2. The planet's existence had been deduced earlier than that by other astronomers, based on Uranus's orbit.

3. One of those astronomers, Frenchman Urbain Le Verrier, later tried to name the planet after himself. He did not succeed.

4. You can't see Neptune without a good telescope.

5. It's dimmer than Jupiter's larger moons.

6. Neptune takes 165 Earth-years to get around the sun once.

7. It takes more than four hours for the sun's light to reach Neptune.

8. The light that reaches it is very dim.

9. Neptune appears blue because its atmosphere contains methane gas.

10. The blue of Uranus and the blue of Neptune are different shades, so Neptune may have something else in its atmosphere interacting with the methane.

11. Methane, along with water and ammonia, form the planet's substance.

12. Scientists dub Neptune an "ice" planet for that reason, though the fluid mixture is actually hot and dense.

13. The planet likely has a core of iron and nickel.

14. Neptune is the windiest planet in the solar system.

15. Winds on Neptune can reach 1,200 miles per hour.

16. At least one storm on its surface, titled the "Great Dark Spot," was large enough to contain Earth.

17. It has since dissipated.

18. Neptune has 14 known, named moons as of January 2023.

19. Its main moon, Triton (a little smaller than our own), is doomed.

20. In as little as 10 million years, when Triton falls from the sky and spirals toward Neptune, the planet's gravity will crumble it into a huge ring.

21. Like Uranus, Neptune has only been visited by a *Voyager 2* flyby in 1989.

22. Neptune is about 2.8 billion miles away from the sun, or 30 AU.

23. The planet has faint rings.

24. Neptune, like Earth, has seasons.

25. Since it takes so long to orbit the sun, each season on Neptune lasts decades.

26. Scientists think Neptune was closer to the sun when it was formed and later moved to the edge of the solar system.

27. Neptune's astrological symbol resembles a trident, like one carried by the god Neptune.

35 Dwarf Planet Facts

1. Some objects aren't big enough to be classified as planets, so they're called dwarf planets.

2. There are thousands in the universe.

3. Pluto is the most famous dwarf planet.

4. Clyde Tombaugh (1906–97) discovered Pluto in 1930.

5. He laboriously flipped through photographic plates of regions of the sky captured at different dates that showed the special wandering that denotes a planet.

6. In more recent years, researchers have found many problematic bodies in our solar system.

7. Scientists were faced with either assigning planetary status to many more bodies or reclassifying Pluto.

8. In 2006, the International Astronomical Union stipulated that planets have to fulfill the following criteria: 1) the body must orbit the sun; 2) the body must be big enough for gravity to quash it into a round ball; 3) the body must have cleared all other debris (including asteroids and comets) out of its orbit.

9. Pluto was reclassified as a dwarf planet because it did not fulfill the third criterion.

10. Because Pluto's orbit around the sun takes approximately 247.9 Earth-years, it didn't even get to celebrate its first anniversary of being a planet.

11. Many scientists contend that Earth, Mars, Neptune, and Jupiter have not cleared their orbits either.

12. Pluto actually ventures into Neptune's orbit at times, and there are 10,000 asteroids in Earth's orbit and 100,000 in Jupiter's.

13. Pluto's orbit is off kilter—it's out of line with the solar system and elliptical enough that for periods of 20 Earth-years at a time, Pluto is closer than Neptune.

14. Pluto's moon, Charon, is about half its size.

15. On January 19, 2006, NASA launched an unmanned probe called *New Horizons* that was bound for Pluto.

16. It carried some of the ashes of Clyde Tombaugh.

17. Officially a dwarf planet, Eris is a little bigger than Pluto and has one dinky moon that we know of.

18. Eris is one of the non-planets that got Pluto kicked out of the Planet Club.

19. Given the argument that ensued, how fitting that Eris was named for a goddess of strife and discord.

20. Eris is way out there. A year on Eris is more than twice as long (557 Earth-years) as a year on Pluto (248 Earth-years).

21. Eris has a small moon called Dysnomia, named after Eris's mythological daughter.

22. Ceres is a dwarf planet found in the Main Asteroid Belt.

23. It's the only dwarf planet found there, as most are found in the outer solar system, in the Kuiper Belt.

24. Ceres has water vapor in its atmosphere.

25. Ceres was explored by the spacecraft *Dawn* in 2015.

26. The dwarf planet Makemake takes about 305 Earth-years to orbit the sun.

27. Makemake is the name of the Rapa Nui god of fertility. (The Rapa Nui are the people of Easter Island.)

28. Slightly smaller than Pluto, Makemake is the second-brightest object in the Kuiper Belt as seen from Earth.

29. NASA's Hubble Space Telescope spotted a small, dark moon orbiting Makemake in 2016.

30. Haumea has a very fast rotation period, making a day on the dwarf planet four hours long.

31. The fast spin distorts the dwarf planet's shape, making Haumea look more like a football or egg than a sphere.

32. It takes Haumea 285 Earth-years to make one trip around the sun.

33. Haumea is named after a Hawaiian goddess of childbirth.

34. Its two moons, Namaka and Hi'iaka, are named after Haumea's mythological daughters.

35. Haumea is the first known Kuiper Belt Object to have rings.

5 Myths About Space

1. **Myth:** There is no gravity in space. **Truth:** There is a difference between "weightlessness" and "zero-g" force. Astronauts may effortlessly float inside a space shuttle, but they are still under the grasp of approximately 10 percent of Earth's gravity.

2. **Myth:** Gravitational forces are powerful enough to distort a person's features. **Truth:** This popular notion can be traced to the fertile minds of Hollywood filmmakers. Although gravitational forces press against the astronauts, they are perfectly capable of performing routine tasks, and their faces do not resemble Halloween masks.

3. **Myth:** An ill-suited astronaut will explode. **Truth:** Another Hollywood invention. The human body is too tough to distort

in a complete vacuum. The astronaut would double over in pain and eventually suffocate, but it wouldn't explode.

4. **Myth:** Stranded space travelers will be asphyxiated. **Truth:** Although the danger of being stranded in space is very real (the Apollo 13 mission comes to mind), astronauts in such a situation would not die from lack of oxygen. Carbon dioxide in a disabled spacecraft could build up to life-threatening levels long before the oxygen ran out.

5. **Myth:** The world watched the *Challenger* disaster. **Truth:** Stories tell of the millions of horrified viewers who watched as the spacecraft and its solid-rocket boosters broke apart on live television. Except for cable network CNN, however, the major networks had ceased their coverage of the launch. However, many schoolchildren were watching to see Christa McAuliffe, the first teacher in space.

14 Facts About Humans & Space

1. Soviet cosmonaut Valentina Tereshkova became the first woman in space in 1963.

2. The 26-year-old Tereshkova did skydiving as a hobby.

3. During her flight, she orbited Earth 48 times.

4. The first "all woman" spacewalk took place in 2019, with astronauts Christina Koch and Jessica Meir.

5. The world's first space station was the Soviet Union's Salyut 1, which launched in 1971.

6. Skylab, the first American space station, fell to Earth in thousands of pieces in 1979.

7. Thankfully, most of them landed in the ocean.

8. The International Space Station was launched in 1998.

9. It has a length of close to 240 feet, a width of close to 358 feet, and a weight of more than 925,000 pounds.

10. The Very Large Array is a not-so-creatively named radio astronomy telescope group in New Mexico.

11. There are 27 antennas that work together to make images of space visible only within radio wavelengths.

12. Since 1959, more than 6,000 pieces of "space junk" (abandoned rocket and satellite parts) have fallen out of orbit, and many of these have hit Earth's surface.

13. U.S. astronaut Peggy Whitson, who was a commander of the International Space Station (twice, in fact), has spent 665 total days in space. That's a record for NASA astronauts.

14. U.S. astronauts can vote from space! They've been able to do so since 1997.

37 Pulsar Discovery Facts

Named for their distinctive pulses of electromagnetic radiation, pulsars are sometimes called the "lighthouses" of the universe. And like a lighthouse, these neutron stars rotate, so their beams of radiation are only detectible when pointing directly at Earth.

✳   ✳   ✳   ✳

1. A pulsar is formed when a massive star dies and detonates as a supernova.

2. The outer layers of the star are blasted off into space, but the inner core is compacted by gravity.

3. Now a smaller size, this tiny, dense object begins to rotate quickly, emitting radiation along its magnetic field lines.

4. On Earth, these beams of radiation can be detected each time the pulsar rotates, which can be several times a second.

5. The existence of these celestial objects was unknown until 1967.

6. It all began with the study of quasars, the bright nuclei at the center of galaxies.

7. Antony Hewish, professor of radio astronomy at the University of Cambridge, designed a radio telescope to observe quasars.

8. Hewish enlisted astronomy graduate student Jocelyn Bell to help build the new telescope, which would cover about 4½ acres.

9. Bell and several other students spent two years constructing the huge telescope, which consisted of 2,000 dipole antennas and 120 miles of wire and cable.

10. The telescope became operational in July 1967.

11. Bell was in charge of analyzing the data, which required her to pore through 100 feet of paper readouts every day.

12. This taught Bell what was typically expected in the data and how it related to the quasars she was studying.

13. A few weeks into her research, Bell noticed what she called "a bit of scruff" in the data.

14. It didn't appear to be man-made interference, and it wasn't the usual data she was used to seeing.

15. Instead, it was a consistent, regular signal, coming from the same place in the sky.

16. Bell shared her finding with Hewish, who was just as confused.

17. The data indicated a series of sharp pulses that occurred every 1.3 seconds, something that had never been observed in the skies before.

18. The two began searching for what could possibly be causing such a signal, ruling out various sources of interference such as television signals, orbiting satellites, or radar reflected off the moon.

19. They even used another telescope to confirm the signal, ruling out a defect with their own equipment.

20. Bell and Hewish half-jokingly considered the possibility of an extraterrestrial message.

21. They playfully nicknamed the signal "LGM-1," for "Little Green Men."

22. Even the extraterrestrial theory was ruled out when Bell discovered another signal emanating from a different part of the sky.

23. Since it would be quite unlikely that two groups of aliens from completely different parts of the galaxy would attempt to contact Earth at the same time, Bell and Hewish realized they were dealing with something natural, yet unknown.

24. By the end of 1967, Bell had found a total of four similar signals, and the unknown sources were dubbed "pulsars," combining "pulse" and "quasar."

25. In February 1968, Hewish held a seminar to announce the unusual discovery that had been made, even though neither he nor Bell was yet certain what they'd found.

26. The press ran with the "extraterrestrial" story, but astronomers took the discovery more seriously, searching the skies for more unusual signals.

27. By the end of 1968, dozens of pulsars had been detected.

28. Astrophysicist Thomas Gold of Cornell University suggested that the signals might be from rapidly rotating neutron stars, which, due to this rotation and their strong magnetic fields, would emit consistent pulses of radiation like a lighthouse beacon.

29. Later in 1968, the discovery of the Crab Pulsar confirmed Gold's theory.

30. The detection of the Crab Pulsar, located in the center of the Crab Nebula, which was already known to be the remnant of a supernova, confirmed that these signals were, indeed, from neutron stars.

31. In 1974, Hewish was awarded the Nobel Prize in Physics for his role in discovering pulsars.

32. While Bell, who originally found the "bit of scruff" that turned out to be an amazing new find, was overlooked, she held no bitterness over the decision.

33. Today, around 1,600 pulsars are known to be shining their beams throughout the universe.

34. The fastest pulsars emit 716 pulses per second.

35. The slowest emit only one pulse every 23.5 seconds.

36. Pulsars are so incredibly regular in their rotations that they have been known to rival even atomic clocks in keeping precise time.

37. Some scientists have even suggested that these "lighthouses" could be used to help spacecraft navigate around the universe.

19 Facts About Being an Astronaut

1. Skylab astronauts grew 1½ to 2¼ inches due to spinal lengthening and straightening as a result of zero gravity.

2. Because of the lack of laundry facilities, astronauts on missions change their socks, shirts, and underwear just every two days.

3. They only change their pants once a week.

4. It's sponge-bath only for space crews.

5. Water droplets could escape and float out, posing a danger to expensive electronics.

6. To be safe, the astronauts step in a cylindrical stall, where they have about a one-gallon ration of water to use.

7. The dirty water is sucked up by a vacuum and stored in special trash tanks.

8. Water isn't part of the tooth-brushing regime at all.

9. NASA developed a unique kind of toothpaste that astronauts swish around without the need for liquid.

10. Contrary to common perception, most garbage is sealed in bags to be brought back to Earth for disposal—it isn't just tossed into orbit.

11. As for the other, shall we say, "waste" produced onboard, astronauts use toilets similar to those back home.

12. Their facilities, though, have no water.

13. Instead, the space toilet uses a constant vacuum-like airflow in the bowl to keep things from floating back up in the zero gravity.

14. Crew members also have to strap in their feet and thighs to keep themselves from floating away mid-act.

15. To sleep, astronauts zip themselves into specially designed sleeping bags that attach to their lockers.

16. Some crew members may also rest in removable bunk beds.

17. Because of objects' tendency to float in space, magnets are a common commodity.

18. Meal trays are magnetic and designed to keep forks, spoons, and knives stuck down.

19. The food packages are adhered to the trays with strips of Velcro.

14 Facts About Katherine Johnson

Who was NASA mathematician Katherine Johnson? Read on to find out.

1. Johnson was born Creola Katherine Coleman in West Virginia in 1918.

2. Her mother was a teacher.

3. Johnson enrolled in high school when she was only ten years old.

4. She enrolled in college when she was only 14.

5. Johnson graduated college summa cum laude at the age of 18.

6. She did some graduate coursework at West Virginia University, one of three Black students to integrate the graduate school there.

7. In 1952, she joined an all-Black computer section at NACA, the predecessor to NASA, at their Langley facility.

8. The head of that group was Dorothy Vaughan (played by Octavia Spencer in the 2016 *Hidden Figures* movie).

9. Johnson was portrayed by Taraji P. Henson in *Hidden Figures*.

10. Johnson did trajectory analysis for the first human spaceflight in 1961.

11. She was the co-author of a report in 1960 and was credited for it, the first woman in the division to get that kind of credit.

12. In 1962, Johnson worked on the equations for John Glenn's orbital mission, by Glenn's specific request.

13. Johnson worked at Langley for 33 years before retiring.

14. She passed away in 2020 at the age of 101.

29 Largest Objects in the Solar System

1. Sun
2. Jupiter
3. Saturn
4. Uranus
5. Neptune
6. Earth
7. Venus
8. Mars
9. Ganymede (moon of Jupiter)
10. Titan (moon of Saturn)
11. Mercury
12. Callisto (moon of Jupiter)
13. Io (moon of Jupiter)
14. Moon (Earth's moon)
15. Europa (moon of Jupiter)
16. Triton (moon of Neptune)
17. Pluto (dwarf planet)
18. Eris (dwarf planet)
19. Haumea (dwarf planet)
20. Titania (moon of Uranus)
21. Rhea (moon of Saturn)
22. Oberon (moon of Uranus)
23. Iapetus (moon of Saturn)
24. Makemake (dwarf planet)
25. Gonggong (dwarf planet)
26. Charon (moon of Pluto)
27. Umbriel (moon of Uranus)
28. Ariel (moon of Uranus)
29. Dione (moon of Saturn)

Chapter 8

Arts and Culture

49 Facts About Artists

1. Michelangelo (1475–1564) considered himself a sculptor when the pope commissioned him to paint the ceiling of the Sistine Chapel.

2. The huge painting (141 feet long by 43 feet wide) contained nine scenes from the Bible.

3. No two of the 300 people painted on the Sistine Chapel ceiling look alike.

4. Leonardo da Vinci (1452–1519) was an artist, scientist, and inventor during the Italian Renaissance.

5. Vincent van Gogh (1853–90) was not commercially successful during his lifetime.

6. However, the Dutch post-impressionist painter posthumously became one of the most famous and influential artists in history.

7. The impressionism movement was named after *Impression: Sunrise*, a painting by Claude Monet (1840–1926).

8. Critics derided the painting's "unfinished" appearance.

9. Mary Cassatt (1844–1926) was the only official American member of the Impressionists.

10. The 13-foot-tall *David* statue is Michelangelo's most famous work and considered nearly perfect by many art experts.

11. Unlike some artists, Leonardo da Vinci was famous for his paintings while he was still alive.

12. Van Gogh's suicide at age 37 came after years of mental illness, depression, and poverty.

13. His brother Theo died six months afterward and was buried next to him.

14. Grant Wood (1891–1942) used his sister, Nan, and his 62-year-old dentist as models for his iconic *American Gothic* painting.

15. Monet's parents called him Oscar to distinguish him from his father, who also was named Claude.

16. Known for his water lily paintings, Monet supposedly hired six full-time employees to tend to his gardens.

17. Pablo Picasso (1881–1973) is considered one of the inventors of Cubism, a style that reduces subjects to geometric forms.

18. Georgia O'Keeffe (1887–1986) is known for her close-up artworks of flowers.

19. But flowers only make up approximately 200 of O'Keeffe's more than 2,000 paintings.

20. Salvador Dalí (1904–89) believed that he was a reincarnation of his older brother who had died nine months before he was born.

21. All of the watches in Dalí's *The Persistence of Memory* tell different times.

22. In 1984, Mexico declared the works of Frida Kahlo (1907–54) part of the country's national cultural heritage.

23. Her paintings often featured aspects of Aztec mythology and Mexican folklore.

24. Picasso loved pets and owned a mouse, a turtle, a monkey, and many cats and dogs.

25. Picasso's full name is Pablo Diego José Francisco de Paula Juan Nepomuceno María de los Remedios Cipriano de la Santísima Trinidad Ruiz y Picasso.

26. O'Keeffe customized her car so that she should paint inside her Model-A Ford instead of in the desert sun.

27. She rarely signed her paintings, but sometimes wrote on the back.

28. Dalí's famous long, curled moustache was most likely inspired by the Spanish Golden Age painter Diego Velázquez.

29. Van Gogh lopped off his own left earlobe following an argument with painter Paul Gauguin in 1888.

30. He wrapped his severed lobe in a piece of cloth and presented it to a prostitute named Rachel. No word on what she did with the "gift."

31. Kahlo's *The Frame* was the first painting by a Mexican artist acquired by the Louvre.

32. Some of the stories in Dalí's autobiography *The Secret Life of Salvador Dalí* are true, but some are just made up.

33. Picasso's first word was "piz, piz" which is short for *lápis,* or pencil in Spanish.

34. O'Keeffe received the Presidential Medal of Freedom from President Gerald Ford in 1977.

35. Monet's father wanted his son to join the family's grocery and shipping supply company.

36. Picasso completed his famous *Guernica* in just three weeks.

37. The black-and-white painting depicts the destruction of Guernica, Spain, during the Spanish Civil War.

38. Andy Warhol (1928–87) is behind some of the most recognizable art of the 20th century's latter half.

39. In fact, Warhol's pop art helped shape pop(ular) culture in general.

40. Neo-expressionist artist Jean-Michel Basquiat (1960–88) became a junior member of the Brooklyn Museum at age six.

41. Dalí designed the film sets in Alfred Hitchcock's 1945 thriller *Spellbound,* in which a psychiatrist solves a murder by analyzing her patient's dreams.

42. Warhol was the first artist to exhibit video footage as art, essentially creating the "multi-media" in 1965.

43. Warhol regularly taped conversations with others or dictated his ideas into a tape recorder.

44. There are approximately 3,400 of these audiotapes.

45. At a Sotheby's auction in May 2017, *Untitled*, a 1982 painting by Basquiat depicting a black skull with red and yellow rivulets, sold for $110.5 million, becoming one of the most expensive paintings ever purchased.

46. It also set a new record high for an American artist at auction.

47. American digital artist Mike Winkelmann (born 1981) is known professionally as Beeple.

48. Winkelmann sold his jpeg collage of images, *Everydays: the First 5000 Days*, for $69.3 million in March 2021 at Christie's.

49. It was the first purely non-fungible token (NFT) sold by Christie's.

14 Famous First Lines

1. *"Marley was dead, to begin with. There is no doubt whatever about that."* –CHARLES DICKENS, *A CHRISTMAS CAROL*

2. *"It is a truth universally acknowledged, that a single man in possession of a good fortune, must be in want of a wife."* –JANE AUSTEN, *PRIDE AND PREJUDICE*

3. *"You don't know about me without you have read a book by the name of* The Adventures of Tom Sawyer; *but that ain't no matter."* –MARK TWAIN, *THE ADVENTURES OF HUCKLEBERRY FINN*

4. *"It was a pleasure to burn."* –RAY BRADBURY, *FAHRENHEIT 451*

5. *"One morning, when Gregor Samsa woke from troubled dreams, he found himself transformed in his bed into a monstrous vermin."* –FRANZ KAFKA, *METAMORPHOSIS*

6. *"If I am out of my mind, it's all right with me, thought Moses Herzog."* –SAUL BELLOW, *HERZOG*

7. *"Happy families are all alike; every unhappy family is unhappy in its own way."* –LEO TOLSTOY, *ANNA KARENINA*

8. *"In the country of Westphalia, in the castle of the most noble Baron of Thunder-ten-tronckh, lived a youth whom Nature had endowed with a most sweet disposition."* –VOLTAIRE, *CANDIDE*

9. *"At a village of La Mancha, whose name I do not wish to remember, there lived a little while ago one of those gentlemen who are wont to keep a lance in the rack, an old buckler, a lean horse and a swift greyhound."* –MIGUEL DE CERVANTES, *DON QUIXOTE*

10. *"When he was nearly thirteen, my brother Jem got his arm badly broken at the elbow."* –HARPER LEE, *TO KILL A MOCKINGBIRD*

11. *"I was leaning against the bar in a speakeasy on Fifty-second Street, waiting for Nora to finish her Christmas shopping, when a girl got up from the table where she had been sitting with three other people and came over to me."* –DASHIELL HAMMETT, *THE THIN MAN*

12. *"He was an old man who fished alone in a skiff in the Gulf Stream and he had gone 84 days now without taking a fish."* –ERNEST HEMINGWAY, *THE OLD MAN AND THE SEA*

13. *"We are at rest five miles behind the front."* –ERICH MARIA REMARQUE, *ALL QUIET ON THE WESTERN FRONT*

14. *"It was a bright cold day in April, and the clocks were striking thirteen."* –GEORGE ORWELL, *1984*

51 Unusual World Customs

1. In much of the Far East, belching is considered a compliment to the chef and a sign that you have eaten well and enjoyed your meal.

2. In Chile, pounding your left palm with your right fist is considered vulgar.

3. The people of Ottery St. Mary, England, celebrate Guy Fawkes Day by racing through the streets of their town with barrels of flaming tar strapped to their backs.

4. In ancient Rome, urine was commonly used as a tooth whitener.

5. In most of the Middle and Far East, it's considered an insult to point your feet (particularly the soles) at another person, or to display them in any way, for example, by resting with your feet up.

6. For almost 500 years, a form of conflict resolution known as "dueling" took place in western Europe and the United States.

7. The highly ritualistic tradition began with the offended party throwing down his glove at the foot of another and ended with a sword or pistol fight—often to the death.

8. It's considered impolite to use silverware to eat chicken in Turkey.

9. In Japan and Korea, tipping is considered an insult, rather than a compliment; for them, accepting tips is akin to begging.

10. Yawning in public is considered gauche in Ecuador.

11. While a bone-crushing handshake is seen as admirable in the U.S. and UK, in much of the East, particularly the Philippines, it is seen as a sign of aggression.

12. In Italy, eat spaghetti as the Romans do: with a fork only.

13. Using a spoon to help collect the pasta is considered uncouth.

14. Keep your right elbow off the table when eating in Chile.

15. Until being banned in 1912, foot binding was common in China.

16. The practice, which involved breaking girls' toes and wrapping them tightly in cloth, prevented women's feet from growing normally.

17. Small, dainty feet were considered a symbol of status.

18. In Sweden, the Christmas season begins on Santa Lucia's Day (December 13), when the eldest daughter of a household, clad in white and wearing a wreath holding seven lit candles on her head, serves her family breakfast in bed.

19. In most Asian countries, a business card is seen as an extension of the person it represents.

20. Therefore, to disrespect a card—by folding it, writing on it, or just shoving it into your pocket without looking at it—is to disrespect the person who gave it to you.

21. In Germany and most of South America, the "okay" sign (thumb and forefinger touching to make a circle) is an insult, similar to giving someone the finger in the United States.

22. In Turkey, the same "okay" sign is a derogatory gesture used to imply that someone is homosexual.

23. Up until the 19th century, long fingernails were considered a symbol of gentility and wealth among the Chinese aristocracy.

24. Wealthy Chinese often sported fingernails several inches in length and protected them by wearing special coverings made of gold.

25. Chewing gum might be good for dental hygiene, but in many parts of the world, particularly Luxembourg, Switzerland, and France, public gum-chewing is considered vulgar.

26. In parts of India, some women still perform sati, an ancient custom in which a widow throws herself on the funeral pyre of her deceased husband to commit suicide.

27. In Greece, any signal that involves showing your open palm is extremely offensive.

28. Such gestures include waving, as well as making a "stop" sign.

29. If you do wish to wave goodbye to someone in Greece, you need to do so with your palm facing in.

30. Tipping is uncommon, and even rude, in many Asian countries.

31. When dining in China, never force yourself to clear your plate out of politeness—it would be bad manners for your host not to keep refilling it.

32. Instead, you should leave some food on your plate at each course as an acknowledgment of your host's generosity.

33. Orthodox Jews won't shake hands with someone of the opposite sex.

34. A strict Muslim woman will not shake hands with a man.

35. Although, to confuse matters, a Muslim man *will* shake hands with a non-Muslim woman.

36. People in these cultures generally avoid touching people of the opposite sex who are not family members.

37. When conversing in Quebec, keep your hands visible.

38. Talking with your hands in your pockets is considered rude.

39. In most Arab countries, the left hand is considered unclean, and it is extremely rude to offer it for a handshake or to wave a greeting.

40. Similarly, it is impolite to pass food or eat with the left hand.

41. Historically, people living in deserts didn't have access to toilet paper, so the left hand was used for "hygienic functions," then cleaned by rubbing it in the sand.

42. In the UK, when the two-fingered "V for victory" or "peace" salute is given with the palm facing inward, it is considered extremely rude, having a meaning similar to raising the middle finger to someone in the United States.

43. In the U.S., one should never butter an entire piece of bread before eating it.

44. The proper way to eat bread is to pull off a small bit from your larger piece and butter it before popping it into your mouth.

45. In many countries, particularly in Asia and South America, it is essential to remove your shoes when entering someone's home.

46. In most of Europe it's polite to ask your host whether they would prefer you to do so.

47. In many Latin American cultures, a girl's 15th birthday is considered one of the most important days of her life.

48. Known as the quinceañera, the celebration can be as elaborate as a wedding.

49. In Thailand, using one's foot to move an object or gesture toward somebody is considered the height of rudeness.

50. Similarly, one should never cross their legs when in the presence of elders.

51. In Singapore, most gums have been illegal since 1992 when residents grew tired of scraping the sticky stuff off their sidewalks.

5 Fabulous Fads of the 1950s

1. Poodle Skirts
2. Sock Hops
3. 3-D Movies
4. The Conical Bra
5. Beatniks

30 *Mona Lisa* Facts

1. The *Mona Lisa* was painted by Italian artist Leonardo da Vinci.

2. The Renaissance genius began the portrait in 1503 when he was about 51 years old.

3. Although it looms large in cultural influence, the *Mona Lisa* is rather small—just 30 inches by 21 inches.

4. The *Mona Lisa* weighs 18 pounds.

5. The *Mona Lisa* is painted on three pieces of wood, which are about 1.5 inches thick.

6. Leonardo da Vinci used a process of brushwork he called "sfumato" (from the Italian *fumo*, meaning smoke).

7. He said the painting was composed "without lines or borders, in the manner of smoke or beyond the focus plane."

8. The government of France owns the *Mona Lisa*.

9. The world-famous painting has been on permanent display at the Louvre museum in Paris since 1797.

10. More than 80 percent of museum visitors go directly to the *Mona Lisa*, skipping the Louvre's 6,000+ other paintings.

11. The subject of the painting is Lisa del Giocondo, an Italian noblewoman and a member of the Gherardini family.

12. She was the wife of wealthy Florentine silk merchant Francesco del Giocondo.

13. Lisa was Francesco's third wife.

14. She married him when she was 16 and he was 30.

15. Lisa likely sat for the portrait soon after the birth of her third child, when she was about 24.

16. Although Francesco del Giocondo commissioned da Vinci to paint the *Mona Lisa*, for some reason, da Vinci did not give it to him.

17. "Mona" is simply a contraction of *ma donna*, or "my lady," in Italian.

18. The title is the equivalent of "Madam Lisa" in English.

19. While it appears that the *Mona Lisa* does not have eyelashes or eyebrows, a 2007 study revealed that she originally was painted with them.

20. They gradually disappeared, most likely due to overcleaning.

21. The *Mona Lisa* is believed to have been painted from 1503–1506.

22. But da Vinci may have continued working on it as late as 1517.

23. When da Vinci died in 1519, the *Mona Lisa* was inherited by his student and assistant Salaì.

24. King Francis I of France probably bought the *Mona Lisa* from Salaì, which is how a painting by an Italian artist became the property of France.

25. Over 8 million people visit the *Mona Lisa* at the Louvre each year.

26. They each spend about 15 seconds looking at her.

27. Studies using x-ray fluorescence spectroscopy have revealed that da Vinci was constantly testing new methods.

28. He did not always use glaze, for example.

29. While working on the *Mona Lisa*, he mixed manganese oxide in with his paint.

30. Ongoing studies are sure to reveal even more of da Vinci's masterful techniques.

16 Popular TV Shows of the 1950s

1. I Love Lucy
2. You Bet Your Life
3. Gunsmoke
4. I've Got a Secret
5. Dragnet
6. The Jack Benny Program

7. The Adventures of Ozzie and Harriet

8. The Milton Berle Show

9. The Honeymooners

10. Leave It to Beaver

11. The Ed Sullivan Show

12. The Donna Reed Show

13. Father Knows Best

14. Make Room for Daddy (The Danny Thomas Show)

15. Have Gun, Will Travel

16. The Lone Ranger

48 Facts to Dance About

1. The first archeological proof of dance comes from 10,000-year-old rock and cave paintings.

2. David Meenan holds the record for greatest distance tap danced.

3. He tap danced 32 miles in 7 hours and 35 minutes at track in New Jersey in October 2001.

4. The Greek word *chora*, meaning "source of joy," is related to the word *choros*, the Greek word for "dance."

5. According to Plato, dance "gave the body its just proportions."

6. In ancient Greek mythology, nine muses were associated with each one of the arts. Terpsichore was the muse of dancing.

7. Ballet began in Italy around 1500.

8. When Catherine de Medici of Italy married the French King Henry II, she introduced early dance styles into court life in France.

9. From Italian roots, ballets in France and Russia developed their own stylistic character.

10. Dancing en pointe (on toe) became popular during the early part of the 19th century.

11. Every time a ballerina jumps en pointe, three times her body weight falls on the tip of her big toe.

12. As a teenager, rapper Tupac Shakur was a ballet dancer.

13. He was a member of the 127th Street Ensemble, a Harlem-based theater company.

14. Salsa is derived from the Spanish word for "sauce," and refers to the hot and spicy flavor of the dance.

15. Germany suffered a plague of uncontrollable dancing in 1518.

16. "The Dancing Plagues," or dancing manias, occurred sporadically throughout Western Europe from the 14th to the 17th centuries.

17. Victims were seized by an uncontrollable urge to dance.

18. For unknown reasons, people danced until they collapsed from exhaustion or died.

19. The film *Footloose* is based on a true story.

20. An Oklahoma town outlawed dancing until local high schoolers successful challenged the law in 1979.

21. Lipizzaner stallions are trained to dance a Grande Quadrille, a ballet that involves six to eight horses (and riders) all in step.

22. In 1913, Kaiser Wilhelm II banned German military officers in uniform from dancing the tango.

23. James Brown's 1972 hit "Get the Good Foot" started a new dance craze called "The Good Foot."

24. Roy Castle achieved one million taps in 23 hours and 44 minutes at the Guinness World of Records exhibition in London in 1985.

25. Hula is the state dance of Hawaii.

26. In the Korean buchaechum (fan dance), dancers use large fans painted with blossoms to create formations that represent birds, butterflies, dragons, flowers, and waves.

27. The greatest number of consecutive fouettés en tournant (extremely fast turns on one foot) is 166.

28. It was achieved by Delia Gray at the Harlow Ballet School's summer workshop on June 2, 1991.

29. Onikenbai (devil's sword dance) is popular in northeastern Japan.

30. The dance is so named because the dancers wear face masks which look like demons (oni).

31. During one of Elvis Presley's early performances, cameras would only show him from the waist up.

32. His gyrating hips were considered shocking.

33. The world's longest conga line consisted of 119,986 people.

34. The record was set in Miami, Florida, on March 13, 1988.

35. Square dance is the official state dance of Massachusetts, Mississippi, Illinois, Idaho, Oklahoma, and Oregon.

36. The most popular ballet in the world is Tchaikovsky's *The Nutcracker*, which was first performed in 1892.

37. Brazil's national dance is the samba.

38. Originally, "rumba" was used as a synonym for "party" in northern Cuba.

39. In 1996, an estimated 72,000 people took part in the world's largest chicken dance in history at the Canfield fair in Ohio.

40. Michael Flatley of *Riverdance* and *Lord of the Dance* was once the world's highest-paid dancer, earning $1.6 million a week at his prime.

41. Rudolf Nureyev and Margot Fanteyn received a record 89 curtain calls after performing *Swan Lake* in Vienna in 1964.

42. The waltz is a ballroom and folk dance in ¾ time.

43. Johann Strauss II, who composed over 500 waltzes, did much to popularize the dance, but its origins trace back to the 16th century.

44. Bharata natyam is a classical Indian dance drama.

45. The world's largest Charleston dance consisted of 1,096 participants.

46. It was achieved in Shrewsbury, UK, on September 22, 2018.

47. Dancer, choreographer, and company director Martha Graham (1894–1991) is considered one of the pioneering founders of American modern dance.

48. In 2015, Misty Copeland became the first Black woman promoted to principal dancer for American Ballet Theatre, one of the three leading classical ballet companies in the U.S.

17 Unusual Book Titles

1. *How to Avoid Huge Ships* by John W. Trimmer

2. *Scouts in Bondage* by Michael Bell

3. *Be Bold with Bananas* by Crescent Books

4. *Fancy Coffins to Make Yourself* by Dale L. Power

5. *The Flat-Footed Flies of Europe* by Peter J. Chandler

6. *101 Uses for an Old Farm Tractor* by Michael Dregni

7. *Across Europe by Kangaroo* by Joseph R. Barry

8. *101 Super Uses for Tampon Applicators* by Lori Katz and Barbara Meyer

9. *Suture Self* by Mary Daheim

10. *The Making of a Moron* by Niall Brennan

11. *How to Make Love While Conscious* by Guy Kettelhack

12. *Underwater Acoustics Handbook* by Vernon Martin Albers

13. *Superfluous Hair and Its Removal* by A. F. Niemoeller

14. *Lightweight Sandwich Construction* by J. M. Davies

15. *The Devil's Cloth: A History of Stripes* by Michel Pastoureaut

16. *How to Be a Pope: What to Do and Where to Go Once You're in the Vatican* by Piers Marchant

17. *How to Read a Book* by Mortimer J. Adler and Charles Van Doren

48 Fun Facts About the Movies

1. Thomas Edison's "Black Maria" in West Orange, New Jersey, was the first film studio, completed in 1893.

2. Edison made short vaudeville-act films there.

3. The oldest continuously operating cinema theater is the State Theatre in Washington, Iowa, which opened on May 14, 1897.

4. The first feature-length film was an Australian Western titled *The Story of the Kelly Gang* from 1906.

5. This tale of Australia's most famous bandit, Ned Kelly, was reportedly about an hour long.

6. The first film ever made in Hollywood was D.W. Griffith's 1910 *In Old California*.

7. *In Old California* was a biograph melodrama about a Spanish maiden (Marion Leonard) who has an illegitimate son with a man who later becomes governor of California.

8. The first movie to be filmed in Technicolor was *Becky Sharp* (1934).

9. The first building designed specifically to show movies was The Nickelodeon in Pittsburgh.

10. It opened in June 1905 and showed three or four one-reel films continuously throughout the day.

11. Within three years, thousands of similar operations, called nickelodeons, had opened around the country.

12. Contrary to popular belief, *The Jazz Singer* (1927) was not the first film with sync sound.

13. Sync sound aligns the picture and the soundtrack so that an action and its corresponding sound coincide.

14. The first feature-length film with sync sound was *Don Juan* (1926).

15. Movies hit the friendly skies for the first time in 1925.

16. That's when a British Imperial Airways flight screened *The Lost World* for passengers.

17. One pair of ruby slippers worn by actress Judy Garland in *The Wizard of Oz* (1939) are on display at the National Museum of American History.

18. The three lead actors in *Rebel Without a Cause* (1955) all met an untimely death.

19. James Dean died in a car crash, Natalie Wood drowned, and Sal Mineo was stabbed to death.

20. Alfred Hitchcock's *Psycho* (1960) was the first American film to show a toilet being flushed on screen.

21. Julie Andrew learned how to play the guitar for *The Sound of Music* (1965).

22. The now ubiquitous F-word was first uttered in a film in 1968, in *I'll Never Forget Whatshisname*.

23. Actress Marianne Faithfull shocked audiences when she spoke the four-letter word.

24. For comparison, *Casino* (1995) reputedly had nearly 400 uses of it—but who's counting?

25. The scream of horror you hear in *The Godfather* (1972) when Jack Woltz discovers the severed horse head in bed is real.

26. Actor John Marley (who played Jack Woltz) was not told in advance that the horse head would be real.

27. You won't find his name in the credits, but George Lucas worked on a montage of crime scene photos and newspaper headlines for *The Godfather*.

28. The beach on fictitious Amity Island where sunbathers and waders get a scare in the 1975 film *Jaws* is actually Joseph A. Sylvia State Beach on Martha's Vineyard in Massachusetts.

29. The mechanical shark used in *Jaws* was nicknamed "Bruce."

30. R2-D2 of *Star Wars* fame was sometimes operated by remote

control and other times by an actor who would crouch inside and move the droid.

31. Francis Ford Coppola shot his 1979 war epic *Apocalypse Now* in the Philippines instead of Vietnam.

32. The Philippine government allowed Coppola to use its military helicopters, only to divert them to fight insurgents several times during the shoot.

33. *E.T. the Extra-Terrestrial* (1982) was originally intended for the horror genre.

34. The original title for *Ghostbusters* (1984) was *Ghost Smashers*.

35. O.J. Simpson was considered for the lead in 1984's *The Terminator*.

36. Producers passed on Simpson because they thought he was too "nice" and "innocent" to be believable playing a cyborg killer.

37. It took a group of over 120 people three years to complete the stop-motion animated musical *The Nightmare Before Christmas* (1993).

38. For one second of film, up to 12 stop-motion moves had to be made.

39. Some of the dinosaur sounds in *Jurassic Park* (1993) were actually tortoises mating.

40. The first feature film created solely with Computer Generated Imagery (CGI) was *Toy Story* (1995).

41. *Titanic* (1997) came out when theaters still showed movies from film reels.

42. Due to the popular film's long theatrical run, Paramount had to send replacement reels to theaters that actually literally wore theirs out.

43. The first movie to gross over $2 billion was James Cameron's *Avatar* (2009).

44. During one of Leonardo DiCaprio's big scenes in *Django Unchained* (2012), he really did cut his hand on broken glass and begin to bleed.

45. He stayed in character and continued the scene.

46. Director Quentin Tarantino was so impressed that he used this take in the final film.

47. The largest film stunt explosion was achieved during filming for Bond movie *Spectre* in June 2015.

48. The explosion was the result of detonating 2,224 gallons of kerosene with 73 pounds of powder explosives, and lasted for over 7.5 seconds.

25 of The Beatles' Top Singles

1. "I Want to Hold Your Hand" (1963)

2. "She Loves You" (1963)

3. "From Me to You" (1963)

4. "Twist and Shout" (1964)

5. "Can't Buy Me Love" (1964)

6. "A Hard Day's Night" (1964)

7. "I Feel Fine" (1964)

8. "Eight Days a Week" (1965)

9. "Ticket to Ride" (1965)

10. "Help!" (1965)

11. "Yesterday" (1965)

12. "We Can Work It Out" (1965)

13. "Day Tripper" (1965)

14. "Nowhere Man" (1966)

15. "Paperback Writer" (1966)

16. "Yellow Submarine" (1966)

17. "Eleanor Rigby" (1966)

18. "Hello, Goodbye" (1967)

19. "With a Little Help from My Friends" (1967)

20. "Lady Madonna" (1968)

21. "Hey Jude" (1968)

22. "Come Together" (1969)

23. "Get Back" (1969)

24. "Let It Be" (1970)

25. "The Long and Winding Road" (1970)

Origins of 7 Common Phrases

1. **Blue Bloods:** In the Middle Ages, the veins of the fair-complexioned people of Spain appeared blue. To distinguish them as untainted by the Moors, they referred to themselves as blue-blooded.

2. **Rob Peter to Pay Paul:** In the mid-1550s, estates in St. Peter's, Westminster, were appropriated to pay for the new St. Paul's Cathedral. This process revived a phrase that preacher John Wycliffe had used 170 years before in *Select English Works*.

3. **Humble Pie:** While medieval lords and ladies dined on the finest foods, servants had to utilize leftovers (the "umbles," or offal) when preparing their meals. To eat humble pie means to exercise humility or self-effacement.

4. **Men of Straw:** In medieval times, men would hang around English courts of law, eager to be hired as false witnesses. They identified themselves with a straw in their shoe.

5. **White Elephant:** Once upon a time in Siam, rare albino elephants were to receive nothing but the best from their owners. Therefore, no one wanted to own one.

6. **Touch and Go:** English ships in the 18th century would often hit bottom in shallow water, only to be released with the next wave. The phrase indicated that they had narrowly averted danger.

7. **By Hook or By Crook:** This phrase describes a feudal custom that allowed tenants to gather as much wood from their lord's

land as they could rake from the undergrowth or pull down from the trees with a crook.

8 Groovy Fads of the 1960s

1. Hippies
2. Go-go Boots and Minidresses
3. Fallout Shelters
4. Surfing
5. Peace Symbol
6. The Twist
7. Tie-Dye
8. Lava Lamps

35 Mighty Mouse Facts

1. Mighty Mouse was almost a housefly. When Terrytoons studio writer Izzy Klein first came up with the idea of a tiny superpowered creature, he designed a character called Superfly.

2. Thankfully, cooler heads prevailed: Studio head Paul Terry changed our hero into a mouse. Good thing, too; no one likes flies. (As opposed to mice. We all love having a mouse for a houseguest, don't we?)

3. Even after he became a mouse, our hero still wasn't "Mighty."

4. In his first two cartoon appearances, the flying mouse went by the name Super Mouse.

5. Terrytoons changed the name to Mighty Mouse only after learning that a new comic book, Coo Coo Comics, had introduced its own Super Mouse character in 1943.

6. Terrytoons renamed their mouse "Mighty Mouse" to avoid competition.

7. The studio later altered Mighty Mouse's first two adventures upon re-release to reflect the change.

8. The first Mighty Mouse cartoon debuted in 1942.

9. It was titled "The Mouse of Tomorrow."

10. Mighty Mouse's most famous villain is Oil Can Harry, a nasty cat.

11. Oil Can, though, made his debut long before the flying mouse, starring in Terrytoons' 1933 cartoon "The Banker's Daughter."

12. In that cartoon, he was a villainous human out to steal the virtuous Fanny Zilch from her stalwart lover, J. Leffingwell Strongheart.

13. During his early adventures, Mighty Mouse's girlfriend was the sweet Pearl Pureheart.

14. Pearl usually managed to get captured by Oil Can Harry.

15. The first Mighty Mouse cartoons were actually parodies of the silent-film serials that often featured heroines in peril with the requisite strong-chinned heroes rushing to their rescue.

16. Many Mighty Mouse cartoons started with the mouse and Pearl already in some trap devised by the nefarious Oil Can—say, tied to railroad tracks as a locomotive steamed toward them.

17. Mighty Mouse is known for his operatic cartoons, in which characters sing their lines.

18. The first operatic Mighty Mouse episode titled "Mighty Mouse and the Pirates" aired in January 1945.

19. Mighty Mouse has mostly been slighted by the Academy.

20. Only one of his cartoons was nominated for an Oscar in 1945. It didn't win.

21. Some of the original Mighty Mouse cartoons, especially the operatic cartoons, were rather violent for kiddie fare.

22. Mighty Mouse would often pummel his enemies, mostly cats of some sort, until they fled the scene.

23. In some cartoons, Mighty Mouse boasts telekinetic powers, making objects fly through the air.

24. In one adventure, he even managed to turn back time.

25. Mighty Mouse cartoons first aired as animated shorts that ran before feature films in movie theaters.

26. In 1955, CBS brought the character and his old adventures to television screens.

27. The network then replayed the old shorts for 12 years on the "Mighty Mouse Playhouse."

28. Famed animator Ralph Bakshi—who created the adult "Fritz the Cat"—created his own version of Mighty Mouse in the 1980s.

29. Called "Mighty Mouse: The New Adventures," the cartoon shared elements of the flying mouse's backstory with viewers.

30. For instance, when he wasn't fighting crime, Mighty Mouse posed as an ordinary mouse named Mike Mouse.

31. He also gained a sidekick in this version, Scrappy Mouse.

32. The Bakshi version of the cartoon lasted only two seasons, but that was long enough for it to generate controversy.

33. In one cartoon, Mighty Mouse sniffs a white powder.

34. Reverend Donald Wildmon, founder of the American Family Association, claimed the mouse was snorting cocaine.

35. Bakshi denied this, saying the mouse was actually sniffing his lucky cheese. Later, Bakshi said Mighty Mouse was sniffing crushed flowers.

Birth Names of 26 Vintage Film Stars

1. Fred Astaire	Frederic Austerlitz
2. Jack Benny	Benjamin Kubelsky
3. Milton Berle	Milton Berlinger
4. George Burns	Nathan Birnbaum
5. Eddie Cantor	Edward Israel Iskowitz
6. Gary Cooper	Frank James Cooper
7. Joan Crawford	Lucille Fay LeSueur

8. Bing Crosby	Harry Lillis Crosby
9. Tony Curtis	Bernard Schwartz
10. Bette Davis	Ruth Elizabeth Davis
11. Doris Day	Doris Kappelhoff
12. Dale Evans	Frances Smith
13. Judy Garland	Frances Gumm
14. Cary Grant	Archibald Leach
15. Rita Hayworth	Margarita Cansino
16. William Holden	William Beedle
17. Al Jolson	Asa Yoelson
18. Boris Karloff	William Henry Pratt
19. Jerry Lewis	Joseph Levitch
20. Bela Lugosi	Béla Blaskó
21. Dean Martin	Dino Crocetti
22. Marilyn Monroe	Norma Jean Baker
23. Roy Rogers	Leonard Slye
24. Barbara Stanwyck	Ruby Stevens
25. Lana Turner	Julia Jean Turner
26. John Wayne	Marion Morrison

26 *Star Wars* Facts

1. Released in 1977, the first *Star Wars* movie cost just over $11 million to make, compared to $113 million for *Star Wars: Episode III—Revenge of the Sith* (2005) produced nearly 30 years later.

2. In the late 1990s, remastering and reediting *Star Wars* for its 20th anniversary edition cost nearly as much as it did to make the original movie.

3. The first U.S. theater run for the original *Star Wars* pulled in $215 million.

4. The first-ever *Star Wars* trailer began showing a full six months before the movie came out.

5. Vague taglines such as "the story of a boy, a girl, and a universe" and "a billion years in the making" were meant to build up buzz before the film's debut.

6. The first *Star Wars* film was originally going to be titled *The Star Wars*, but George Lucas decided to drop the introductory article.

7. The movie's full title, *Star Wars: Episode IV: A New Hope*, was not used on posters, promotions, or publicity until the film was rereleased in 1981.

8. George Lucas was initially set to receive about $165,000 for the making of *Star Wars*.

9. But when production costs rose, he waived his fee in exchange for 40 percent of the box-office returns.

10. That, combined with a lucrative merchandising deal and his foresight into snagging creative control and full merchandising rights for the sequels, made him millions.

11. The filmmaker who directed *Scarface* (1983) cowrote the opening crawl text at the beginning of *Star Wars*.

12. Brian De Palma helped pen the words: "It is a period of civil war. Rebel spaceships, striking from a hidden base, have won their first victory against the evil Galactic Empire..."

13. The actors who played C-3PO and R2-D2—Anthony Daniels and Kenny Baker, respectively—are the only two people credited with appearing in all six of the *Star Wars* movies.

14. Daniels's C-3PO costume was precision engineered and fit very tightly.

15. If he moved too much, the pieces of the suit cut into him.

16. On the first day of shooting, walking in the costume resulted in a great many scrapes, cuts, and abrasions—a persistent problem throughout the shoot.

17. After Harrison Ford tested for the part of Han Solo, he had the edge for the role, though early in the project, Lucas had decided he didn't want to use anyone he had directed in the past.

18. The potential candidate list included Kurt Russell and Christopher Walken, among others.

19. Jodie Foster and Cindy Williams auditioned for the role of Princess Leia, along with Linda Purl, Terry Nunn, and many others.

20. Foster and Nunn were rejected because they were under 18.

21. The character of Luke Skywalker went through many incarnations during script development.

22. In the original treatment for the *Star Wars* script, which was written in 1973, Luke Skywalker was a general assigned to protect a rebel princess.

23. In a later version, his name was Kane Starkiller, and he was a half-man, half-machine character who is friends with Han Solo.

24. Later, he evolved into the young man closer to the character we are familiar with, but his name was Luke Starkiller.

25. Just before shooting began, Lucas decided that the name might remind people of murderers like Charles Manson (who was literally responsible for killing stars), so the name was changed to Luke Skywalker.

26. The stormtroopers' weapons in *Star Wars* were retooled military weapons from the 1940s.

51 Facts About Music

1. An ancient bone flute, found at a Neanderthal cave site in Slovenia, is thought to be at least 50,000 years old, making it the oldest known musical instrument.

2. It's carved from the femur of a cave bear.

3. There are eight notes on the musical scale.

4. Elvis Presley's "Don't Be Cruel/Hound Dog" was the best-selling single in the 1950s.

5. Simon and Garfunkel first formed under the name Tom and Jerry, after the cartoon characters.

6. Buddy Holly's real name was Charles "Buddy" Holley.

7. Decca Records misspelled his last name on his original recording contract, and he decided just to stick with the e-free version.

8. Ludwig van Beethoven (1770–1827) is one of the most influential composers of all time.

9. He ultimately lost his hearing, but continued to compose, conduct, and perform regardless.

10. The Beatles' "I Want to Hold Your Hand" was the top-selling single of the 1960s.

11. Harry Belafonte was the first Black male singer to top the British music charts.

12. Thomas Silkman changed 226 guitar strings in a single hour on April 20, 2018, setting the world record.

13. Singer Farrah Franklin, best known for her stint in the hit group Destiny's Child, actually has the middle name Destiny.

14. Pink Floyd has put out many albums, but only one band member has played on every single one: drummer Nick Mason.

15. Bagpipes are thought to have been first used in ancient Egypt.

16. The Beatles' classic "Yesterday" initially had the words "scrambled eggs" in its chorus until Paul McCartney came up with the magic word.

17. Ever wonder who the dude Steven Tyler is referring to in Aerosmith's "Dude Looks Like a Lady"? The song was written about Mötley Crüe frontman Vince Neil.

18. The inspiration for Duran Duran's "Hungry like the Wolf" came from Little Red Riding Hood.

19. The rock 'n' roll classic "Johnny B. Goode," written by Chuck Berry, originally contained the controversial lyric "that little colored boy can play."

20. It was later changed to "that little country boy can play."

21. Michael Jackson's 1982 *Thriller* is the best-selling album of all time with 70 million copies sold.

22. Marvin Gaye's hit single, "I Heard It Through the Grapevine," is a cover—Smokey Robinson and the Miracles recorded it first.

23. Katy Perry made her debut with gospel album *Katy Hudson*.

24. Pop group ABBA's name is merely an acronym formed from the first letters of each member's first name: Agnetha Fältskog, Björn Ulvaeus, Benny Andersson, and Anni-Frid Lyngstad.

25. "I Will Survive" was originally slated to be a B-side for Gloria Gaynor's single "Substitute."

26. Chumbawamba released an album with a 156-word title in March 2008, earning the record for the world's longest music album title.

27. In 1992, child rappers Kriss Kross wore their clothes backward to get attention for their single, "Jump."

28. It must have worked, because the song went to No. 1.

29. The late conductor Georg Solti holds the record for the most Grammy Awards in any genre with 31.

30. Bruce Springsteen's 1984 album *Born in the U.S.A.* was, rather fittingly, the first compact disc ever made in the United States.

31. Think Creedence Clearwater Revival is a strange name? The band was performing as The Golliwogs before that.

32. Baroque composer Johann Sebastian Bach (1685–1750) had eye surgery at age 65.

33. Traveling surgeon John Taylor performed the cataract couching procedure.

34. Post op, Taylor gave the composer eyedrops containing pigeon blood, mercury, and pulverized sugar. Bach died shortly after.

35. Taylor went on to botch eye surgery on composer George Frideric Handel (1685–1759).

36. Freddie Mercury was the lead singer of Queen and the otherworldly voice behind such iconic hits as "Bohemian Rhapsody" and "Somebody to Love."

37. Mercury announced to the press that he had AIDS just one day before he died of the disease in 1991.

38. ABBA turned down a $1 billion offer to reunite.

39. Singer and pianist Tori Amos was expelled from the prestigious Peabody Conservatory music school in Baltimore when she was 11 years old.

40. Apparently she hated to read sheet music.

41. The catchy six-note riff that serves as the melody of Missy "Misdemeanor" Elliott's hit "Get Ur Freak On" was played on a tumbi, an unusual single-stringed guitar.

42. French romantic composer Hector Berlioz (1803–69) contributed greatly to the importance of the modern orchestra.

43. He sometimes wrote music for a thousand musicians.

44. The best-selling pop duo in history is Hall & Oates.

45. Karaoke means empty orchestra in Japanese.

46. Elvis got an unimpressive C grade in his junior high music class.

47. The castrati were male sopranos and alto-sopranos whose manhood was intentionally removed before puberty to keep their voices "sweet."

48. The Beatles' anthem "A Day In The Life" ends with a high-pitched whistle that no human could possibly hear (although a dog could).

49. In 2021, Taylor Swift became the first female artist to win the Album of the Year Grammy three times.

50. "Wolfie" was the nickname of composer Wolfgang Amadeus Mozart (1756–91).

51. A rare musical genius, Mozart could listen to music just once and then write it down from memory without any mistakes.

20 Popular TV Shows of the 1960s

1. Candid Camera
2. The Andy Griffith Show
3. Bonanza
4. The Red Skelton Show
5. Gunsmoke
6. The Lucy Show
7. Gomer Pyle U.S.M.C.
8. The Beverly Hillbillies
9. The Dick Van Dyke Show
10. Gilligan's Island
11. My Three Sons
12. Lassie
13. The Munsters
14. Bewitched
15. Green Acres
16. Get Smart
17. Batman
18. Family Affair
19. The Twilight Zone
20. I Dream of Jeannie

42 Shakespearean Facts

1. William Shakespeare was born in Stratford-upon-Avon, England, in April 1564.

2. At age 18, Shakespeare married Anne Hathaway, who was eight years his elder.

3. She was three months pregnant with their first child, Susanna.

4. Twins Hamnet and Judith were born two years later.

5. Shakespeare disappears from the historical record from 1585 to 1592.

6. Sometime in the late 1580s, Shakespeare left his family in Stratford-upon-Avon to seek his fortune in London.

7. Shakespeare joined a theater company in London called the Lord Chamberlain's Men.

8. He began his career as an actor and playwright.

9. The Bard resurfaced in 1592 when records show that several of his plays were on the London stage.

10. After 1594, Shakespeare's plays were performed exclusively by the Lord Chamberlain's Men.

11. Shakespeare's plays were performed for Queen Elizabeth I and King James I.

12. This was a time when the arts flourished.

13. Many people loved to watch plays, and London had the country's only theaters.

14. Outside the capital, plays were sometimes performed at the homes of aristocrats and at public inns.

15. Only men were allowed to act on stage in Shakespeare's day.

16. Men and boys played all the female parts.

17. London's theaters were closed in 1593 because of an outbreak of bubonic plague.

18. In 1599, a partnership of members from Shakespeare's theater company built their own theater on the south bank of the River Thames, which they named the Globe Theatre.

19. The Globe Theatre could house up to 3,000 spectators.

20. It cost just one penny to watch a play, as long as you didn't mind standing.

21. Standing audience members were known as "groundlings."

22. Seats, and cushions to sit on, cost more.

23. People who could afford it sat in the galleries, which were tiers of wooden seats covered by a thatched roof.

24. After the death of Queen Elizabeth in 1603, the company

became part of the new King James I's court, and changed its name to the King's Men.

25. After 1608, they performed at the indoor Blackfriars Theatre during the winter and the Globe during the summer.

26. The original Globe Theatre was destroyed in 1613 when a real cannon used for special effects during a production of Shakespeare's *Henry VIII* set fire to the thatched roof, burning the building to the ground.

27. The Globe was rebuilt in 1614 and finally demolished in 1644.

28. Many of Shakespeare's works, such as *Hamlet* and *King Lear*, were based on writings of former playwrights or even of Shakespeare's contemporaries—a common practice of the time.

29. Shakespearean comedies include *Twelfth Night* and *As You Like It*.

30. The lead actor and star of many of Shakespeare's plays was Richard Burbage.

31. Shakespeare's most famous tragedies include *Hamlet* and *Romeo & Juliet*.

32. Fourteen characters die in the tragedy *Titus Andronicus*.

33. Shakespeare retired in 1613, returning to his hometown with some wealth.

34. He died in 1616 and was laid to rest in the Holy Trinity Church of Stratford-upon-Avon.

35. Shakespeare left behind scant personal correspondence, but he gave the world 38 plays, 154 sonnets, and two narrative poems.

36. Thirty-six of the plays were published seven years after his death in what is now called the *First Folio*.

37. Almost nothing is known about when Shakespeare's 154 sonnets were written, to whom they were addressed, or whether they are assembled in the correct order.

38. Shakespeare invented over 1,700 words that are still used in English today, like zany, fashionable, and puking.

39. Shakespeare was the first to use "elbow" as a verb in *King Lear* Act IV Scene III.

40. These commonly-used phrases first appeared in Shakespeare's works: "break the ice" (*The Taming of the Shrew*); "it's Greek to me" (*Julius Caesar*); and "a laughing stock" (*The Merry Wives of Windsor*).

41. The Broadway-phenom-turned-movie-musical "West Side Story" retells Shakespeare's classic *Romeo & Juliet* love story on the streets of New York City.

42. However, instead of the Montagues and the Capulets, it's the Jets versus the Sharks.

21 One-Hit Wonders

1. "Tequila" by The Champs (1958)

2. "Monster Mash" by Bobby Pickett (1962)

3. "Spirit in the Sky" by Norman Greenbaum (1969)

4. "O-o-h Child" by Five Stairsteps (1970)

5. "Disco Duck" by Rick Dees and His Cast of Idiots (1976)

6. "Don't Give Up on Us" by David Soul (1976)

7. "You Light Up My Life" by Debby Boone (1977)

8. "My Sharona" by The Knack (1979)

9. "Funkytown" by Lipps, Inc. (1980)

10. "Turning Japanese" by The Vapors (1980)

11. "Mickey" by Toni Basil (1981)

12. "Come on Eileen" by Dexys Midnight Runners (1982)

13. "99 Luftballons" by Nena (1983)

14. "Take on Me" by A-ha (1984)

15. "Don't Worry, Be Happy" by Bobby McFerrin (1988)

16. "She's Like the Wind" by Patrick Swayze (1988)

17. "I'm Too Sexy" by Right Said Fred (1992)

18. "Macarena" by Los del Rio (1996)

19. "Tubthumping" by Chumbawamba (1997)

20. "Butterfly" by Crazy Town (2001)

21. "Bad Day" by Daniel Powter (2006)

28 *Godfather* Facts

1. Marlon Brando wasn't the only guy up for the role of Don Vito Corleone.

2. Paramount Pictures actually wanted Ernest Borgnine.

3. Trade magazines also claimed at the time that George C. Scott and Laurence Olivier were considered for the role, and Burt Lancaster is said to have wanted the part as well.

4. During a screen test, Brando actually stuffed his cheeks with cotton or tissues to make his character look "like a bulldog."

5. To achieve the look of an older, jowly man, Brando wore a dental device during his scenes.

6. The piece is now displayed at the Museum of the Moving Image in Queens, New York.

7. Producer Robert Evans wasn't pleased with director Francis Ford Coppola's preliminary work, and from the beginning, he and Coppola had butted heads.

8. Evans talked about having Elia Kazan on standby in case he decided to fire Coppola.

9. Kazan had worked previously with Brando on *A Streetcar Named Desire* (1951) and *On the Waterfront* (1954), and, therefore, might have been able to handle the actor's temperament better than Coppola.

10. Martin Sheen, Warren Beatty, Jack Nicholson, Dustin Hoffman, Ryan O'Neal, and James Caan all read for the role of Michael Corleone before Al Pacino got the part.

11. Pacino has said that studio execs weren't impressed with his early scenes and that they considered firing him.

12. But the scene in which his character shoots Sollozzo and McCluskey in the restaurant convinced the studio to keep him onboard.

13. Brando never learned the majority of his lines.

14. Instead, he read from cue cards while filming.

15. This is typical of the way Brando worked because he liked to improvise the exact wording of his lines as a way of internalizing his characters.

16. Sylvester Stallone tried out for the part of Paulie, which went to John Martino.

17. Did you know that *The Godfather* spawned a board game? It hit store shelves in the early 1970s.

18. In one of the most famous programming events in television history, *The Godfather Saga* was broadcast in November 1977.

19. The four-night, nine-hour broadcast was edited by Coppola and film editor Walter Murch, and it integrated footage from both *The Godfather* and *The Godfather: Part II* (1974) in chronological order.

20. *The Godfather Saga* contained additional footage not included in either *Godfather* theatrical release, bridging the two stories and fleshing out certain subplots.

21. *TV Guide's* "50 Greatest Movies on TV and Video" ranked *The Godfather: Part II* at No. 1 and the original at No. 7.

22. The cat Brando holds in the opening scene of *The Godfather* wasn't part of the script.

23. Coppola found the stray wandering around the Paramount lot and, knowing the actor's talent with props, he plopped it in Brando's lap just before shooting the scene.

24. The score for *The Godfather* created a minor scandal.

25. Composer Nino Rota was nominated for an Oscar, but then someone realized that he had simply taken his music from *Fortunella* (1958) and changed it around a bit.

26. Subsequently, Rota's Oscar nomination was withdrawn.

27. Paramount considered shooting *The Godfather* in Kansas City, Missouri, because executives had experienced problems with the unions when shooting in New York City previously.

28. The actors playing Brando's sons in the movie ranged from 6 to 16 years younger than him in real life.

8 Funky Fads of the 1970s

1. Disco
2. Afros
3. Roller Skates
4. Pet Rock
5. Leisure Suits
6. Mood Rings
7. CB Radio
8. Punk Rock

20 Audrey Hepburn Films

Movie—Character

1. *Roman Holiday* (1953)—Princess Ann
2. *Sabrina* (1954)—Sabrina Fairchild
3. *War and Peace* (1956)—Natasha Rostov
4. *Funny Face* (1957)—Jo Stockton
5. *Love in the Afternoon* (1957)—Ariane Chavasse
6. *Green Mansions* (1959)—Rima
7. *The Nun's Story* (1959)—Sister Luke
8. *The Unforgiven* (1960)—Rachel Zachary
9. *Breakfast at Tiffany's* (1961)—Holly Golightly
10. *The Children's Hour* (1961)—Karen Wright
11. *Charade* (1963)—Reggie Lampert

12. *Paris—When It Sizzles* (1964)—Gabrielle Simpson

13. *My Fair Lady* (1964)—Eliza Doolittle

14. *How to Steal a Million* (1966)—Nicole Bonnet

15. *Two for the Road* (1967)—Joanna Wallace

16. *Wait Until Dark* (1967)—Susy Hendrix

17. *Robin and Marian* (1976)—Lady Marian

18. *Bloodline* (1979)—Elizabeth Roffe

19. *They All Laughed* (1981)—Angela Niotes

20. *Always* (1989)—Hap

53 Slang Terms by Decade

1920s

1. 23 skiddoo—to get going; move along; leave; or scram

2. The cat's pajamas—the best; the height of excellence

3. Gams—legs

4. The real McCoy—sincere; genuine; the real thing

5. Hotsy-totsy—perfect

6. Moll—a female companion of a gangster

7. Speakeasy—a place where alcohol was illegally sold and drunk during Prohibition

8. The bee's knees—excellent; outstanding

1930s

9. I'll be a monkey's uncle—sign of disbelief; I don't believe it!

10. Gig—a job

11. Girl Friday—a secretary or female assistant

12. Juke joint—a casual and inexpensive establishment with drinking, dancing, and blues music, typically in the southeastern United States

13. Skivvies—men's underwear

1940s

14. Blockbuster—a huge success

15. Keeping up with the Joneses—competing to have a lifestyle or socioeconomic status comparable to one's neighbors

16. Cool—excellent; clever; sophisticated; fashionable; or enjoyable

17. Sitting in the hot seat—in a highly uncomfortable or embarrassing situation

18. Smooch—kiss

1950s

19. Big brother is watching you—someone of authority is monitoring your actions

20. Boo-boo—a mistake; a wound

21. Hi-fi—high fidelity; a record player or turntable

22. Hipster—an innovative and trendy person

1960s

23. Daddy-o—a man; used to address a hipster or beatnik

24. Groovy—cool; hip; excellent

25. Hippie—derived from hipster; a young adult who rebelled against established institutions, criticized middle-class values, opposed the Vietnam War, and promoted sexual freedom

26. The Man—a person of authority; a group in power

1970s

27. Catch you on the flip side—see you later

28. Dig it—to like or understand something

29. Get down/Boogie—dance

30. Mind-blowing—unbelievable; originally an expression for the effects of hallucinogenic drugs

31. Pump iron—lift weights

32. Workaholic—a person who works too much or is addicted to his or her job

1980s

33. Bodacious—beautiful

34. Chillin'—relaxing

35. Dweeb—a nerd; someone who is not cool

36. Fly—cool; very hip

37. Gag me with a spoon—disgusting

38. Gnarly—exceptional; very cool

39. Preppy—one who dresses in designer clothing and has a neat, clean-cut appearance

40. Wicked—excellent; great

41. Yuppie—Young Urban Professional; a college-educated person with a well-paying job who lives near a big city; often associated with a materialistic and superficial personality

1990s

42. Diss—show disrespect

43. Get jiggy—dance; flirt

44. Homey/Homeboy—a friend or buddy

45. My bad—my mistake

46. Phat—cool or hip; highly attractive; hot

47. Wassup?—What's up?; How are you?

48. Word—yes; I agree

2000s

49. Barney Bag—a gigantic purse

50. Newbie—a newcomer; someone who is inexperienced

51. Peeps—friends; people

52. Rents—parents

53. Sweet—beyond cool

31 Big Band Leader Facts

1. Chick Webb (1905–39) became one of the giants of the big band era.

2. "A-tisket, A-tasket" signifies Webb's biggest hit.

3. With mega-hits including "In the Mood," "Chattanooga Choo Choo," and "Moonlight Serenade," Iowan Glenn Miller (1904–44) achieved big-band supremacy.

4. One-half of the famed Dorsey Brothers, younger brother Tommy Dorsey (1905–56) found shared success with such songs as "I Get a Kick out of You" and "Lullaby of Broadway."

5. He later broke off on his own and scored again with "I'm Getting Sentimental over You" and "I'll Never Smile Again."

6. The older half of the Dorsey Brothers, Jimmy Dorsey (1904–57) was content to remain with the band while his brother took the lead role.

7. Top-ten hits including "What a Difference a Day Made" and "I Believe in Miracles" stand as testament to the fruitful pairing.

8. Paul Whiteman (1890–1967) was progressive in his musical outlook. He proclaimed jazz "the folk music of the industrial age."

9. Fittingly, Whiteman was a well-oiled machine when it came to churning out hits.

10. "Let's Fall in Love," "Together," and "Rhapsody in Blue" kept his "factory" hummin.'

11. Gene Krupa (1909–73) is considered one of the most influential drummers of the 20th century.

12. He invented the pairing of drums that would come to be known as the standard "kit" and raised drum solos to an art form.

13. "It Don't Mean a Thing if It Ain't Got that Swing" was one of

his most famous numbers, and swing Duke Ellington (1899–1974) certainly did.

14. "Don't You Know I Care (Or Don't You Care to Know)" and "Perdido" continued such hip, "to and fro" swaying.

15. A onetime contortionist from a traveling circus, Harry James (1916–83) twisted himself into a successful bandleader.

16. His swinging version of "You Made Me Love You" turned appreciative teens into quivering masses of flesh.

17. Count Basie (1904–84) was bestowed with the honor of "Count" beside the regal likes of fellow bandleaders Benny "The King of Swing" Goodman and Edward "Duke" Ellington.

18. Basie's band featured energetic ensemble work and generous soloing.

19. "One O' Clock Jump" and "Jumpin' at the Woodside" kept toes a-tappin.'

20. Benny Goodman (1909–86), the "King of Swing," scored more than 100 hits during his career.

21. "Sing, Sing, Sing," "Blue Moon," "Moonglow," and "Jersey Bounce" were among them.

22. Considered by many to be the greatest drummer to ever pick up sticks, Buddy Rich (1917–87) had a career that spanned seven decades.

23. His caustic humor, as finely honed as his precision drum licks, made him a talk-show favorite.

24. John Birks "Dizzy" Gillespie (1917–93) was known for his comical antics—hence his nickname—in addition to his world-class trumpeting.

25. With his upturned horn, Gillespie ushered in the bebop era.

26. Responsible for such ditties as "Hi-De-Ho" and "Minnie the Moocher," former law school student Cab Calloway (1907–94) was known for his energetic "scat" singing and frequent appearances at the Cotton Club.

27. Known as "King of the Vibes," percussionist/bandleader Lionel Hampton (1908–2002) elevated the vibraphone to first-class status.

28. "On the Sunny Side of the Street" and "Hot Mallets" rank as two of his most popular songs.

29. Born Arthur Arshawsky, Artie Shaw (1910–2004) was noted for such hits as "Begin the Beguine" and "Everything's Jumping."

30. He was equally noted for his eight marriages. Actresses Lana Turner and Ava Gardner were but two "victims" who famously took the plunge with Shaw.

31. The last living great big band leader, Artie Shaw died at age 94 on December 30, 2004.

21 Popular TV Shows of the 1970s

1. M*A*S*H

2. All in the Family

3. Sanford and Son

4. Hawaii Five-O

5. Happy Days

6. Maude

7. Laverne & Shirley

8. Mary Tyler Moore

9. The Brady Bunch

10. The Jeffersons

11. Three's Company

12. Alice

13. One Day at a Time

14. Good Times

15. Little House on the Prairie

16. Starsky and Hutch

17. The Waltons

18. The Partridge Family

19. The Carol Burnett Show

20. Welcome Back, Kotter

21. Charlie's Angels

17 Coco Chanel Facts

1. "Coco" was the nickname for this revolutionary fashion designer.

2. Her actual first name was Gabrielle.

3. Chanel opened her first shop in Paris in 1909, where she sold hats.

4. In 1910, she moved to the Rue Cambon, where the House of Chanel remains to this day.

5. Before Chanel, black was a color of mourning only. After she began working with it, black became synonymous with chic.

6. The concept we have of the classic little black dress is credited to Coco herself.

7. So are sweater sets, pleated skirts, triangular scarves, and fake pearls.

8. Chanel couldn't actually sew, but she did possess a great skill for draping and pinning.

9. She would pin the garment the way she wanted it to look and then hand it to her workers to create.

10. When Coco returned from vacation with a dark tan in the 1920s, tanning quickly became a symbol of wealth and leisure.

11. Prior to this, no proper woman would expose her delicate, pale skin to the sun.

12. Chanel also produced the first artificial suntan lotion.

13. Chanel No. 5 perfume was introduced in 1921.

14. The number was chosen because it was the fifth sample presented to Chanel and the one she liked best.

15. In 1945, Chanel lived in exile in Switzerland after she became romantically involved with a Nazi officer.

16. People said of Chanel, "She knows what a woman wants to wear before the woman knows it herself."

17. The haute couture House of Chanel produced clothing designed by Karl Lagerfeld for Chanel from 1983 until his death in 2019.

28 Movie Prop Facts

1. The record for the greatest number of props used in a single movie belongs to *Gone with the Wind* (1939).

2. *Gone with the Wind* boasted more than 1,250,000 items.

3. The telephone that Harpo Marx ate in *The Cocoanuts* (1929) was made of chocolate, and the bottle of ink he drank was cola.

4. Similarly, the old shoe Charlie Chaplin ate in *The Gold Rush* (1925) was made of licorice.

5. The gun used by Johnny Mack Brown in the 1930 version of *Billy the Kid* was William "Billy the Kid" Bonney's actual firearm.

6. Wild Bill Hickok's pocket Derringer was used as a prop in the 1924 Western *The Iron Horse*.

7. The golden spike that united the Union Pacific and Central Pacific Railroads at Promontory Point, Utah, in 1869, was used to reenact that remarkable event in the 1939 flick *Union Pacific*.

8. The severed horse's head that made such an impact in *The Godfather* (1972) was real. It came from a butcher's shop.

9. For the 1916 war drama *The Crisis*, the U.S. government loaned filmmakers the dispatch box (a box for holding official papers and transporting them from place to place) used by Abraham Lincoln during his presidency.

10. Jimmy Stewart kept every hat he wore in a movie starting with his debut in *The Murder Man* in 1935.

11. The simple, tattered hat worn by Henry Fonda in *On Golden Pond* (1981) was once owned by Spencer Tracy, who had received it as a gift from the great director John Ford.

12. It was given to Fonda as a gift by his costar, Katharine Hepburn, during their first day on the set.

13. In *Legal Eagles* (1986), starring Robert Redford, $10 million worth of original paintings and sculptures were used as props.

14. Among them were works by Pablo Picasso and Roy Lichtenstein.

15. One of the most expensive props ever featured in a movie was the full-size replica of a Spanish galleon constructed for Roman Polanski's *Pirates* (1986).

16. It cost more than $10 million to make.

17. The armor worn by Geraldine Farrar in the 1916 movie *Joan the Woman* was crafted of pure silver.

18. This wasn't because of extravagance, but for necessity—prior to the widespread use of aluminum, silver was the lightest durable metal available.

19. Custard pies were a staple of early slapstick comedies, but directors found that real custard pies fell apart when tossed.

20. A California pastry shop solved the dilemma by creating pies with double-thick crusts and a filling of flour, water, and whipped cream.

21. Today, shaving cream is also used.

22. Milk was added to the water used to create the raindrops in *Singin' in the Rain* (1952) so they would show up better on film.

23. The swarm of locusts in *The Good Earth* (1937) was created by filming coffee grounds settling in water, then reversing the film.

24. Perhaps the most amazing effect of all is the parting of the Red Sea in Cecil B. DeMille's *The Ten Commandments* (1923).

25. This effect was created by pouring water over two giant blocks of clear gelatin, then reversing the film.

26. In *The Maltese Falcon* (1941), the aforementioned bird symbolizes "the stuff that dreams are made of."

27. Supposedly, two heavy lead falcons were made for the film, each weighing more than 40 pounds.

28. Sources claim that the second was made because the first was dented when it was dropped on the film's star, Humphrey Bogart.

35 Facts About Mimes

1. Miming began as an ancient Greco-Roman tradition.

2. "Pantomime" means "an imitator of nature"—derived from Pan, the Greek god of nature, and mimos, meaning "an imitator."

3. The first record of pantomime performed as entertainment comes from ancient Greece, where mimes performed at religious festivals honoring Greek gods.

4. As early as 581 B.C., Aristotle wrote of seeing mimes perform.

5. From religious festivals, Greek mime made its way to the stage: Actors performed panto-mimic scenes as "overtures" to the tragedies that depicted the moral lesson of the play to follow.

6. Greek settlers brought mime to Italy, where it flourished during the Roman Empire and spread throughout Europe as the empire expanded.

7. The ancient Romans distinguished between the pantomime and mime: Pantomimes were tragic actors who performed in complete silence, while mimes were comedic and often used speech in their acts.

8. The Roman Empire brought pantomime and mime to England around 52 B.C., but with the fall of the Empire in the fifth century and the progress of Christianity, both were banished as forms of paganism.

9. Pantomime and mime weren't really gone, though: The sacred religious dramas of the Middle Ages were acted as "dumb shows" (no words were used), and historians believe that comedic mime was used by court jesters, who included humorous imitations in their acts.

10. After the Middle Ages, mime resurged during the Renaissance and swept through Europe as part of the Italian theater called *Commedia dell'arte*, in which comedic characters performed in masks and incorporated mime, pantomime, music, and dance.

11. The first silent mime appeared on the English stage in 1702, in John Weaver's Tavern Bilkers at the Drury Lane Theatre.

12. It was really more of a "silent ballet" than silent acting.

13. British actor John Rich is credited with adapting pantomime as an acting style for the English stage in 1717.

14. His "Italian Mimic Scenes" combined elements of both Commedia dell'arte and John Weaver's ballet.

15. Meanwhile, mime flourished as a silent art in 18th-century France, when Napoleon forbade the use of dialogue in stage performance for fear something slanderous might be said.

16. The classic white-faced/black-dressed mime was introduced and popularized in the 19th-century French circus by Jean-Gaspard Deburau, who was deemed too clumsy to participate in his family's aerial and acrobatics act.

17. Mime started to fade in popularity at the beginning of the 1900s but was revitalized with the birth of silent films.

18. Stars such as Charlie Chaplin and Buster Keaton relied on elements of pantomime.

19. In the 1920s, French performer Etienne Decroux declared mime an independent art form—different from the circus form introduced by Deburau—and launched the era of modern mime.

20. In 1952, French-trained American mime Paul Curtis founded the American Mime Theatre in New York.

21. In 1957, Etienne Decroux traveled to New York to teach a workshop at the Actors Studio, which inspired him to open a mime school in the city.

22. Decroux's most famous student, Marcel Marceau, expanded modern mime's influence in the 1960s by touring the United States and inviting mimes to train with him.

23. When he was growing up, Marceau had been greatly influenced by the actor Charlie Chaplin.

24. In fact, Marceau's alter ego, "Bip" the clown, was inspired by Chaplin's own "Little Tramp" character.

25. The San Francisco Mime Troupe (SFMT), one of the most powerful political theaters in the U.S., began as a silent mime company in 1959.

26. It was founded by Ronnie Davis, who had previously performed with the American Mime Theatre.

27. Future concert promoter Bill Graham was so moved by an SFMT performance in 1965 that he left his corporate job to manage the group.

28. That led to his career as the legendary promoter of The Rolling Stones, the Grateful Dead, and Janis Joplin, among others, in the 1960s and '70s.

29. Robert Shields, a former student of Marceau's, developed the "street mime" form in the 1970s.

30. He performed in San Francisco's Union Square, where he occasionally received traffic citations, landed in jail, and was beaten up by people for imitating them!

31. Shields married fellow mime Lorene Yarnell in a mime wedding in Union Square.

32. Shields and his wife brought Marceau's mime technique to TV in the late 1970s with the Emmy-award-winning show Shields and Yarnell.

33. Though the popularity of mime in the U.S. declined after the 1970s, it still influences aspects of current culture.

34. Urban street dances, including break dancing, incorporate aspects of mime.

35. Most notable is the evolution of the moonwalk, universalized by Michael Jackson, who was inspired by Marceau.

8 Awesome Fads of the '80s

1. "Valspeak"
2. The Walkman
3. Atari
4. Break Dancing
5. Parachute Pants
6. Swatch Watches
7. Hair Bands
8. Preppies

Former Names of 17 Famous Bands

1. Cheap Trick—Fuse

2. U2—Feedback, The Hype

3. The Beatles—The Quarrymen, Johnny and the Moondogs

4. Styx—The Tradewinds

5. Queen—Smile

6. Led Zeppelin—The New Yardbirds

7. The Beach Boys—The Pendletones

8. Green Day—Sweet Children

9. KISS—Wicked Lester

10. The Who—The Detours, The High Numbers

11. Def Leppard—Atomic Mass

12. Pink Floyd—Tea Set

13. Boyz II Men—Unique Attraction

14. Blondie—Angel and the Snake

15. Simon and Garfunkel—Tom and Jerry

16. Journey—Golden Gate Rhythm Section

17. Pearl Jam—Mookie Blaylock

31 Disney Facts

1. Everyone's favorite rodent, Mickey Mouse was introduced in the cartoon short *Steamboat Willie* (1928), which was released just after the sync-sound revolution.

2. The cartoon's use of synchronized music and sound effects helped make it a critical success.

3. With a voice provided by Walt Disney himself, Mickey Mouse quickly became a hit in movie theaters.

4. Disney's cartoons won every Animated Short Subject Academy Award during the 1930s.

5. Disney's *Snow White and the Seven Dwarfs* (1937) was the first full-length animated feature produced in the United States.

6. The movie debuted two decades after Argentina created the first-ever full-length animated movie, *El Apóstol*.

7. Walt Disney called his nine animators the "Nine Old Men," though many were quite young when they started working for Disney Studios.

8. All nine remained with Disney from *Snow White and the Seven Dwarfs* through *The Rescuers* (1977).

9. The influence of the Nine Old Men—Ollie Johnston, Milt Kahl, Les Clark, Frank Thomas, Wolfgang Reitherman, John Lounsbery, Eric Larson, Ward Kimball, and Mark Davis—on commercial Hollywood animation cannot be underestimated.

10. *Alice in Wonderland* (1951) was based on two of Lewis Carroll's books: *Alice's Adventures in Wonderland* and *Through the Looking Glass*.

11. But the animated adaptation had one character that Carroll didn't create: the doorknob.

12. *Beauty and the Beast* (1991) holds the honor of being the first full-length animated feature to receive a Best Picture Oscar nomination.

13. It ended up losing to *The Silence of the Lambs*, but it still made history.

14. The movie took home the Golden Globe for Best Picture, becoming the first animated film to do so.

15. *Dumbo* is the only Disney title character who never speaks in his movie.

16. The last film that bore the personal stamp of Walt Disney himself was *The Jungle Book* (1967).

17. Although Kathleen Turner voiced all of Jessica Rabbit's speaking parts in *Who Framed Roger Rabbit* (1988), she was never credited.

18. Actress Amy Irving did the character's singing.

19. Dan Castellaneta—best known as the voice of Homer Simpson—stepped in to voice the genie in *The Return of Jafar* (1994), the sequel to *Aladdin* (1992), after Robin Williams turned down the chance to reprise his role.

20. Williams later came back for a third film, the direct-to-video *Aladdin and the King of Thieves* (1995).

21. Castellaneta had already voiced the film, but Disney discarded his voice track when Williams agreed to take the role.

22. Castellaneta did not lose out entirely, however, because he voiced the genie for Disney's *Aladdin* cartoon television series.

23. Singer Ricky Martin voiced the main character in the Spanish version of Disney's *Hercules* (1997).

24. Tate Donovan did the voice in the original English flick.

25. *The Little Mermaid* (1989) was the last animated Disney film to use hand-painted cels shot on analog film.

26. Artists created more than a million drawings for the movie.

27. The idea for *The Little Mermaid* was actually born in the 1930s.

26. Disney art director Kay Nielsen created pastel and watercolor sketches for the project, but it never got off the ground during her tenure.

27. Nielsen's drawings were pulled from Disney's archives and were used by the modern-day artists working on the movie.

28. Nielsen was credited for "visual development" for her contribution, six decades after she actually did the work.

29. Animators studied the movement of skateboarder Tony Hawk while making *Tarzan* (1999).

30. The way Tarzan slid down a log was based on Hawk's movements while riding his board.

31. Pumbaa, the warthog from *The Lion King*, is thought to be the first character to pass gas in an animated Disney movie.

9 Foreign Slang Terms

1. **Backpfeifengesicht (German):** a face that's just begging for someone to slap it—a familiar concept to anyone fond of daytime TV.

2. **Bakku-shan (Japanese):** a girl who looks pretty from the back but not the front. This loanword would in fact be a loanword regifted, since it's already a combination of the English word "back" with the German word schoen, meaning "beautiful."

3. **Kummerspeck (German):** literally this means "grief bacon"—excess weight gained from overeating during emotionally trying times.

4. **Ølfrygt (Viking Danish):** the fear of a lack of beer. Often sets in during trips away from one's hometown, with its familiar watering holes.

5. **Drachenfutter (German):** literally "dragon fodder"—a makeup gift bought in advance. Traditionally used to denote offerings made by a man to his wife when he knows he's guilty of something.

6. **Bol (Mayan):** for the Mayans of South Mexico and Honduras, the word bol pulls double duty, meaning both "in-laws" as well as "stupidity."

7. **Uitwaaien (Dutch):** walking in windy weather for the sheer fun of it.

8. **Blechlawine (German):** literally "sheet metal avalanche"—the endless lineup of cars stuck in a traffic jam on the highway.

9. **Karelu (Tulu, south of India):** the mark left on the skin by wearing anything tight.

27 Popular TV Shows of the 1980s

1. Diff'rent Strokes
2. Dallas
3. The Cosby Show
4. Cheers
5. The Golden Girls
6. Dynasty
7. Murder, She Wrote
8. Who's the Boss?
9. Simon & Simon
10. Falcon Crest
11. Family Ties
12. A Different World
13. The Wonder Years
14. Facts of Life
15. Magnum P.I.
16. The A-Team
17. Knots Landing
18. Growing Pains
19. Moonlighting
20. Alf
21. L.A. Law
22. Trapper John, M.D.
23. Matlock
24. Miami Vice
25. Silver Spoons
26. thirtysomething
27. Mama's Family

31 Facts About Leading Men

1. Humphrey Bogart is buried with a small, gold whistle.

2. His wife Lauren Bacall placed it there in remembrance of her famous line to him in *To Have and Have Not* (1944): "You know how to whistle, don't you, Steve? You just put your lips together and blow."

3. The two met on the set of that film, and Bogart gave the whistle to Bacall before they married.

4. Executives at Paramount Studios gave Cary Grant his stage name.

5. They apparently thought the name Cary Grant would resonate with the public, though they first tried to get their new discovery to accept the name Cary Lockwood.

6. Leading man Jimmy Stewart held a degree in architecture from Princeton University.

7. He graduated in 1932, three years before his first credited film appearance.

8. Jack Nicholson began his Hollywood career as a messenger boy for MGM.

9. After he finished high school in the mid-1950s, Nicholson worked in the studio's cartoon department and mailroom.

10. Paul Newman graduated from Ohio's Kenyon College in 1949.

11. He worked in his family's sporting goods store, then sold copies of *Encyclopaedia Britannica* to pay for his tuition at the Yale Drama School.

12. Robert Redford almost snagged the lead role of Ben Braddock in *The Graduate* (1967).

13. But director Mike Nichols thought it would be unrealistic that Redford would have such a hard time getting the girl.

14. James Dean died in a car accident shortly after getting a speeding ticket.

15. While filming *Giant*, Dean appeared in a PSA for auto safety, ironically ending with the line: "The life you save may be...mine."

16. Filmmakers had to get creative when working with Marlon Brando, who rarely memorized his lines.

17. Cue cards were often employed, and in *Superman* (1978), Brando's lines were supposedly written on the baby Kal-El's diaper.

18. In 1964, Sidney Poitier became the first African American man to win an Academy Award when he received the honor for his role in *Lilies of the Field* (1963).

19. He later became the first Black actor to leave his hand- and foot-prints at Grauman's Chinese Theatre.

20. While under contract to MGM, Clark Gable balked at playing gangsters and villains.

21. Studio head Louis B. Mayer decided to punish him by lending him to the much smaller Columbia Pictures to play the lead in a wacky little comedy called *It Happened One Night* (1934).

22. Gable won an Oscar for the role.

23. During the late 1960s, Harrison Ford worked as a carpenter in LA.

24. He was apparently known as one of the finest cabinetmakers in the city, and he still does carpentry as a hobby.

25. Clint Eastwood wrote the scores for some of his films, including *Million Dollar Baby* (2004) and *Changeling* (2008).

26. Occasionally, his musician son, Kyle, scores his films, keeping it all in the family.

27. George Clooney's first steady acting job was on a mid-1980s medical sitcom called *E/R*, which also starred Elliott Gould and Jason Alexander.

28. A decade later, Clooney landed the role of Dr. Doug Ross in the hit medical drama *ER*.

29. He's the only actor to star in two separate fictional series with the same name.

30. Brad Pitt played plenty of sports in high school: He was on the golf, tennis, and swimming teams at Kickapoo High in Springfield, Missouri.

31. Pitt also belonged to the Forensics Club and the Key Club, a service organization for high school students.

7 Fantastic Fads of the 1990s

1. Grunge
2. Hypercolor T-shirts
3. The Macarena
4. The Waif Look
5. Tattoos and Piercing
6. Hip-Hop Fashion
7. Tags

20 Films by the Master of Suspense

Nicknamed "the Master of Suspense," Alfred Hitchcock made more than 65 full-length movies. Here are some of the best.

✳   ✳   ✳   ✳

1. *The Birds* (1963)
2. *To Catch a Thief* (1955)
3. *Dial "M" for Murder* (1954)
4. *Family Plot* (1976)
5. *Frenzy* (1972)
6. *Lifeboat* (1944)
7. *The Man Who Knew Too Much* (1956)
8. *Marnie* (1964)
9. *Mr. & Mrs. Smith* (1941)
10. *North by Northwest* (1959)
11. *Psycho* (1960)
12. *Rear Window* (1954)
13. *Rebecca* (1940)
14. *Rope* (1948)
15. *Saboteur* (1942)
16. *Shadow of a Doubt* (1943)
17. *Topaz* (1969)
18. *Torn Curtain* (1966)
19. *The Trouble with Harry* (1955)
20. *Vertigo* (1958)

27 *Madame Bovary* Facts

1. Gustave Flaubert's novel *Madame Bovary* was charged with offending national sensibility, morality, and religion.

2. *Madame Bovary* was first published serially in the *Revue de Paris* in 1851.

3. The *Revue's* editor, Leon Laurent-Pichat; the work's author, Gustave Flaubert; and the publisher, Auguste-Alexis Pillet, were charged with "offenses to public morality and religion" by the conservative Restoration Government of Napoleon III.

4. Many, including Flaubert, believed that his work was singled out because of the regime's distaste for the notoriously liberal *Revue*.

5. The prosecutor at trial, Ernest Pinard, based his case upon the premise that adultery must always be condemned as an affront to the sanctity of marriage and society at large.

6. The novel tells the tale of Emma Bovary's gradual but inevitable acceptance of her need for sexual satisfaction outside the confines of a provincial marriage.

7. *Madame Bovary* conspicuously lacks any voice reminding the reader that adultery is reprehensible.

8. Other works of the period, notably the popular plays of Alexandre Dumas, commonly featured adulterous characters.

9. But in Dumas's plays, a voice of reason reminded the audience that the character's actions were wrong and worthy of punishment.

10. That *Madame Bovary* lacked such perspective was certainly unprecedented and—according to the government—worthy of censure.

11. The defending attorney, Maitre Jules Sénard, was a close friend of Flaubert's and one of the people to whom the work was dedicated.

12. Sénard argued that literature must always be considered art for art's sake and that Flaubert, in particular, was a consummate artist whose intentions had nothing to do with affecting society at large.

13. Whether or not Flaubert intended to undermine any aspect of French society is debatable.

14. As the son of a wealthy family, Flaubert could afford to sit in his ivory tower and decry what he perceived as the petty hypocrisies of the emerging middle class.

15. Certainly, Gustave Flaubert was a perfectionist who spent weeks reworking single pages of prose.

16. In *Madame Bovary*, he sought to create a novel that was stylistically beautiful above all else.

17. To test his craft, Flaubert would shout passages out loud to try their rhythm.

18. It took the author five years of solitary toil to complete the work.

19. The literary elite, notably Sainte-Beuve, Victor Hugo, and Charles-Pierre Baudelaire, immediately recognized the novel's genius.

20. But the general public largely ignored *Madame Bovary* when it was first published.

21. In the end, the judges agreed with Sénard and acquitted all of the accused.

22. The sensational trial sparked public interest in a work that might otherwise have gone unnoticed by the very society—the emerging middle class of France's provinces—the trial was meant to protect.

23. *Madame Bovary* was Flaubert's first published novel, and when it appeared in full, he received a commensurate fee: 800 francs.

24. The morality trial, however, boosted sales.

25. Flaubert estimated that he missed out on some 40,000 francs worth of income in the deal.

26. Although it survived the trial, the influential *Revue de Paris* finally capitulated to political and financial pressure and stopped publication a year later.

27. *Madame Bovary* was first published in book form in 1857.

16 Popular TV Shows of the 1990s

1. Beverly Hills, 90210
2. Home Improvement
3. Seinfeld
4. Friends
5. NYPD Blue
6. Frasier
7. ER
8. The Simpsons
9. Roseanne
10. Mad About You
11. Married with Children
12. Murphy Brown
13. Full House
14. Saved by the Bell
15. Sisters
16. The X Files

30 King of Ragtime Facts

1. Scott Joplin was the most famous ragtime composer of all time.

2. Despite his enduring musical legacy, the details of much of Scott Joplin's early life remain imprecise.

3. Joplin's birth date is often recorded as November 24, 1868, although it may have been as much as a year before that.

4. He was born near Linden, Texas, where Jiles Joplin, his father and a former slave, worked as a laborer.

5. By the time that he was seven years old, Scott Joplin was already an experienced banjo player.

6. When his family moved to the Texas–Arkansas border town of Texarkana, he was introduced to the instrument that would make him famous.

7. His mother, Florence, did domestic work for a neighboring attorney, who allowed young Scott to experiment on his piano.

8. The elder Joplin saved enough money to buy his son a used piano, and Scott's talent quickly blossomed.

9. At age 11, Joplin began taking free lessons from Julius Weiss, a German-born piano teacher who helped shape Joplin's musical influences, including European opera.

10. As a teen, Joplin formed a vocal quartet and performed in the dance halls of Texarkana before venturing out as a pianist on the saloon and honky-tonk circuit.

11. In St. Louis, Joplin encountered a style of music that featured abbreviated melody lines called "ragged time," or "ragtime" for short.

12. Joplin adopted the principles of ragtime into longer musical forms including a ballet—*The Ragtime Dance*, written in 1899—and two operas—*The Guest of Honor* in 1903 and *Treemonisha* in 1910.

13. While the orchestration scores for both operas were sadly lost during the copyright process, a piano-vocal score for *Treemonisha* was later published.

14. But it was Joplin's shorter compositions that earned him the title "The King of Ragtime."

15. One of his first compositions to be published, "Maple Leaf Rag" in 1899, went on to sell more than one million copies of sheet music.

16. The piece was named after one of the music clubs Joplin enjoyed playing in Sedalia, Missouri, the Maple Leaf Club.

17. Joplin occasionally returned to Texarkana to perform, but by 1907 he was living in New York City.

18. It was in New York City that Joplin wrote his instructional manual, *The School of Ragtime*.

19. In 1916, Joplin's health deteriorated in part due to syphilis, which he'd contracted a few years before.

20. His playing became inconsistent, and he was eventually forced to enter the Manhattan State Hospital.

21. Joplin died there on April 1, 1917.

22. Joplin was married and divorced twice.

23. His only child, a daughter, died in infancy.

24. It wasn't until years after his death that Scott Joplin achieved the full recognition his work deserved.

25. In 1971, the New York Public Library published Joplin's collected works.

26. His music found a whole new audience with the release of the popular 1973 Paul Newman and Robert Redford movie, *The Sting*.

27. Joplin's work, adapted by Marvin Hamlisch, was featured heavily in the film's score, which won an Academy Award.

28. "The Entertainer," released as a single from the movie, became a bona fide top ten hit.

29. Joplin himself was posthumously awarded a Pulitzer Prize in

1976 for *Treemonisha*, which has been recognized as the first grand opera by an African American composer.

30. Today, a large mural on Texarkana's Main Street depicts the life and accomplishments of one of the town's most famous sons.

27 American Terms and Their British Equivalents

American Term—British Term

1. ballpoint pen—biro
2. toilet paper—bog roll
3. umbrella—brolly
4. fanny pack—bum bag
5. cotton candy—candy floss
6. french fry—chip
7. plastic wrap—clingfilm
8. zucchini—courgette
9. potato chip—crisp
10. checkers—draughts
11. thumbtack—drawing pin
12. busy signal—engaged tone
13. soccer—football
14. astonished—gobsmacked
15. sweater—jumper
16. elevator—lift
17. restroom—loo
18. truck—lorry
19. ground beef—mince
20. diaper—nappy

21. mailbox—pillar box

22. bandage (Band-Aid)—plaster

23. baby carriage/stroller—pram

24. collect call—reverse-charge call

25. aluminum can—tin

26. to go drastically wrong—to go pear-shaped

27. complain—whinge

64 Facts About the Academy Awards

1. The inaugural Academy Awards ceremony took place on May 16, 1929, in the Blossom Room of the Hollywood Roosevelt Hotel.

2. The 270 invited guests, all of whom were members of the industry, each paid $5 to attend.

3. Over 9,000 people (including many actors) vote on the Academy Award winners.

4. All are members of the Academy of Motion Picture Arts and Sciences (AMPAS).

5. Fifteen Oscars were awarded in 1929, the first year the ceremony was held.

6. The Best Actress award was the only honor bestowed upon a woman that year.

7. Janet Gaynor took the honor for her roles in *7th Heaven* (1927), *Sunrise* (1927), and *Street Angel* (1928).

8. After the first year, nominees were selected for only one film.

9. You can't buy a ticket to the Academy Awards—it's an invitation-only event.

10. The winner of the first Oscar ever presented didn't show up to receive his award.

11. Emil Jannings won the Best Actor award in 1929 for his roles in *The Way of All Flesh* (1927) and *The Last Command* (1928), but he was in Europe during the ceremony, so he received the statue early.

12. Made of gold-plated Britannia metal, Oscar statues are 13.5 inches tall and weigh 8.5 pounds.

13. The original Oscar mold was cast in 1928 at the C.W. Shumway & Sons Foundry in Batavia, Illinois.

14. The statuettes were manufactured by R.S. Owens & Company in Chicago, Illinois, up until 2016.

15. The shortest-ever Oscar ceremony lasted only 15 minutes in 1929, the first year for the event.

16. The Academy Awards ceremony was first broadcast by radio in 1930 and first televised in 1953.

17. The first picture to sweep all five major Academy Awards— Best Picture, Best Actor, Best Actress, Best Director, and Best Screenplay (adaptation)—was Frank Capra's *It Happened One Night* (1934).

18. *One Flew Over the Cuckoo's Nest* (1975) and *The Silence of the Lambs* (1991) later achieved the same feat of winning all five major Oscars.

19. The art deco statuette was first dubbed "Oscar" in 1934 when Margaret Herrick, the Academy's librarian, remarked that the statue bore a striking resemblance to her uncle, Oscar Pierce.

20. The gilded prize officially became known as Oscar in 1939.

21. Accounting firm PricewaterhouseCoopers has been protecting the Oscars' integrity almost since the beginning.

22. The company—then known as Price Waterhouse—signed with the Academy in 1934.

23. At every ceremony, PricewaterhouseCoopers (which collates the results) stations employees in the wings with instructions to immediately correct a presenter if an error is made.

24. In 1934, Bette Davis was nominated for an Oscar through a write-in campaign.

25. She was nominated for Best Actress for her performance in *Of Human Bondage* (1934).

26. The Academy has since prohibited such write-in votes on final Oscar ballots.

27. The first Academy Award winner who refused to accept an Oscar was writer Dudley Nichols, who turned down his statue for the 1935 film *The Informer* because the Writers Guild was on strike at the time.

28. The first Best Supporting Actor and Actress awards were given out in 1937 for films made in 1936.

29. Walter Brennan won the former for his role in *Come and Get It*, while Gale Sondergaard took the latter honors for her part in *Anthony Adverse*.

30. In 1938, *A Star Is Born* became the first all-color movie to receive a Best Picture nomination.

31. Two years later, *Gone with the Wind* became the first color movie to win the award.

32. The first African American Oscar winner was Hattie McDaniel who was awarded the 1939 Best Supporting Actress for her role as Mammy in *Gone with the Wind*.

33. Twenty-four years would lapse before another African American would win: Sidney Poitier for *Lilies of the Field* (1963).

34. At the 1943 Academy Awards ceremony, Greer Garson made a six-minute acceptance Best Actress Oscar speech for her role in *Mrs. Miniver* (1942).

35. Later, the Academy created a 45-second rule to avoid such lengthy speeches.

36. From 1942 until 1944, Oscars were made of plaster due to a metal shortage during World War II.

37. Later, the recipients were allowed to exchange their awards for gold-plated ones.

38. *All About Eve* (1950), *Titanic* (1997) and *La La Land* (2016) are tied for the most total nominations with 14 each.

39. In 1968, the Oscars were pushed back two days following the assassination of Dr. Martin Luther King Jr.

40. During the April 1974 Oscar broadcast, actor David Niven was innocently introducing Best Picture presenter Elizabeth Taylor when a streaker ran across the stage totally nude, throwing the two-finger "peace" sign. Live television caught it all.

41. In 1981, producers delayed the ceremony by 24 hours after John Hinckley Jr.'s attempt on President Ronald Reagan's life.

42. Actor George C. Scott called the Oscars "a two-hour meat parade" and swore that if he won for *Patton* (1970), he wouldn't be there to collect.

43. Scott *was* named the winner that year and refused the "honor," becoming the first actor to do so.

44. The legendary Katharine Hepburn won a record four Best Actress Oscars.

45. Meryl Streep holds the record for most Best Actress nominations with 17.

46. Three-time Academy Award recipient Jack Nicholson reportedly uses one of his Oscars as a hat stand.

47. In 2001, *Shrek* won the first-ever Academy Award for Best Animated Feature Film.

48. John C. Reilly appeared in three of the five films up for Best Picture at the 2002 Academy Awards: *Chicago, Gangs of New York*, and *The Hours*.

49. Ang Lee became the first Asian to win the Best Director Academy Award for *Brokeback Mountain* (2005).

50. Lee won again for *Life of Pi* (2012).

51. Kathryn Bigelow became the first woman to win the Best Director Oscar in 2009 for *The Hurt Locker*.

52. *Wings* (1927) was the only silent film to win the Oscar for Best Picture until 2012, when *The Artist* took home the gold statuette.

53. *The Lord of the Rings: The Return of the King* (2003) won in every category for which it was nominated—11 Oscars total.

54. Two other films have won 11 Academy Awards: *Ben-Hur* (1959) and *Titanic* (1997).

55. Alfonso Cuarón became the first Mexican director to win Best Director for *Gravity* (2013).

56. He won again for *Roma* (2018).

57. At the 2017 Academy Awards ceremony, Warren Beatty and Faye Dunaway accidentally announced the wrong Best Picture winner—*La La Land* rather than *Moonlight*.

58. It turns out a PricewaterhouseCoopers accountant handed the presenters the wrong envelope backstage.

59. *Parasite* (2019) was the first film to win both the Best International Feature Film and the Best Picture Oscar.

60. The 93rd Academy Awards marked the first time multiple women were nominated for Best Director: Chloé Zhao for *Nomadland* (2020) and Emerald Fennell for *Promising Young Woman* (2020).

61. Chloé Zhao became the first woman of color to win the award.

62. Jane Campion became just the third woman to win Best Director at the 94th Academy Awards for *The Power of the Dog* (2021).

63. Michelle Yeoh became the first Asian woman to win Best Actress for her role in *Everything Everywhere All At Once* (2022).

64. At the 95th Academy Awards, A24 became the first studio to win all top six Oscars for *Everything Everywhere All At Once* (2022) and *The Whale* (2022).

5 Cyber-Chic Fads of the 2000s

1. Reality TV
2. Text messaging
3. Sudoku
4. YouTube and MySpace
5. Speed Dating

32 Comic Book Facts

1. *The Yellow Kid in McFadden's Flats* is widely regarded as the first comic book.

2. Published in 1897, the phrase "comic book" was written on its back cover.

3. *Funnies on Parade* and other American comic books emerged in the 1930s when newspaper strips were reprinted in the now-standard comic book format.

4. During World War II, superheroes and talking animals were the two most popular comic genres.

5. While comics can be the work of a single creator, the process of making a comic often involves a number of specialists.

6. The comic writer comes up with a story idea or concept, plots the storyline, then writes the script.

7. The penciller, working exclusively in pencils, interprets the script and lays down the panels and artwork on the page.

8. These will be the backbone for the inker.

9. The inker, working in ink, outlines and finishes the penciller's artwork.

10. Some comic artists hold the role of both penciller and inker.

11. The colorist adds color to the pages to help control the comic's mood and style.

12. Lastly, the letterer adds captions and speech bubbles from the writer's script.

13. Superman debuted in 1938's *Action Comics #1*.

14. The success of Superman prompted editors at National Comics Publications, corporate predecessor of DC Comics, to request more superheroes.

15. Bob Kane and Bill Finger created *Detective Comics #27*'s Batman in May of 1939.

16. One year later, Robin joined Batman's side in *Detective Comics #38*.

17. In 1941, H. G. Peter and William Moultin Marston created the female superhero Wonder Woman.

18. When *Pep Comics* introduced "Archie" in 1942, the teen-humor comic became the most popular of any MLJ properties, leading the publisher to rename itself *Archie Comics*.

19. The Golden Age of Comic Books extends from Superman's first appearance to the 1956 introduction of The Flash.

20. Popular heroes of the time include Superman, Batman, Captain Marvel, Captain America, and Wonder Woman.

21. After World War II, superhero stories began to fade out of popularity.

22. The industry expanded to other genres such as horror, crime, science fiction, and romance.

23. The Comics Code Authority (CCA) was formed in 1954 in response to concern over depictions of horror and violence.

24. Many publishers adhered to the code until 2011, when it was finally rendered defunct.

25. The underground comix movement began at the end of the Silver Age of Comic Books in response to CCA restrictions.

26. Marvel's Stan Lee revolutionized superhero comics by introducing heroes that would appeal to older audiences.

27. These included *The Fantastic Four #1* and *The Amazing Spider-Man* titles.

28. In 1988, DC Comics left the life of Jason Todd, the second Robin, up to the mercy of *A Death in the Family* readers.

29. After an incredibly close poll (won only by a margin of 72 votes!), Jason's time as Robin came to a tragic end.

30. The darker tones of *Batman: The Dark Knight Returns* and *Watchmen*, both published by DC Comics, had a profound impact on American comic book readers and ultimately led to what fans would nickname the "grim-and-gritty" era of the 1990s.

31. American comic books are one of three major comic book schools.

32. The other two are Japanese manga and Franco-Belgian comic books.

23 Grandma Moses Facts

1. Anna Mary Robertson Moses (1860–1961) is known to the world as "Grandma" Moses.

2. Grandma Moses was one of the most successful, famous artists in the United States, and possibly the best-known American artist in Europe.

3. Featured on radio, on television, and in magazines, Moses was the first artist to become a media superstar.

4. The influence of her unique, folk-art style continues to this day.

5. It all started in 1938. During a drive through Moses's hometown of Hoosick Falls, New York, art collector Louis J. Caldor spied her paintings in a drugstore.

6. He bought them all, then found her home and snapped up more of her work.

7. The following year, three of Moses's paintings were included in an exhibition of contemporary unknown painters at the Museum of Modern Art in New York.

8. Her first solo exhibition, "What a Farmwife Painted," opened at New York's Galerie St. Etienne in 1940.

9. Grandma Moses began to paint seriously when she was 76 years old (due to arthritis, which made her forfeit her beloved embroidery).

10. Her subject matter included ordinary farm activities, such as maple sugaring or making candles, soap, and apple butter.

11. These paintings were later reproduced on everything from ceramic plates to greeting cards.

12. Grandma Moses was, indeed, a farmwife, with all that entailed, and a "hired girl" before that.

13. At 12, she left school and home, where she had four sisters and five brothers, to go to work.

14. At 27, she married the hired hand on the farm where she did the housework.

15. The couple had ten children, five of whom died as infants.

16. Grandma Moses eventually outlived her husband and all of her children.

17. But she did enjoy her 9 grandchildren and more than 30 great-grandchildren.

18. Grandma Moses painted more than 1,000 pictures, including 25 after her 100th birthday.

19. In 2001, the Galerie St. Etienne created a traveling exhibition entitled "Grandma Moses in the 21st Century."

20. It earned rave reviews from the public and critics.

21. The *New York Observer*'s notoriously tough Hilton Kramer gushed: "Grandma Moses is back, and she's enchanting."

22. Grandma Moses and Norman Rockwell were friends who lived just across the New York–Vermont border from one another.

23. In November 2006, her 1943 work *Sugaring Off* became Moses's highest-selling artwork at $1.2 million.

24 Unusual Museums in the U.S.

1. Circus World Museum, Baraboo, Wisconsin

2. Wooden Nickel Museum, San Antonio, Texas

3. National Museum of Dentistry, Baltimore, Maryland

4. Sing Sing Prison Museum, Ossining, New York

5. Lizzie Borden House, Fall River, Massachusetts

6. National Museum of Health and Medicine, Washington, D.C.

7. Toilet Seat Museum, Alamo Heights, Texas

8. Mardi Gras World, New Orleans, Louisiana

9. Idaho Potato Museum, Blackfoot, Idaho

10. Mütter Museum at The College of Physicians of Philadelphia, PA

11. Computer History Museum, Mountain View, California

12. International UFO Museum & Research Center, Roswell, NM

13. Devil's Rope Museum, McLean, Texas

14. Leila's Hair Museum, Independence, Missouri

15. National Mustard Museum, Middleton, Wisconsin

16. Carrousel Factory Museum, North Towanda, New York

17. Neon Museum, Las Vegas, Nevada

18. SPAM Museum, Austin, Minnesota

19. National Cowboy & Western Heritage Museum, Oklahoma, OK

20. International Spy Museum, Washington, D.C.

21. Mobile Carnival Museum, Mobile, Alabama

22. Glore Psychiatric Museum, St. Joseph, Missouri

23. National Museum of Funeral History, Houston, Texas

24. Museum of Glass, Tacoma, Washington

43 Leading Ladies Facts

1. Elizabeth Taylor was the first leading lady to get a million-dollar paycheck.

2. She scored the big bucks for her role in *Cleopatra* (1963).

3. Grace Kelly gave up her glamorous acting career to become a real-life princess when she married the Prince of Monaco in 1956. The couple had three children.

4. There is a rose named after Ingrid Bergman.

5. The Ingrid Bergman rose is a highly fragrant, dark red tea rose.

6. Katharine Hepburn was actually born on May 12, 1907, though for years, she gave November 8, her deceased older brother's birth date, as a way to ensure that he was not forgotten.

7. Julia Roberts, the soon-to-become "pretty woman," played clarinet in her high school band.

8. Bette Davis was born Ruth Elizabeth Davis.

9. When Davis started her acting career, studio executives wanted her to take the name Bettina Dawes.

10. But Davis refused to adopt the name, saying that it sounded too much like "between the drawers."

11. By the time she was 30, Meryl Streep had won a Tony, an Emmy, and an Oscar.

12. Ava Gardner did her own singing in *The Killers* (1946). However, most other films dubbed the vocals for her songs.

13. In *Showboat* (1951), Gardner's singing voice wasn't used in the film, but it made it onto the soundtrack album.

14. Grace Kelly starred in only 11 feature films in her career, and three of them were for Alfred Hitchcock.

15. That means more than a quarter of her film work was for "The Master of Suspense."

16. Ingrid Bergman's 5-foot-10-inch frame made things challenging for some of her shorter male costars.

17. Some men, including 5'8" Humphrey Bogart, had to use lifts in their shoes so they wouldn't appear too short next to her.

18. Katharine Hepburn's mother was a renowned suffragette who championed birth control before women even had the right to vote.

19. Her father was a urologist who spoke out against venereal disease in an era when few people acknowledged that it was a problem.

20. Elizabeth Taylor is one of two women who, in the same year, won an Oscar for playing a prostitute.

21. In 1961, Taylor received the Best Actress award for her role in *Butterfield 8*.

22. That same year, Shirley Jones won Best Supporting Actress for her role in *Elmer Gantry*.

23. Meryl Streep was both a cheerleader and a homecoming queen during her high school years.

24. An entire museum exists solely to showcase Ava Gardner memorabilia.

25. The Ava Gardner Museum is located in Smithfield, North Carolina.

26. In 1982, Bette Davis told *Playboy* magazine that she had worked as a nude model in her younger years.

27. She stated that an artist had created a statue of her that was featured in a public place in Boston. However, the statue has never been located or identified.

28. Katharine Hepburn may have been shy about revealing her age, but she wasn't shy about her body.

29. On one occasion, she allegedly walked around a movie studio in her underwear (in the modest 1930s era) when someone from the costume department took her slacks away.

30. Hepburn refused to put on any other clothes until her pants were returned.

31. Ingrid Bergman battled to keep her birth name after becoming an actress.

32. Early in her career, producers looked at re-branding her Ingrid Berriman or having her use her married name, Ingrid Lindstrom.

33. In 1993, Grace Kelly became the first American actress ever depicted on a U.S. postage stamp.

34. That same year, Monaco released a stamp bearing her likeness.

35. In the United States, she was listed as Grace Kelly, while in Monaco, the stamp called her Princess Grace.

36. Ava Gardner may have been known for her beauty, but she hardly had a ladylike reputation.

37. A reporter once described her language as being "like a sailor and a truck driver...having a competition."

38. He also said Gardner threw a champagne glass at him during an interview.

39. Meryl Streep almost went to law school.

40. She applied, but when she overslept and missed her interview, she took it as a sign that law school wasn't where she belonged.

41. Julia Roberts was reportedly offered the leading role in the 1992 thriller *Basic Instinct*, but she turned it down, thus allowing Sharon Stone to make it her own.

42. Liz Taylor, who suffered from numerous health problems during her life, was even pronounced dead once.

43. Doctors made the mistaken declaration while Taylor was sick with pneumonia during the filming of *Butterfield 8* (1960).

Top 30 Songs of Elvis

1. "Heartbreak Hotel" (1956)

2. "Don't Be Cruel" (1956)

3. "Hound Dog" (1956)

4. "Love Me Tender" (1956)

5. "Too Much" (1957)

6. "All Shook Up" (1957)

7. "(Let Me Be Your) Teddy Bear" (1957)

8. "Jailhouse Rock" (1957)

9. "Don't" (1958)

10. "Hard Headed Woman" (1958)

11. "One Night" (1958)

12. "(Now and Then There's) A Fool Such as I" (1959)

13. "A Big Hunk o' Love" (1959)

14. "Stuck on You" (1960)

15. "It's Now or Never" (1960)

16. "Are You Lonesome Tonight?" (1960)

17. "Surrender" (1961)

18. "(Marie's the Name) His Latest Flame" (1961)

19. "Can't Help Falling in Love" (1961)

20. "Good Luck Charm" (1962)

21. "She's Not You" (1962)

22. "Return to Sender" (1962)

23. "(You're the) Devil in Disguise" (1963)

24. "Wooden Heart" (1964)

25. "Crying in the Chapel" (1965)

26. "In the Ghetto" (1969)

27. "Suspicious Minds" (1969)

28. "The Wonder of You" (1970)

29. "Burning Love" (1972)

30. "Way Down" (1977)

22 Popular TV Shows of the 2000s

1. Grey's Anatomy
2. CSI
3. Law & Order
4. Who Wants to Be a Millionaire
5. Survivor
6. The Apprentice
7. Desperate Housewives
8. Everybody Loves Raymond
9. American Idol
10. Will & Grace
11. The Sopranos
12. Sex and the City
13. King of Queens
14. The West Wing
15. Ally McBeal
16. Lost
17. Dancing with the Stars
18. 24
19. Scrubs
20. The Amazing Race
21. The Office
22. Fear Factor

17 Cajun vs. Creole Facts

1. Cajun culture—its language, its accordion-heavy music, and its crawfish étouffée—is an integral part of the romance of New Orleans, Louisiana.

2. The history of Louisiana Cajuns goes back to the French and Indian War of the mid-18th century.

3. England and France battled over large swaths of colonial land, including what was then known as Acadia (now part of Nova Scotia, Canada).

4. Though Acadia was part of a British colony then, it was populated mostly by French settlers.

5. Wary of having a colony full of French people during an impending war with France, the Brits kicked out everyone of French descent.

6. These displaced settlers scattered all over North America, but a large percentage of them headed down to another French colony, Louisiana.

7. Though New Orleans was a thriving port community at that time, Acadians instead settled in the surrounding swampy, alligator-infested bayou regions.

8. Through the years, the Acadians, or Cajuns, as they came to be known, developed a close-knit, if isolated, community with its own dialect, music, and folk wisdom.

9. Technically, only people who are descended from the communities settled by those original displaced Acadians are considered Cajun.

10. Creole, on the other hand, can refer to any number of things.

11. Originally the term Creole, which dates back to the Spanish conquest of Latin America, meant any person descended from colonial settlers.

12. Eventually, any people of mixed race who were native to the colonies became known as Creoles.

13. To add further confusion to the definition, there is something called a Creole language, which is most often born of the contact between a colonial language and a native one.

14. In Louisiana, Creole refers to people of any race born in Louisiana who descended from the original French settlers of the colony.

15. These folks differ from Cajuns in that they came from places other than Acadia.

16. Louisianan Creoles, too, have their own language—which differs from Cajun—that blends French, West African, and Native American languages.

17. Louisianan Creoles also have their own music (such as zydeco) and cuisine.

15 Car Art Facts

1. Crafted in 1987, *Carhenge* stands as an exact proportional replica to the famed Stonehenge in England.

2. Car for car to stone for stone, the measurements match up with precision.

3. A total of 38 automobiles make up this tributary structure near Alliance, Nebraska.

4. Some cars are positioned upright, their trunks buried deep underground, and others settled into various contorted angles.

5. This Stonehenge clone is complete with its coating of gray paint.

6. *Cadillac Ranch* is located just off Route 66 in Amarillo, Texas.

7. It features a row of ten Golden Age Cadillacs buried in the ground with only their back halves sticking up.

8. What can be seen of the cars above ground is spray-painted in various ever-changing shades of graffiti.

9. The cars are periodically repainted by the sculpture's keepers.

10. But visitors are strongly encouraged to add their paint contributions to *Cadillac Ranch*.

11. In 1989, a 40-foot spike was erected in a parking lot in Berwyn, Illinois.

12. Eight cars were threaded onto the spike in a towering vertical row.

13. The *Spindle*, sometimes referred to as the car-kabob, was dismantled in May 2008 to make way for the construction of a new building.

14. For those lovers of car art who want function with their form, there are also annual celebrations of art cars.

15. One such event, *Cartopia*, held in Berwyn, Illinois, provides a showcase for the country's coolest art cars, complete with a parade.

19 EGOT Winners

An EGOT-winner is someone who has won all four of the major American entertainment awards—the Emmy, Grammy, Oscar, and Tony—during their career. Below is a list of 19 EGOT-winners, along with the year they achieved it:

✳   ✳   ✳   ✳

1. Richard Rogers, composer (1962)

2. Helen Hayes, actress (1977)

3. Rita Moreno, actress (1977)

4. John Gielgud, actor (1991)

5. Audrey Hepburn, actress (1994) *posthumously

6. Marvin Hamlisch, composer (1995)

7. Jonathan Tunick, music director and composer (1997)

8. Mel Brooks, performer, writer, and director (2001)

9. Mike Nichols, performer, director, and producer (2001)

10. Whoopi Goldberg, performer and producer (2002)

11. Scott Rudin, producer (2012)

12. Robert Lopez, composer (2014)

13. John Legend, songwriter and producer (2018)

14. Andrew Lloyd Webber, composer and producer (2018)

15. Tim Rice, lyricist and producer (2018)

16. John Legend, singer, composer, and producer (2018)

17. Alan Menken, composer and producer (2020)

18. Jennifer Hudson, singer, actress, and producer (2022)

19. Viola Davis, actress and producer (2023)

Chapter 9

Bible

29 Facts by the Numbers

1. The Bible was written by at least 40 different authors.

2. It took nearly 1,600 years to write the Good Book.

3. There are a combined total of 66 books in the Bible.

4. The Bible contains 1,189 chapters and 31,071 verses.

5. Goliath stood "six cubits and a span," which is over nine feet tall.

6. David had five stones in his shepherd's bag when he faced Goliath.

7. The word "amen" appears 773,692 times in the King James Version (KJV) of the Bible.

8. The Old Testament contains 39 books, 929 chapters, and 23,114 verses.

9. The Bible includes 3,294 questions.

10. The King James Version contains more than 8,670 different words in Hebrew, 5,624 in Greek, and 12,143 in English.

11. The New Testament has 27 books, 260 chapters, and 7,957 verses.

12. The Bible includes 1,260 promises.

13. God is mentioned by name a total of 3,358 times in the Bible.

14. The word "Lord" appears 7,736 times.

15. The Old Testament contains 17 historical books, 5 poetical books, and 17 prophetic books.

16. The Bible includes 6,468 commands.

17. The University of Göttingen in Germany houses a Bible written on 2,470 palm leaves.

18. The Bible has been translated into more than 1,200 languages—including Klingon.

19. More than 8,000 predictions appear in the Bible.

20. The word "Christian" appears only three times in some versions.

21. The phrase "Do not be afraid" is repeated in the Bible 365 times.

22. The number seven is of great significance in the Bible, often used as a symbol of completion or perfection.

23. For example, God rested on the seventh day after Creation; Psalm 12:6 says the Lord's words are perfect like silver refined "seven times."; Jesus performed seven healing miracles on the seventh day; and there are seven seals, trumpets, angels, and bowls in Revelation.

24. The number 40 is also frequently mentioned in the Bible, usually associated with a new beginning.

25. Jesus fasted for 40 days in the desert; it rained for 40 days and nights during the Great Flood; and Jonah warned Nineveh they had 40 days before God would punish them.

26. Twelve is another prominent number in the Bible: 12 tribes of Israel, 12 disciples of Christ, 12 apostles, 12 gates to heaven, etc.

27. The name "Jesus" appears 700 times in the Gospels, which tell the story of his life.

28. The beast's number is 666 (Revelation 13:18).

29. It would take about 70 hours to read the entire Bible aloud.

42 Biblical Kings & Queens Facts

1. Melchizedek was not only a king of Salem, but also a "priest of the most high God" (Genesis 14:18).

2. Abraham paid homage to this mysterious king.

3. Esther, a Jewish girl, became queen of Persia by essentially winning a beauty contest.

4. King Ahab was dominated by Jezebel, a foreign queen who tried to spread false worship in Israel.

5. Hezekiah and Josiah were among Judah's greatest kings.

6. Josiah was eight years old when he began his reign.

7. King Hezekiah got bad news and asked God to rescind it.

8. He learned that he was about to die, but God gave him 15 more years to live.

9. Eglon, king of Moab, was a very fat man.

10. When Ehud thrust a blade into Eglon's belly, the fat closed upon the blade and it couldn't be removed.

11. Ramses II was one of Egypt's greatest pharaohs, reigning for 67 years in the 13th century B.C.

12. Second Kings 17 tells of Assyrian king Shalmaneser's conquest of Samaria in 722 B.C.

13. However, in the Assyrian annals, King Sargon II, Shalmaneser's successor, claims the credit.

14. Ahaziah, king of Judah, was 22 years old when he began his reign.

15. When Ahaziah was killed, his wicked mother Athaliah slaughtered all remaining royal family members she could find to assure the throne for herself.

16. But King Ahaziah's infant son Joash was saved by his aunt before he could be killed.

17. The high priest Jehoiada hid Joash in the temple for six years while Athaliah reigned.

18. When Joash was seven years old, Jehoiada engineered a plan to crown the boy king.

19. The coup was successful, Athaliah was assassinated, and Joash was crowned king of Judah.

20. Tiberius Caesar ruled the Roman Empire from A.D. 14–37.

21. Samuel anointed Saul king of Israel.

22. When Saul was tormented by an evil spirit, David played the harp to soothe him.

23. After being wounded by Philistine archers, King Saul asked his armor-bearer to kill him.

24. When the armor-bearer refused, Saul fell upon his sword.

25. David went on to become Israel's best king, "a man after God's own heart."

26. Yet King David made some mistakes, including committing adultery with Bathsheba.

27. In a desperate attempt at cover-up, the king arranged for her soldier-husband to be killed in battle.

28. Then David took Bathsheba as his queen.

29. The ruse would have worked—except for a prophet who challenged King David and provoked his repentance (see Psalm 51).

30. Bathsheba later gave birth to famous son Solomon.

31. King David's son Absalom died when he was hung on a tree by his hair.

32. Absalom's hair got tangled in the tree when the mule he was riding split.

33. The book of Daniel portrays Belshazzar as the king of Babylon and son of Nebuchadnezzar.

34. But Belshazzar was actually the son of Nabonidus—one of Nebuchadnezzar's successors.

35. King Nebuchadnezzar threw Shadrach, Meshach, and Abednego into a fiery furnace for their refusal to serve his gods or worship the golden image he set up.

36. Zimri's reign as king of Israel only lasted seven days.

37. King Cyrus of Persia allowed the Jews to return to Jerusalem.

38. Herod the Great (a.k.a. Herod I) ruled Judea from about 37–4 B.C.

39. This is the Herod known for great building projects, including renovating the Second Temple in Jerusalem.

40. According to the Gospel of Matthew, Herod the Great ordered the Massacre of the Innocents around the time of Jesus' birth.

41. As a prisoner, the Apostle Paul spoke before the Jewish king Agrippa and his sister/queen Bernice, telling them all about his faith.

42. Agrippa and Bernice belonged to the ruling Herodian dynasty.

22 Facts About Noah's Ark

1. God saw only wickedness and violence, so he decided to wipe the earth clean with a global flood.

2. Not everyone was wicked, however.

3. God found one righteous man—Noah—and instructed him to build a massive ship out of wood.

4. He was then to fill it with a mated pair of every kind of bird and animal.

5. Noah was also instructed to stock food for the animals as well as for himself, his wife, his three sons, and their wives.

6. God told Noah to build the ark 300 cubits long, 50 cubits wide, and 30 cubits high.

7. A cubit was a common measurement, based on the distance from a grown man's elbow to the tip of his middle finger.

8. The standard cubit was approximately 17.5–18 inches long, which would make Noah's ark about 438 feet by 73 feet and nearly three stories high.

9. The dimensions of Noah's ark make it just over half the size of the *Titanic*.

10. God even told Noah how to waterproof the ark by sealing it inside and out with pitch.

11. Building the ark wasn't some weekend project.

12. In fact, Noah spent around 120 years constructing it.

13. He was 480 years old when God first spoke to him and 600 when the flood finally occurred.

14. Not all of the animals entered Noah's ark two by two.

15. Some boarded the boat in parties of 14.

16. God produced torrential rains for 40 days and 40 nights, enough to flood the entire planet and kill everyone and everything on it.

17. After the rain stopped, the floodwaters were still high.

18. As the waters receded, the ark came to rest "upon the mountains of Ararat" (Genesis 8:4).

19. Noah sent out a dove to assess the situation, and it brought back an olive leaf, indicating that some trees were above water.

20. When Noah finally exited the ark, he built an altar and worshipped God with burnt offerings.

21. God was pleased, and he promised never to destroy the entire earth by flood again.

22. Originally, humans were vegetarians, but after the Great Flood, God gave animals for food as well.

43 Dreams and Visions Facts

God often communicated with Bible characters in strange ways.

✳   ✳   ✳   ✳

Marry Mary (Matthew 1:20–21)

1. Joseph was engaged to Mary when he learned she was pregnant.

2. He was planning to avoid scandal by quietly breaking the engagement.

3. But the Lord spoke to him in a dream: "fear not to take unto thee Mary thy wife."

4. Joseph did so, taking on the role of stepfather to the Savior.

Take a Bow (Genesis 37:5–11)

5. In Genesis 37:5–11, a different Joseph told his 11 brothers what he dreamt: their stars bowing down to his; their bundles of grain paying homage to his.

6. This didn't make the boy popular.

7. Joseph's jealous brothers staged his death and sold him as a slave.

Have a Cow (Genesis 41)

8. As a slave in Egypt, Joseph continued his saga of survival.

9. Falsely accused of rape, he was tossed in prison—where he interpreted dreams for fellow prisoners.

10. When the pharaoh had a strange dream about fat cows and skinny cows, he was told of Joseph's skill at explaining dreams.

11. Brought out of jail to face the pharaoh, Joseph told of upcoming years of plenty and then famine.

12. For this, Joseph was immediately hired to manage the nation's response.

Up on the Roof (Acts 10)

13. Peter was meditating on a rooftop when the Lord gave him a vision of nonkosher animals.

14. A voice said, "Kill and eat."

15. When he protested, the voice reminded him that God decides what's kosher.

16. Just then there were messengers at the door, asking him to preach in the home of a gentile—a "nonkosher" person.

17. Peter got the message, and he went with them.

Holy, Holy, Holy (Isaiah 6)

18. The prophet Isaiah had a vision of the Lord sitting on a high

throne, surrounded by seraphim—fiery angels that sang, "Holy, holy, holy is the Lord of hosts: the whole earth is full of his glory."

19. This was Isaiah's calling as a prophet.

20. When he worried about his "unclean lips," an angel took a coal from the altar and touched his lips with it, cleansing them.

21. "Who will go for us?" the Lord asked, and Isaiah offered a response that many have echoed since: "Here am I; send me."

Word to the Wise (1 Kings 3:3–15)

22. Solomon didn't feel ready to succeed David as king.

23. God appeared to him in a dream, offering him what he wanted.

24. Acknowledging that he was "but a little child: I know not how to go out or come in," Solomon asked for wisdom. "Give therefore thy servant an understanding heart to judge thy people."

Come and Help Us (Acts 16:9–10)

25. Paul was on his second mission trip, revisiting areas throughout Asia Minor (modern Turkey).

26. Suddenly he and his team didn't know where to go next.

27. It seemed that the Lord was thwarting all their plans.

28. But then, in a dream, Paul saw a man saying, "Come over into Macedonia, and help us."

29. This meant crossing into the European continent, but Paul boldly did so, continuing his successful ministry.

Dry Bones (Ezekiel 37:1–14)

30. The prophet Ezekiel did some crazy stuff, and he saw some odd visions, too.

31. One of his most dramatic visions was a valley of dry bones.

32. As he watched, the bones reassembled into skeletons and added sinews and flesh.

33. He prayed that God would breathe new life into them, and that's what happened.

34. "I will open your graves," God said, "and cause you to come up out of your graves, and bring you into the land of Israel."

Feet of Clay (Daniel 2)

35. Daniel had a Joseph-like experience as a Jewish exile in Babylon.

36. King Nebuchadnezzar had a troubling dream, but he couldn't remember what it was.

37. Of all his advisors, only Daniel could reveal what he dreamt and what it meant.

38. It was a vision of a huge statue, head of gold, chest of silver, and inferior metals down to the feet of clay and iron.

39. Daniel announced that this foretold a series of kingdoms, progressively inferior, that would follow Babylon.

Sevens Are Wild (Revelation)

40. John was exiled to the island of Patmos when the Lord gave him the remarkable vision that we know as the book of Revelation.

41. It starts as a series of divine messages to seven churches in Asia Minor but continues with a vivid depiction of events in heaven and on Earth—seven seals sealing the scroll, seven angels proclaiming events, with seven trumpets.

42. The images are highly symbolic, allowing for all sorts of interpretations.

43. The key fact, through all these supernatural struggles, is that God *wins*!

24 Tribes of Israel Facts

1. God promised Jacob (later renamed Israel) that an entire nation would spring from him.

2. Jacob had 12 sons by four different women—Leah, Rachel, Zilpah, and Bilhah.

3. Jacob's first wife, Leah, was the mother of Reuben, Simeon, Levi, Judah, Zebulun, and Issachar.

4. Rachel, Jacob's favorite wife, was the mother of Joseph and Benjamin.

5. Jacob favored Joseph over all his children, as he was the "son of his old age" (Genesis 37:1–3).

6. Rachel died giving birth to Benjamin.

7. As he was born, Rachel named him *Benoni* which means "son of my sorrow," but he was renamed Benjamin ("son of the right hand") by his father (Genesis 35).

8. Gad and Asher were the sons of Zilpah, Leah's maid.

9. The sons of Bilhah, Rachel's maid, were Dan and Naphtali.

10. The 12 tribes of Israel were named after the sons or grandsons of Jacob: Benjamin, Ephraim, Manasseh, Naphtali, Dan, Asher, Issachar, Judah, Zebulon, Simeon, Reuben, and Gad.

11. While no tribe was named after Jacob's son Joseph, two tribes were named after Joseph's sons Manasseh and Ephraim.

12. Little is known of most of Jacob's 12 sons, but the tribes made up of their descendants played vital roles in Israelite history.

13. First, Moses appointed a leader for each tribe, and these leaders guided the people during the 40-year sojourn in the wilderness.

14. When the Israelites entered Canaan, each tribe—except the tribe of Levi—was given its own land.

15. The Levites, who served as priests for all the tribes, were supported by the tithes of the people.

16. However, to keep the number of tribal lands at 12, the descendants of Joseph were split in two tribes, descended from Joseph's sons Ephraim and Manasseh.

17. Under David and Solomon, all 12 tribes were brought together in a single great kingdom.

18. After Solomon's death, however, the kingdom split.

19. Ten of the tribes formed the northern kingdom of Israel, while

the tribes of Judah and Benjamin formed the southern kingdom of Judah.

20. In 721 B.C., the Assyrians invaded the northern kingdom and permanently dispersed the people.

21. Consequently, they are often referred to as the Ten Lost Tribes of Israel.

22. Only Judah and Benjamin remained in the south, and Benjamin was slowly absorbed by Judah and foreign powers.

23. Most Jews today trace their ancestry to the tribe of Judah.

24. In fact, the word *Jew* is taken from the name of this tribe.

10 Plagues of Egypt

1. Water into blood
2. Frogs
3. Gnats/lice
4. Flies
5. Pestilence of livestock
6. Boils
7. Hailstorm
8. Locusts
9. Darkness
10. Death of the firstborn

20 Facts About King David's Men

King David of Israel needed not only great advisors but also fearless champions: men of loyalty, might, and leadership.

✳ ✳ ✳ ✳

1. King David had 37 such valiant men about him, called the Thirty (not sure why).

2. Three of these men were especially dedicated: Eleazar, Shammah, and Josheb-Basshebeth.

3. When David commented he'd like some water from Philistine-occupied Bethlehem, these three champions fought their way in, got the water, and brought it to him.

4. Josheb-Basshebeth the Tachmonite was chief among these three captains.

5. This legendary spearman was not a person to mess with.

6. Josheb-Basshebeth once killed 800 enemies in a single battle with his spear.

7. That should have taught someone a lesson.

8. Son of Agee the Hararite, Shammah earned renown in a battle against King David's favorite enemies: the Philistines.

9. Second Samuel 23 tells of a battle in a lentil field from which the entire Israelite fighting force fled in evident disorder.

10. Except Shammah, that is, who stayed behind in the lentils clobbering the Philistines with God's help.

11. Shammah obviously feared nothing and no one, which says a great deal for his personal faith.

12. Eleazar was the son of Dodo (no, not the bird).

13. Before one battle, King David and Eleazar gave the gathered Philistines some biblical trash talk.

14. Perhaps they wanted to provoke rash action; maybe they just enjoyed insulting the Philistines.

15. Whatever the motive, it worked too well.

16. When the Philistines charged, Eleazar was the only Israelite who stood fast at King David's side.

17. Eleazar fought until his hand froze claw-like about his sword, but he won a great victory.

18. The Bible says Abishai commanded the Thirty without being a member.

19. Put a spear in Abishai's hand and it was a bad day for the enemies of King David.

20. Abishai once killed 300 enemies with a spear.

48 Animals of the Bible Facts

1. A serpent is featured in Genesis, the first book of the Bible.

2. It is also noted in Revelation, the final book.

3. The animal mentioned most often in the Bible is the sheep.

4. Hebrew has some 12 different words for sheep, such as ram, ewe, and flock.

5. This undoubtedly reflects the important place sheep had in Israelite life in Old Testament times.

6. In the Bible, sheep are symbolically seen as innocent, sacrificial animals.

7. It's unlikely that Jonah was swallowed by a whale, as is commonly believed by many.

8. Whales are rarely sighted in the Mediterranean Sea today and were probably unknown in biblical times.

9. Many Bibles don't mention a whale at all, noting instead that Jonah was swallowed by a "great fish" or "large sea monster."

10. The dove is best remembered as the bird that informed Noah that the waters of the Great Flood were receding.

11. However, the dove wasn't the first animal released from the ark.

12. That honor goes to the raven, which angered Noah by flying back and forth rather than doing the job it was assigned.

13. The one domesticated animal that is *not* mentioned in the Bible is the cat.

14. Dogs are mentioned several times, often in unflattering terms.

15. God often used insects as weapons.

16. For example, he sent swarms of hornets into the land of Canaan ahead of the Israelites to drive out their enemies.

17. In many biblical lands, priests and prophets practiced

divination of omens by examining animal entrails and the markings of their livers.

18. The liver was so important for this that detailed clay models were made, with various lobes and lines marked and sometimes even inscribed with omens and magical formulas.

19. Leviticus 11:29–30 mentions several animals that are unclean, some of them reptiles.

20. They include the tortoise, the chameleon, and the lizard.

21. These all were among the animals "that creep upon the earth."

22. Lizards were (and are) common in Palestine.

23. The ostrich was common in biblical lands, but today it is extinct there.

24. Ostriches are described in considerable detail in Job 39:13–18.

25. Unicorns are mentioned several times throughout the KJV Bible.

26. But rather than magical, single-horned horses, the animals in question were actually oryx, a horse-like creature with two long, straight horns.

27. These exotic animals were nearly hunted to extinction in the 19th century, but they are slowly making a comeback.

28. The Bible mentions bees a number of times.

29. Twice the swarming habits of honeybees are mentioned: Psalm 118:12 says that the nations "compassed me about like bees," while Deuteronomy 1:44 mentions the Amorites chasing the Israelites like bees.

30. The Hebrew word for "bee" is *deborah*.

31. Horses were common in biblical lands.

32. They are mentioned more than 140 times in scripture.

33. In Old Testament times, royalty held horses, and they were a symbol of human power.

34. In Israel, kings were not to accumulate horses, but to leave military matters to God.

35. David kept a few horses after one battle, but his son Absalom was able to capture them during his revolt against his father.

36. Did you know that the word *scapegoat* comes from the Bible?

37. Leviticus 16 describes a ritual involving two goats.

38. One was sacrificed as a sin offering.

39. The other was released into the desert after Israel's sins were symbolically transferred onto its head.

40. The Bible calls this animal the "scapegoat" (Leviticus 16:8).

41. Eventually, the word came to refer to anyone who takes the blame for others.

42. The Hebrew word *behemoth* is a rare term for "beasts."

43. In Job 40:15, behemoth refers to a large animal, probably the hippopotamus.

44. The modern-day usage of behemoth as something large takes its meaning from this passage.

45. Eagles are mentioned more often than any other bird of prey in the Bible.

46. They are impressive for their great size, strength, speed, and soaring abilities.

47. Eagles are usually mentioned using stirring words, like in Isaiah 40:31: "They that wait upon the Lord shall renew their strength; they shall mount up with wings as eagles."

48. Jesus rode a donkey colt into Jerusalem.

12 Disciples of Christ

1. Simon Peter, brother of Andrew

2. Andrew, brother of Peter

3. James the elder (or greater), son of Zebedee, brother of John

4. John, son of Zebedee, brother of James

5. Bartholomew (or Nathanael)

6. James the younger (or lesser), son of Alpheus

7. Thaddeus (or Judas/Jude), son of James

8. Thomas

9. Philip

10. Matthew (or Levi)

11. Simon the Zealot

12. Judas Iscariot

35 Choice Proverbs

Attributed to Solomon, Proverbs offers so much in the way of life guidance. Here are some of the choicest morsels.

Anger/Quarreling

1. Do not quarrel with anyone without cause, when no harm has been done to you (3:30).

2. A soft answer turns away wrath, but a harsh word stirs up anger (15:1).

3. One who is slow to anger is better than the mighty, and one whose temper is controlled than one who captures a city (16:32).

4. Like a city breached, without walls, is one who lacks self-control (25:28).

5. Like somebody who takes a passing dog by the ears is one who meddles in the quarrel of another (26:17).

Fools/Foolishness

6. The fear of the Lord is the beginning of knowledge; fools despise wisdom and instruction (1:7).

7. In vain is the net baited while the bird is looking on (1:17).

8. The lips of the righteous feed many, but fools die for lack of sense (10:21).

9. A rebuke strikes deeper into a discerning person than a hundred blows into a fool (17:10).

10. Even fools who keep silent are considered wise; when they close their lips, they are deemed intelligent (17:28).

11. Do not answer fools according to their folly, or you will be a fool yourself (26:4).

12. The legs of a disabled person hang limp; so does a proverb in the mouth of a fool (26:7).

Love/Friendship

13. Hatred stirs up strife, but love covers all sins (10:12).

14. Better is a dinner of vegetables where love is than a fatted ox and hatred with it (15:17).

15. A friend loves at all times, and a brother is born for adversity (17:17).

16. Some friends play at friendship but a true friend sticks closer than one's nearest kin (18:24).

17. Iron sharpens iron, so a man sharpens his friend (27:17).

Wealth/Money

18. Wealth hastily gotten will dwindle, but those who gather little by little will increase it (13:11).

19. Better is a little with righteousness than large income with injustice (16:8).

20. Whoever is kind to the poor lends to the Lord, and will be repaid in full (19:17).

21. It is better to be poor than a liar (19:22).

22. The rich rule over the poor, and the borrower is the slave of the lender (22:7).

Wickedness

23. Do not enter the path of the wicked, and do not walk in the way of evildoers (4:14).

24. There are six things that the Lord hates, seven that are an abomination to him: haughty eyes, a lying tongue, hands that shed innocent blood, a heart that devises wicked plans, feet that hurry to run to evil, a lying witness who testifies falsely, and one who sows discord in a family (6:16–19).

25. The light of the righteous rejoices, but the lamp of the wicked goes out (13:9).

26. The wicked flee when no one pursues, but the righteous are as bold as a lion (28:1).

Wisdom

27. Trust in the Lord with all your heart and lean not on your own understanding (3:5).

28. Blessed are those who find wisdom, those who gain understanding (3:13).

29. A scoffer who is rebuked will only hate you; the wise, when rebuked, will love you (9:8).

Wine/Drink

30. Wine is a mocker, strong drink a brawler, and whoever is led astray by it is not wise (20:1).

31. Do not look at wine when it is red, when it sparkles in the cup and goes down smoothly. At the last it bites like a serpent, and stings like an adder. Your eyes will see strange things, and your mind utter perverse things (23:31–33).

Women/Mothers

32. Like a gold ring in a pig's snout is a beautiful woman without good sense (11:22).

33. A wife of noble character is her husband's crown, but a disgraceful wife is like decay in his bones (12:4).

34. The rod and reproof give wisdom, but a mother is disgraced by a neglected child (29:15).

35. Who can find a virtuous woman? She is worth far more than rubies (31:10).

14 Long and Short Facts

1. The longest name in the Bible is Mahershalalhashbaz (Isaiah 8:1). That's 18 letters long!

2. The Bible's shortest verse is just two words: "Jesus wept" (John 11:35).

3. The longest verse in the Bible is Esther 8:9.

4. The number of words Esther 8:9 contains varies depending on the version, but it's a loooong sentence.

5. The shortest book in the Old Testament is Obadiah.

6. The longest book in the Bible is Psalms, with an impressive 150 chapters.

7. The book of Jeremiah (with a mere 52 chapters) has the most words.

8. The shortest book, by number of words, is 3 John.

9. The longest chapter is Psalm 119 (a dizzying 176 verses).

10. The Bible's shortest chapter, Psalm 117, is only two verses long.

11. The shortest chapter comes just two chapters before the Bible's longest chapter.

12. Five books share the distinction of being only one chapter long: Obadiah, Philemon, 2 John, 3 John, and Jude.

13. No short person's height is actually given, but Zacchaeus was noted for his short stature.

14. The book of Revelation is the longest example of apocalyptic literature in the Bible.

58 Facts About Bible Places

1. The first battle mentioned in the Bible took place at the vale of Siddim, near the Dead Sea.

2. Jesus performed his first miracle—turning water into wine—at a wedding in Cana.

3. Abraham was called to sacrifice his own son Isaac in the mountainous region of Moriah.

4. The ancient city of Ur was built on the bank of the Euphrates River.

5. The river has since changed course. Ur's ruins are in a desert.

6. The land of Shinar, mentioned in Genesis 10:10, might be another name for Sumer.

7. Sumer dominated the Tigris and Euphrates region in the third millennium B.C.

8. The Sumerians were the world's first literate people. They invented cuneiform (triangular symbols pressed into wet clay with a reed stylus).

9. Antioch was the first city in which Jesus' followers were referred to as Christians.

10. After killing an Egyptian, Moses fled to Midian, a region located in the northwestern Arabian Desert.

11. Babylon—home of the famous Hanging Gardens—was located on the Euphrates River about 55 miles south of Baghdad.

12. David slew Goliath in the valley of Elah.

13. Jonah took a ship to Tarshish rather than going to Nineveh as God commanded.

14. Tarshish was either part of the island of Sardinia (off the coast of Italy) or a region in far-off Spain.

15. Moses was allowed to see the Promised Land from Mount Nebo.

16. Perhaps best known as Jesus' birthplace, Bethlehem was already ancient by his time.

17. Bethlehem was David's city and the burial place of Jacob's wife Rachel.

18. Nonbiblical sources reference Bethlehem as early as the Amarna letters (c.1400 B.C.).

19. The Samaritans (not the good ones, obviously) sacked Bethlehem in A.D. 529. Islamic troops captured it in A.D. 637.

20. Modern Bethlehem (Hebrew, *Bait-lechem*, "house of bread"; Arabic, *Bayt lam*, "house of meat") stands about six miles south of Jerusalem in the West Bank.

21. Bethlehem's population is about 25,000, probably far more than in biblical times.

22. The Apostle Paul was born in Tarsus, a city of Cilicia.

23. Cilicia was located in southern Turkey, along the coast of the Mediterranean Sea.

24. Abraham bought the Cave of Machpelah, near Hebron, to bury his wife Sarah in.

25. Abraham's purchase was significant, since it represented the first clear title to the land that God had promised him.

26. Abraham, Isaac, and Jacob were also buried in the Cave of Machpelah.

27. Macedonia was the first place in Europe that the Gospel was preached.

28. Lot's wife was turned to a pillar of salt while fleeing the destruction of Sodom and Gomorrah.

29. The Ten Commandments were given at Mount Sinai.

30. Moses was buried in a valley in the land of Moab.

31. The ancient city of Corinth was located on the narrow isthmus that connects the Peloponnese peninsula to the mainland of Greece.

32. Galatia was located in the central and northern interior highlands of modern-day Turkey.

33. In Christ's time, Nazareth was a small farming town; it could even be considered, in our lingo, somewhat podunk.

34. Luke describes Nazareth as the site of the Annunciation.

35. Nazareth had a significant Jewish population until about A.D. 630, when the Eastern Roman Empire ran them out.

36. Islamic forces captured the town soon thereafter.

37. Modern Nazareth (Hebrew name *Natzrat* or *Natzeret*; Arabic name *An-Nasira*), the "Arab capital of Israel," is some 16 miles west southwest of the Sea of Galilee.

38. Today, Nazareth's population is over 77,000, predominantly Arabs, of whom 69 percent are Muslim and 30.9 percent are Christian.

39. Nazareth's chief modern attraction is the Church of the Annunciation, built over the spot where tradition says Gabriel appeared to Mary.

40. Did you know that Hell exists right here on Earth? In the northwest Grand Cayman town of West Bay, to be precise.

41. But don't worry—there are no tormented souls there.

42. This Hell is actually just a group of short, black, limestone formations covering a spit of land about half the size of a football field.

43. There is also a Hell in Norway, which freezes over in winter!

44. Paul's epistle addresses the inhabitants of Thessalonica, which falls in Greece's northeastern region of Macedonia.

45. Biblical Capernaum (or Capharnaum) was a relatively young city in Jesus' day, established sometime around 200 B.C.

46. Capernaum lasted over a thousand years before dying out, ironically enough, just before the Crusader era (roughly A.D. 1050).

47. After Christ got fed up with the Nazarenes, he moved to Capernaum.

48. Jesus lived in Capernaum for some time, and it was the site of several healing miracles.

49. Some scholars believe Jesus gave his famous Sermon on the Mount atop a hill on the north end of the Sea of Galilee that acts as a natural amphitheater.

50. Voices are naturally amplified there, so Jesus, speaking in a normal voice, could easily have been heard as far as 200 yards away.

51. Before his crucifixion, Jesus went to the Garden of Gethsemane to pray.

52. *Gethsemane* comes from an Aramaic word meaning "oil press."

53. The Garden of Gethsemane today still has many very old olive trees.

54. Jesus was crucified at a place called Golgotha.

55. The name Golgotha means place of a skull.

56. Saul (later known as the Apostle Paul) had his conversion on the road to Damascus, in present-day Syria.

57. A "Road to Damascus" moment is a turning point or life-changing experience, often a sudden, radical change in attitude, perspective, or belief.

58. While sailing to Italy, Paul was shipwrecked on the island Melita, now known as Malta.

10 Commandments

1. You shall have no other gods before me (Exodus 20:3).

2. You shall not make for yourself any graven image or idol (Exodus 20:4).

3. You shall not take the Lord's name in vain (Exodus 20:7).

4. Remember the sabbath day, and keep it holy (Exodus 20:8).

5. Honor your father and mother (Exodus 20:12).

6. You shall not murder (Exodus 20:13).

7. You shall not commit adultery (Exodus 20:14).

8. You shall not steal (Exodus 20:15).

9. You shall not bear false witness against your neighbor (Exodus 20:16).

10. You shall not covet your neighbor's house, wife, servant, ox, or donkey (Exodus 20:17).

60 Old Testament Prophet Facts

1. Moses' primary complaint was against the pharaoh.

2. The treacherous Egyptian monarch broke deal after deal with the Israelites, deliberately making their forced labor more difficult.

3. Moses conveyed God's prophecies and wishes to the Israelites, leading them out of Egyptian servitude toward the Promised Land and establishing the Jewish faith.

4. Elijah railed against the worship of other gods.

5. He challenged the priests of Baal to a sacrifice-off: Elijah predicted that his God could burn a sacrificial altar at will, and he challenged Baal's priests to do the same.

6. Elijah and God won the challenge (go figure).

7. Later Elijah prophesied a very unpleasant end for King Ahab and Jezebel, his wife, over their foul plot to seize Naboth's vineyard.

8. In Jezebel's case, it was unpleasant indeed—she was chucked out a window and eaten by dogs.

9. Many things got under Isaiah's skin: church and government corruption, rich people exploiting the poor, foreign entanglements that trusted in alliances and deals (rather than God) to protect Israel, and especially worship of other gods.

10. Isaiah foretold the defeat of Sennacherib's Assyrian army.

11. This worked out pretty well for the Israelites when an angel showed up and killed 185,000 Assyrians, causing Sennacherib to hightail it back to Nineveh.

12. Receiving Elijah's prophetic mantle in a literal sense, Elisha was sensitive about his bald head.

13. Head-shaving was probably a ritual for prophets at this time, so they were likely mocking his religion, not his appearance.

14. When a gang of boys mocked his shiny scalp, Elisha cursed them in God's name.

15. Two bears immediately came out of the woods and mauled 42 of the bratty boys. Talk about bad news bears!

16. Elisha had few kind words for Syria.

17. His last prediction was a victory for Israel over Syria.

18. Jeremiah saw that Israel had been unfaithful to God.

19. He foretold that Israel would experience another round of foreign captivity in which to reflect—this time in Babylon, and they'd better learn from it.

20. One of Jeremiah's favorite metaphors was to compare God to a husband and Judah to an unfaithful wife.

21. He made some enemies this way, but what self-respecting prophet didn't?

22. In Lamentations, Jeremiah laments how far the Israelites had fallen away from God.

23. God told the prophet Hosea to marry a prostitute, Gomer.

24. This instruction came with a warning that she would be unfaithful and bear illegitimate children.

25. It was all a life-size object lesson for Israel. The nation had been unfaithful to God, following after idols, but God still loved them madly.

26. Hosea indicted Israel in language that involved prostitution and loose morals.

27. As with most other prophets, Hosea demanded that Israel repent, threatened it with terrible fates otherwise, and predicted deliverance if it did.

28. Joel's theme was disaster and judgment day.

29. If the Israelites didn't return to worship the true God, Joel foretold devastation, pain, earthquakes, gloom, and fire.

30. If they did, however, they would receive many blessings.

31. What sets Joel apart is not so much his theme of repentance, but the exaltation and joy he foretold if Israel heeded his words.

32. A farmer, Amos called for social justice, never a strong point with the Israelite religious hierarchy of his day.

33. He preached against many nations: Damascus, Gaza, Tyre, Edom, Ammon, Moab, Judah, even Israel.

34. It got Amos kicked out of the Bethel religious sanctuary, but he evidently felt the prophetic truth was worth it.

35. Obadiah's rant was specific to the Edomites, Israel's southeastern neighbors and general targets of scorn.

36. Obadiah calls the Edomites overly proud and accuses them of aiding and abetting the Babylonian invasion of Judah.

37. It's amazing how much yelling Obadiah packs into 21 verses, the gist of which is that Edom is *really* going to get it.

38. When God told Jonah to preach to the most wicked city on Earth, he went sailing in the opposite direction.

39. Tossed overboard, swallowed by a big fish, and vomited onto the shore, the prophet finally realized he couldn't outrun God.

40. So Jonah preached in Nineveh, and they repented!

41. It was a smashing success, except Jonah was angry about it.

42. Jonah didn't really want God to forgive the Ninevites.

43. Jonah, it seems, tended to act first and think second.

44. Micah was deeply annoyed with both Judah and Samaria.

45. He compared the latter to a prostitute (a favorite curse of OT prophets) and foretold the former's conquest (the people would have to shave off all their hair).

46. Like many other prophets, Micah saved special dislike for those who oppressed the poor or tried to confiscate others' property.

47. Micah prophesied the ruin of deeply corrupt Jerusalem.

48. Nahum's chosen enemies were the much-despised Assyrians.

49. Nahum went into exact detail about the fall, which actually occurred in 612 B.C.

50. Zephaniah was absolutely disgusted with God's people in Jerusalem.

51. The strong oppressed the weak. As Zephaniah saw it, they dabbled in idolatry, distrusted the Lord, profaned sacred things, and scoffed at God's law.

52. Zephaniah was mad at other nations too, so at least he didn't single out Israel: God would destroy Assyria, for example.

53. Nahum, one supposes, would gladly have hung out with Zeph.

54. Haggai's book is short but memorable.

55. The Jews had not gotten on with rebuilding the temple, and God (via Haggai) wanted them to get their act together.

56. The sooner they got it squared away, the sooner Haggai prophesied that God would bless them.

57. Sometimes God used the stick, other times the carrot.

58. Malachi, whose book appears last (though it may not be the last written), took God's people to task when they questioned their faith in his covenant.

59. In fairness to them, they were stuck in Babylon at the time, but Malachi was deafer to excuses than a Marine gunnery sergeant.

60. Malachi urged people not to divorce and assured the Israelites of the return of Elijah.

15 Facts About Deborah

1. Deborah bucked the trend in ancient Israel's definition of the woman's role.

2. She was the wife (of Lapidoth), a songwriter, a prophetess, and a judge.

3. Deborah was the only female of the 12 judges who guided Israel.

4. She held the position for 40 years.

5. In a male-dominated society, her legal advice in disputes and security and warfare matters were sought and followed.

6. Deborah described herself as "a mother in Israel" (Judges 5:7), referring to her leadership over Israel.

7. Israel had been quietly submitting to its enemies in those days, and it was not until Deborah arose that Israel began to have hope again.

8. She provided the impetus and leadership for action and eventual victory.

9. Deborah is one of only five prophetesses in the Old Testament.

10. As a prophetess, Deborah received orders from God and passed them on to Barak, the commander of Israel's army.

11. When Deborah told Barak to attack the much larger Canaanite army led by Sisera, he responded like a petulant child.

12. Barak declared that he wouldn't go unless Deborah came with him. So Deborah packed up and went along.

13. With God's help, she knew just the right time to attack.

14. Barak listened to her and won an amazing victory—though because of his hissy fit, Deborah got all the credit.

15. After the complete defeat of the enemy, Deborah wrote a song of praise and victory describing God's deliverance (Judges 4—5), which Barak sang.

28 Harsh Punishment Facts

1. The first instance of exile in the Bible involved the first people.

2. In Genesis 3, Adam and Eve get the heave-ho from the Garden of Eden for eating (presumably fruit) from a tree after God said not to.

3. Ezra 7 also describes the penalty of exile, though it's handed down by a far lesser authority: King Artaxerxes of Persia.

4. In Jeremiah 38, Jeremiah was thrown into a cistern (well) with no water, just mud.

5. Nehemiah was so irritated with the Jews for marrying people from Moab, Ashdod, and Ammon, that he started beating up the offenders and plucking their hair out. Ow!

6. Deuteronomy 25 proposes a severe penalty for losing a lawsuit.

7. If the judges decided that the loser was badly in the wrong, in addition to whatever other penalties were imposed, the loser would take a whipping.

8. They established an upper limit of 40 lashes.

9. Adonibezek was a Canaanite king who lost a battle to the armies of Judah and Simeon.

10. Since Adonibezek had cut off the thumbs and big toes of 70 kings (so they couldn't fight), the same was done to him (Judges 1).

11. From his comments on the topic, Adonibezek seems to have expected something like this eventually. As ye sow…

12. In Ezekiel 23, the prophet rails against the corruption of Samaria and Jerusalem by comparing them to loose women.

13. He says that they will have their noses and ears severed, a punishment also known among the Assyrians.

14. In Daniel 2, King Nebuchadnezzar, who leaned toward grotesque penalties, threatens to tear his dream-interpreters into pieces if they can't reveal both his dreams and their meanings.

15. In Genesis 38, Judah was ready to burn Tamar for sleeping around and getting pregnant, but he changed his mind when he learned he was the father.

16. Leviticus 20 makes it quite clear that a man mustn't sleep with a woman and her daughter at the same time, lest the whole three-some be burned.

17. In Daniel 3, the burning punishment backfired against King Nebuchadnezzar of Babylon, who threw Shadrach, Meshach, and Abednego into a furnace for refusing to worship an idol.

18. They didn't burn, and witnesses reported a fourth figure among them. Wonder who that might have been?

19. The most famous example of death by being fed to animals came in Daniel 6, where King Darius threw Daniel into the lions' den.

20. God shut the lions' mouths, and Daniel suffered no harm.

21. Darius, highly impressed, decided instead to throw Daniel's accusers (including their whole families) into the lions' den.

22. This time God didn't shut the animals' mouths.

23. Leviticus 24, Deuteronomy 13, and Acts 7 all mention death by stoning.

24. In Leviticus, God specified stoning to punish blasphemy.

25. In Deuteronomy, for inciting God's people toward worship of other gods.

26. Acts mentions that Stephen was stoned for preaching the Gospel to an angry mob.

27. In 2 Chronicles 25, Judah's army kills 10,000 men of Seir in the Valley of Salt.

28. That left 10,000 captives remaining. Judah's people threw all of them off a steep height so that they broke into pieces.

64 Followers of Jesus Facts

1. Simon (called Peter) and Andrew were the sons of Jonas.

2. They were fishermen, like many of the disciples.

3. Cephas was the Hebrew name of Peter.

4. Peter's name in Greek (*Petros*) means "rock" (*petra*).

5. A climactic point in Jesus' ministry came when Peter confessed that Jesus was the Messiah, the Son of God.

6. Jesus blessed him and said, "Thou art Peter [*Petros*], and upon this rock [*petra*] I will build my church" (Matthew 16:18).

7. Matthew describes Jesus meeting James and John, also fishermen and brothers, very shortly after recruiting Peter and Andrew.

8. James and John were called "sons of thunder," apparently for their explosive tempers.

9. Peter, James, and John formed Jesus' inner circle.

10. Matthew was a tax collector before Jesus called him.

11. Susanna and Joanna, both of whom Jesus healed, helped support Jesus and his disciples.

12. Philip always appears fifth in lists of the twelve apostles.

13. Bartholomew is named Nathanael in the Gospel of John alone.

14. Mary and Martha were Lazarus's sisters and followers of Jesus.

15. Jesus loved all three siblings.

16. Mary Magdalene was prominent among Jesus' women followers.

17. She provided financial support for Jesus' ministry.

18. The disciple Thomas is often referred to as "doubting Thomas," since he insisted on seeing the wounds after Jesus was resurrected.

19. But Thomas was actually the first to understand that Jesus would die—and also volunteered to die with him.

20. Thomas said, "Let us also go, that we may die with him" (John 11:16).

21. James the lesser or younger was the brother of Thaddaeus (Judas).

22. During the days of the early church, Tabitha (a.k.a. Dorcas) helped anyone in need, especially widows.

23. The disciple Simon was a Zealot and a Canaanite.

24. Judas Iscariot, the traitor, is always the last disciple listed in the Bible.

25. James and John, sons of Zebedee, came from a more well-to-do family than most other disciples.

26. James's name never appears apart from his brother John's.

27. Originally a disciple of John the Baptist, Andrew brought his brother Peter to Jesus.

28. Mary Magdalene is the most frequently-mentioned female follower of Christ in the Bible.

29. John was also known as the Beloved Disciple.

30. Peter raised Dorcas (or Tabitha) from the dead.

31. Judas Iscariot was paid 30 pieces of silver for betraying Jesus, but it hardly made him rich.

32. In fact, it was an almost insultingly small sum, equal to what, back then, you would have had to pay if your ox gored a slave.

33. Peter, James, and John witnessed Jesus' transfiguration.

34. John wrote the Gospel of John, 1 John, 2 John, 3 John, and Revelation.

35. Two Pharisees, Nicodemus and Joseph of Arimathea, took personal and political risks to care for Jesus' body after he was crucified.

36. Joseph of Arimathea (a village in Judea) was a wealthy member of the Sanhedrin who came forward and offered his own tomb for Jesus' burial.

37. The first person to whom Jesus appeared after his resurrection was Mary Magdalene.

38. Matthias replaced Judas Iscariot as one of the twelve apostles.

39. Barnabas was a Levite who gave his possessions to be owned in common with other disciples.

54. Stephen was stoned to death, becoming the first Christian martyr (Acts 6:8–8:2).

55. Saul (who later became Paul) consented to and witnessed Stephen's death.

56. After Judas Iscariot's betrayal and death, the other apostles narrowed the choice to Matthias and Joseph (called Barsabas).

57. They cast lots and elected Matthias.

58. James (son of Zebedee) was beheaded by Herod.

59. The Apostle Paul became one of the greatest missionaries.

60. Paul raised Eutychus from the dead.

61. John was banished to the Isle of Patmos.

62. He was later freed and died a natural death.

63. When Peter was martyred, he asked to be crucified upside down.

64. According to tradition, Peter didn't feel worthy to be crucified in the same manner as Jesus.

25 Bible Law Facts

1. There are more than 600 laws in the Bible—not to mention thousands of rules and regulations that were handed down by word of mouth (the so-called Oral Law).

2. Many of the written laws have to do with animal sacrifices and other cultic matters no longer practiced.

3. But others relate to everyday life and range from common courtesy to downright weird.

4. For example, according to the Bible, thou shalt not have a rounded haircut.

5. Leviticus 19:27 says one should not "round the corners" of the hair on your head, nor "mar the corners" of your beard.

6. Bad news for the '60s-era Beatles and their bowl-cut-wearing fans.

7. Playing football is forbidden. Even touching pig skin is a no-no, warns Leviticus 11:8, because swine are unclean animals. So take that, NFL!

8. Leviticus 19:31 tells us not to turn to mediums or spiritualists, lest we face exile.

9. Proverbs 11:13 reminds us that a gossip betrays a confidence, but a trustworthy person keeps a secret.

10. The Bible says tattoos are a no-no.

11. According to Leviticus 19:28, you shall not "print any marks upon you."

12. We know that kosher laws forbid Jews to eat pork, shellfish, and the like.

13. But do we really need laws that forbid us to eat lizards, mice, rats, owls, bats, or vultures (Leviticus 11:26, 29)?

14. Divorce and remarriage is another no-no, according to the Bible.

15. Mark 10:11–12 notes that those who divorce their spouse and marry another commit adultery.

16. The authors of the Bible really dislike mixed things.

17. Leviticus 19:19 says not to mate different breeds of animals, which makes sense.

18. It also warns not to plant your field with two different kinds of seed or wear clothing made of two different materials.

19. That cotton/polyester sweater you love? Ditch it or face the wrath of God.

20. The Bible says to respect your elders.

21. Leviticus 19:32 says to stand up in the presence of the aged to show respect for the elderly and to revere your God.

22. Although this law is more common courtesy, you may want to cite it if you're a senior looking for a seat on public transport.

23. If you see your neighbor's donkey or ox fallen on the road, Deuteronomy 22:4 says to help the owner get it to its feet.

24. Exodus 23:5 tells us to be kind to and help animals, even those of our enemies.

25. According to Deuteronomy 22:11, you shall not wear garments of "woolen and linen together."

36 Units of Measure Facts

Linear Measurements

1. A *cubit* was about 17.5–18 inches, or the length of adult's forearm.

2. It was the ruler of Bible times, the standard measurement of length.

3. A *handbreadth* was 3 fingers' width, about 3 inches.

4. A *span* was a half-cubit, about 9 inches.

5. A *reed* was 6 cubits, or 8 feet 9 inches.

Distance

6. A *fathom* (or orgyia) was 6 feet.

7. A *furlong* (stadium) was 100 fathoms, or 202 yards.

8. A *mile* (milion) was 8 furlongs, or 1,618 yards.

9. Long-distance measurements were given in terms of about how many days' journey it was to the destination.

10. Miles traveled per day: walking, 20; donkey, 20; horse, 25–30; camel, 30 as pack animal but up to 100 with rider.

The Weighting Game

Dry Goods

11. A *cor* was about 6 bushels, the standard for measuring wheat, flour, and the like.

12. An *ephah* was about three-fifths of a bushel.

13. A *cab* was about 1 quart.

14. An *omer* (translated as "measure" in NRSV) was about 2 quarts.

15. A *seah* was about 7 quarts of dry goods.

Liquids

16. A *bath* was 6 gallons, which was a typical amount carried in a jar from a well.

17. A *log* was about one-third quart.

18. A *hin* was about 1 quart.

Keep Current with Currency

Weighing Pure Silver and Gold

19. A *pim* was about one-third ounce.

20. A *shekel* was about two-fifths ounce.

21. To earn 1 shekel, the average person worked about 4 days.

22. A *mina* was 60 shekels, about 1.25 pounds.

23. To earn 1 mina, a person worked about 3 months.

24. A *talent* was 60 minas, weighing approximately 75 pounds.

25. To earn 1 talent, a person worked about 15 years.

26. Using actual silver and gold as currency was the standard until the Persian Empire introduced coins in about 538 B.C.

Coins

27. An *as* was usually bronze, the basic Roman coin (often translated in English as "penny").

28. A *lepta* was small and copper, a Jewish coin worth about one-eighth of an as coin.

29. This was the poor widow's "mite" in Mark 12:42.

30. A *denarius* was silver, worth 16 as coins.

31. The most common Roman coin, a denarius equaled one day's wage.

32. This is the coin Jesus held up when questioned about paying taxes in Luke 20:24.

33. A *drachma* was a silver Greek coin, about equal to the denarius.

34. The drachma appears only in Jesus' parable about the woman who found a lost coin in Luke 15:8–10.

35. A *stater* was worth 4 drachmas.

36. The stater is the Greek coin Peter found in the mouth of a fish in Matthew 17:27.

43 Facts About Bible People

1. Because Adam and Eve were created by God and not born in the traditional way, it's unlikely that they had belly buttons.

2. Adam's name is related to the word for ground (Hebrew *adamah*).

3. Undoubtedly, this involves a play on words, since Adam was formed from "dust of the ground" (Genesis 2:7).

4. Eve's name in Hebrew means "life" (*chavvah*).

5. Adam called his wife Eve "because she was the mother of all living" (Genesis 3:20).

6. Mary (mother of Jesus) and Elizabeth (mother of John the Baptist) were related, possibly cousins.

7. Abraham was the first person to be called a prophet in the Bible.

8. Moses—who was often called upon to talk of weighty matters—was likely a stutterer.

9. Moses claims to be "slow of speech," which suggests some kind of speech impediment.

10. Miriam, sister of Moses, is the first prophetess mentioned by name in the Bible.

11. Only one of the 12 disciples—John the younger—was at the cross when Jesus died.

12. Samson is considered the strongest man in the Bible.

13. The source of Samson's great strength? His hair!

14. His lover Delilah betrayed Samson by having a man cut off his hair while he slept.

15. The wisest man in the Bible? That trophy goes to King Solomon.

16. Haman plotted against the Jews, but Queen Esther saved them.

17. Zacchaeus was so short he had to climb a sycamore tree to see Jesus over the crowd.

18. Cain was the first murderer, killing his brother Abel.

19. Lamech, a descendant of Cain, was history's second killer and first polygamist.

20. His wife Adah gave birth to Jubal, the world's first musician.

21. Lamech's wife Zillah had a son named Tubalcain, the world's first metalworker.

22. Samson once killed 1,000 men with the jawbone of a donkey.

23. King Solomon had 700 wives and more than 300 concubines.

24. The Bible's oldest man was Methuselah, who lived to be 969 years old.

25. Joshua became a spy at age 40.

26. Rahab was a prostitute and resident of Jericho who harbored two Israelite spies.

27. Gideon fought off 135,000 Midianite invaders with just 300 soldiers and 300 trumpets.

28. Abraham's original name was Abram, which means "exalted father" or "the father is exalted."

29. God gave him his new name, Abraham, which is explained as "father of many nations," emphasizing God's promise that he would have many descendants.

30. The angel Gabriel makes his first appearance in the Bible to Daniel.

31. Gabriel later visited Zechariah, father of John the Baptist, and Mary, mother of Jesus.

32. After the death of their husbands, Ruth stayed with her mother-in-law Naomi rather than returning to her own family.

33. Ruth later married Boaz, who provided for both her and Naomi.

34. Jacob followed twin brother Esau out of the womb and took hold of his twin's heel.

35. Jacob and Esau's descendants were the Israelites and the Edomites, respectively, groups that feuded throughout their histories.

36. Lydia was a merchant who sold purple cloth.

37. Balaam, a prophet hired to curse Israel, holds the dubious distinction of being the only biblical character to converse with a donkey.

38. Jacob loved Rachel, but he had to marry her sister Leah first.

39. Heroine Jael killed the Canaanite captain Sisera by driving a tent nail into his temples.

40. The youngest of Jesse's eight sons, David was a shepherd boy who achieved fame by slaying Philistine giant Goliath.

41. David was an ancestor of Jesus.

42. Hannah prayed so fervently to have a son that the priest Eli thought she was drunk.

43. Hannah's son Samuel was the last judge of Israel.

25 Biblical Waters Facts

1. Genesis tells us that a river flowed out of the Garden of Eden and broke into four branches: Pishon, Gihon, Tigris, and Euphrates.

2. The first plague turned Egypt's water sources to blood.

3. God parted the Red Sea so Moses and his followers could flee from the approaching Egyptians.

4. But that wasn't the only time waters were miraculously parted in the Bible.

5. The Jordan River, in fact, parted on three separate occasions, including one instance in which Joshua needed to cross so he could enter the promised Holy Land.

6. After crossing the parted Red Sea, Moses led his people out into the wilderness.

7. When they arrived at Marah, the water was too bitter to drink, so Moses threw a tree into the water in order to make it drinkable.

8. Jacob met future wife Rachel at a well in Haran.

9. When Jesus asked the "woman at the well" for a drink, she was surprised because he was Jewish and she was Samaritan.

10. Three warriors in the service of King David crept through enemy lines to get him a drink of water from the well of Bethlehem.

11. Jochebed hid baby Moses in a basket made of bulrushes in the Nile River.

12. Jesus was baptized near Jericho in the lower Jordan River.

13. The Jordan River runs south from the Sea of Galilee into the Dead Sea, a landlocked salt lake.

14. Jesus met his disciples James and John while they were mending nets in a ship on the Sea of Galilee (Matthew 4:21).

15. The Sea of Galilee was also known as the Lake of Chinnereth, the Sea of Tiberias, and the Lake of Gennesaret.

16. Jesus wasn't the only one to walk on water. Peter walked on water in Matthew 14:29.

17. Another time the Israelites needed water in the wilderness, Moses struck a rock and it gushed forth with water.

18. God was upset because he had told Moses merely to speak to the rock.

19. Jesus famously calmed a storm from a boat on the Sea of Galilee.

20. Elijah had the prophets of Baal put to death at the Kishon brook.

21. Jesus healed a man who had been ill for 38 years at the pool of Bethesda.

22. The Dead Sea, as its name implies, contains almost no lifeforms because its salinity (saltiness) is too high.

23. Jesus crossed the Cedron brook on the night that he was betrayed by Judas Iscariot.

24. Nazareth has only one spring, and it never runs dry.

25. The water flows into a well where women still gather and fill jugs that they balance on their heads, much like Mary would have done.

35 Great Prayers of Jesus Facts

Jesus taught us to pray and then modeled it in several conversations.

✳ ✳ ✳ ✳

The Lord's Prayer (Matthew 6:5–13)

1. After criticizing the way hypocrites prayed (showing off) and the way pagans prayed (babbling), Jesus gave his disciples a template: Our Father in heaven.

2. This prayer includes praise and petition, confession and commitment.

3. If, as a child, you practiced reciting this prayer as quickly as possible, you missed the point. That would be both showing off and babbling.

Hide and Seek (Luke 10:21)

4. Jesus had regular run-ins with the religious intelligentsia, those who knew everything except their own need for God's grace.

5. After sending out his ragtag band of disciples to extend his own preaching-and-healing ministry, he prayed, "I thank thee, O Father, Lord of heaven and earth, that thou hast hid these things from the wise and prudent, and hast revealed them unto babes."

A Very Public Prayer (John 11:41–42)

6. At the raising of Lazarus, Jesus uttered a rather strange prayer for the benefit of the crowd of mourners.

7. Perhaps he had just prayed silently for Lazarus to be resurrected, or maybe that part of the prayer was not recorded.

8. But then Jesus said, "Father, I thank thee that thou hast heard me. And I knew that thou hearest me always: but because of the people which stand by I said it, that they may believe that thou hast sent me."

9. This miracle was a demonstration of Jesus' identity, and this prayer makes that clear.

A Prayer Answered...Audibly (John 12:27–30)

10. Jesus moved from teaching to prayer and back again, as if he had an "always on" connection with his Father.

11. Talking about his impending suffering, he wondered whether he should ask to avoid it but concluded no, it is for this reason that I have come.

12. And then he prayed: "Father, glorify thy name."

13. A voice from heaven replied, "I have both glorified it, and I will glorify it again."

14. Some thought it was just thunder, but Jesus explained, "This voice came not because of me, but for your sakes."

Do Not Sift (Luke 22:31–32)

15. Simon Peter frequently bragged about his loyalty, which is why it was so stunning when Jesus predicted that Peter would deny him.

16. "Simon, Simon, behold," said Jesus, probably at the Last Supper.

17. "Satan hath desired to have you, that he may sift you as wheat: But I have prayed for thee, that thy faith fail not."

A Prayer for Us (John 17)

18. John's Gospel contains a lengthy account of Jesus' teaching and conversation during the Last Supper.

19. He also includes a prayer Jesus prayed at that time.

20. At one point he even prays for us, specifically. "Neither pray I for these alone, but for them also which shall believe on me through their word."

While You Were Sleeping (Matthew 26:36–42)

21. Jesus had warned his disciples about his "cup of suffering."

22. As his arrest neared, he was troubled about this.

23. Taking his disciples to the Garden of Gethsemane, he withdrew from them a bit and prayed while they slept.

24. "O my Father, if it be possible, let this cup pass from me: nevertheless not as I will, but as thou wilt."

25. It remains a model of surrender to God's desires.

Famous Last Words: Forsaken (Matthew 27:46)

26. While on the cross, Jesus cried, "Eli, Eli, lama sabachthani?"

27. This is Hebrew for "My God, my God, why hast thou forsaken me?"

28. Some bystanders thought he was calling for Elijah, but this is the first verse of Psalm 22, which contains an eerily accurate description of the process of crucifixion (though it was written centuries before the Romans invented crucifixion).

Famous Last Words: Forgiven (Luke 23:34)

29. Jesus had been criticized for offering forgiveness when only God could forgive sins, so it's fitting that he forgave his tormentors with one of his dying statements.

30. "Father, forgive them; for they know not what they do."

31. The Romans, well-practiced in crucifixion, knew exactly what they were doing. But they didn't grasp the full impact of this death.

Famous Last Words: Into Your Hands (Luke 23:46)

32. Earlier, Jesus had made the point that no one would take his life from him, but he would "lay it down" (John 10:18).

33. Just before he breathed his last, he did so. "Father, into thy hands I commend my spirit."

34. When the soldiers came by to break his bones (a standard practice to hasten death), they found him already dead (John 19:33).

35. One wonders if anyone remembered the full content of his earlier quote: "I lay down my life, that I might take it again" (John 10:17).

24 Facts About Catholicism

1. Only one nation on Earth can correctly claim to be 100 percent Christian—Vatican City.

2. Vatican City—home of the pope—is the smallest nation in the world, measuring just 110 acres in size.

3. There are more than 2,000 registered saints.

4. Two-thirds of those saints are claimed by Italy and France.

5. Saint Frances Xavier Cabrini, known during her life as Mother Cabrini, was the first American citizen canonized by the Roman Catholic Church.

6. John is the most commonly shared name among popes.

7. The Pontifical Swiss Guard were originally employed as the pope's personal bodyguards.

8. Today, they serve as the military force of the Vatican.

9. On May 6, 1527, The Swiss Guard's 189 members managed a heroic last stand on the steps of St. Peter's Basilica against 20,000 mercenary troops under Emperor Charles V, who wanted Pope Clement VII dead.

10. When the fighting ceased, only 42 of the Guard remained alive, and none had escaped injury.

11. However, Pope Clement VII managed to escape with his life.

12. Can you name the nation with the highest number of Catholics? Nope, it's not Vatican City. (Good guess, though!)

13. It's Brazil. The place is teeming with Catholics.

14. The Vatican's many museums and galleries total about nine miles in length.

15. If you spent just one minute admiring each painting in the Vatican, it would take you nearly four years to complete the tour.

16. Not all popes lived under a vow of celibacy.

17. In fact, several popes fathered children, including Pope Innocent VIII, who had eight offspring. Not so innocent, apparently!

18. Saint Gelasius I was the first pope (A.D. 492–496) to use the title "Vicar of Christ."

19. Though there is some confusion where exactly Gelasius was born, he was one of only three popes of African descent.

20. Shades of Richard Nixon! Saint Pontian, who was pope from A.D. 230–235, was the first pope to resign his office.

21. Most historians consider Pope John XII (real name: Octavian, Count of Tusculum) to be the worst pope ever to hold the title.

22. Over the course of his reign, he allegedly consecrated a 10-year-old boy as bishop of Todi, converted the Lateran Palace into a brothel, raped female pilgrims, stole church offerings, drank toasts to the devil, and tortured anyone who opposed him.

23. Pope John XII was beaten senseless by the husband of a woman with whom he allegedly had been having an affair.

24. The worst pope ever died a few days later, on May 14, 964.

36 Miscellaneous Facts

1. The word "Bible" comes from the Greek word *bibla*, which means books.

2. Grammatical purists may frown, but it's accurate because the Bible is actually comprised of many individual books.

3. The phrase "ashes to ashes, dust to dust" is often recited at funerals.

4. That's a paraphrase of something God said to Adam—"You are dust, and to dust you shall return"—a not-so-subtle reminder of how Adam was created.

5. The star of Bethlehem, which guided the wise men to Jesus' birthplace, was likely not just a regional event.

6. There are records of astronomers in China observing an unusually bright light in the sky at around the same time period, but they thought little of it.

7. The Bible is peppered with a wide variety of prayers, including confession (Psalm 51); praise (1 Chronicles 29:10–13); thanksgiving (Psalm 105:1–7); petition (Acts 1:24–26); confidence (Luke 2:29–32); intercession (Exodus 32:11–13); commitment (1 Kings 8:56–61); and benediction (Numbers 6:24–26).

8. The New Testament was originally written in Greek.

9. Nearly 5,650 handwritten copies survive in the original language and around 10,000 copies in Latin.

10. Baptism is big in the Bible. The very first such act was performed by Moses, who used ox blood to anoint his people.

11. Luckily, water—which is a lot less messy—later took its place.

12. Though scholars aren't 100 percent certain, it is commonly believed that Jesus spoke Aramaic.

13. Christ is not Jesus' last name. It's the English translation of the Greek word for "anointed one," as well as the transliteration of the Hebrew word for "savior."

14. The entrance to heaven is often referred to as the Pearly Gates.

15. According to Revelation 21:12–21, there are twelve gates—three each on the east, west, north, and south walls.

16. The English word "hell" derives from an old Germanic word meaning "hidden place."

17. Other words for hell found in various versions of the Bible include *sheol, hades,* and *gehenna.*

18. Until Johannes Gutenberg invented movable type in the 1450s, Jewish scribes and Christian monks laboriously and meticulously copied Bibles by hand.

19. However, one scribe apparently wasn't paying attention when he copied 1 Samuel 13:1, which in Hebrew reads: "Saul was one year old when he became king, and he reigned two years over Israel."

20. Since we know that Saul was an adult when he became king and that he reigned for longer than two years, it must be assumed that the sleepy copyist missed at least two numbers.

21. You can thank Cardinal Hugo de S. Caro for the concept of biblical chapters, which he introduced in A.D. 1238.

22. Verse notations were added more than 300 years later by Robertus Stephanus, following the invention of the Gutenberg printing press.

23. Blood represented life in the Bible, and shed blood represented death.

24. The sacrifice for sin required the shedding of blood.

25. This was true whether the victim was an innocent animal or the ultimate sacrificial lamb, Jesus.

26. Because of its special significance, Jews were to drain blood from meat before eating it. This is required for a kosher diet.

27. Several verses in the King James Version contain all letters of the alphabet but one. Ezra 7:21, for example, contains every letter except for "J."

28. Second Kings 16:15 and 1 Chronicles 4:10 contain every letter except "X."

29. Galatians 1:14 contains every letter except "K."

30. The Bible is an exercise in dichotomy.

31. For example, various passages talk of Jesus as God, such as John 1:1 and Romans 9:5, while others specifically address his humanity. Among the latter: John 1:14, Luke 2:7, and Hebrews 4:15.

32. Two books in the Bible make no mention of God at all: Song of Solomon and the book of Esther.

33. Our word *Messiah* comes from the Hebrew word *mashiach*, which means "anointed one."

34. In the New Testament the word is used about 500 times to refer to Jesus, who said he came to set up a spiritual kingdom.

35. If you look closely at Leonardo da Vinci's famous painting *The Last Supper*, you'll see oranges.

36. Turns out da Vinci took some creative license. According to horticultural experts, oranges weren't grown in the Holy Land until long after Jesus' time.

Chapter 10

People

50 Royal Dirt Facts

1. Catherine of Aragon was both a widow and a virgin (allegedly) when she married Henry VIII.

2. After his first wife died, King Louis XIV of France secretly married his longtime mistress, Madame de Maintenon.

3. Because of her low social status, she never became queen.

4. Diane de Poitiers, Duchesse de Valentinois (1499–1566) and mistress of King Henri II of France, more or less ran the realm while Henri did his own thing.

5. Her personal emblem looked very much like the modern symbol for a biohazard.

6. Henry VIII and Anne Boleyn, one of his beheaded wives, were the parents of Queen Elizabeth I (reigned 1558–1603), without whom the Elizabethan Era would have been a bore.

7. Queen Elizabeth I was a beer and ale drinker.

8. She also liked mead, but drank very little wine.

9. For Her Majesty, it was all about the hops.

10. Egypt's most famous queen, Cleopatra, was actually Greek.

11. Alice Perrers, notorious mistress of England's Edward III, is said to have stolen the rings off the king's fingers as he died.

12. Elizabeth I lost a fortune in gold and jewel decorations that came loose and fell off her clothes as she went about her daily business.

13. This was a common problem for nobles at the time. Attendants watched for fallen ornaments and discretely retrieved them.

14. France's King Charles VI supposedly sent England's young Henry V a bunch of small balls, saying his age was more suited to games than to battle.

15. Henry V was not impressed and declared that he would soon play a game of ball in the French streets.

16. He invaded France not long after.

17. Jane Seymour, third wife of Henry VIII, picked out her wedding dress on the day her predecessor, Anne Boleyn, was executed.

18. "Divorced, beheaded, died; divorced, beheaded, survived" is a popular rhyme to remember the fates of Henry VIII's six wives.

19. In Tudor England, being beheaded instead of hanged was considered a "favor" granted to condemned nobles.

20. King Louis XV's longtime mistress, Madame de Pompadour, wielded so much power over the king she was known as "the real Queen of France."

21. Egyptian pharaohs often married their siblings.

22. People believed that pharaohs were gods on Earth and thus could marry only other gods.

23. King Henry IV of England was a childhood friend of King Richard II, whom he later deposed.

24. Edward III was the first English king to claim the French throne.

25. His claim was through his mother, Queen Isabella, and had been included as part of her marriage contract to Edward II.

26. The French king's refusal to honor the contract was the spark that started the Hundred Years' War.

27. Abyssinia's Emperor Menelik II ordered three "electric chairs" from America in 1890, even though his country didn't have any electricity.

28. He kept one unplugged death chair as his throne.

29. England's Richard II was so upset over the death of his first wife, Anne of Bohemia, that he had the house she died in destroyed.

30. Edward I grieved the death of his wife, Eleanor of Castile, by building a stone monument at each place her funeral procession stopped en route from Lincoln to London, England.

31. Three of the twelve original structures, known as Eleanor crosses, still stand.

32. Cleopatra occasionally dressed as the goddess Isis to play up the common belief that Egyptian royalty were the incarnations of gods.

33. Henry VI of England suffered periodic mental breakdowns, during which he would respond to no one. One lasted more than a year.

34. Sultan Murad III of the Ottoman Empire had a harem of several hundred women and sired 103 children.

35. Only 20 of his sons outlived him.

36. After taking power, Murad's heir put his 19 brothers to death.

37. The wedding of King Henri III of France was delayed by several hours when Henri insisted on dressing his bride's hair himself.

38. He also designed her wedding gown and a number of her other dresses.

39. Britain's King George V kept the clocks at his Sandringham estate running 30 minutes fast, so he would never be late.

40. Eleanor of Aquitaine is known for having stormy relationships.

41. Shortly after escaping her first marriage to France's King Louis VII, she married Henry II, who was soon to be the king of England.

42. Although Eleanor bore him five sons, Henry constantly flaunted his mistresses, especially Rosamund Clifford.

43. After she was crowned, the first act of Britain's Queen Victoria was to move her bed out of her mother's bedroom.

44. Ancient Egypt's Queen Hatshepsut declared herself pharaoh around 1438 B.C.

45. To look the part, she reportedly wore men's clothes and a fake beard while conducting official business.

46. Bavaria's King Otto often believed he was a sheep, goat, or stork.

47. He was known to stand on one leg in a pond and attempt to pluck fish out of the water while flapping his arms.

48. Although her first language was Greek, Cleopatra was the only Ptolemaic ruler who actually spoke Egyptian.

49. The Ptolemaic pharaohs were Greeks descended from Ptolemy, one of Alexander the Great's generals, so Greek was the dominant language of their court.

50. Early medieval European monarchs had outstanding names, including Charles the Bald (reigned 823–877), Louis the Stammerer (877–879), Charles the Fat (881–887), Louis the Blind (901–905), and Charles the Simple (893–922).

30 Audrey Hepburn Facts

1. Audrey Hepburn was born in Belgium in 1929.

2. She later moved to London, where she practiced ballet and modeling.

3. Hepburn was born Audrey Kathleen Ruston.

4. After WWII, her father discovered documents indicating that he had ancestors named Hepburn, so he legally changed the surname.

5. Audrey became Audrey Kathleen Hepburn-Ruston.

6. Hepburn's first film appearance came in 1948 when a producer offered her a small part in the European *Dutch in Seven Lessons*.

7. In the early 1950s, Hepburn moved to America to play the lead role in the Broadway play *Gigi*.

8. It didn't take long for her to succeed—she snagged the Best Actress Oscar for *Roman Holiday* (1953), her first role as a leading lady.

9. Hepburn was paid $12,500 for her role in *Roman Holiday*.

10. In her final film—Steven Spielberg's *Always* (1989)—she reportedly made $1 million.

11. Audrey Hepburn's best-known movies include *Sabrina* (1954), *Funny Face* (1957), *Love in the Afternoon* (1957), *The Nun's Story* (1959), *Breakfast at Tiffany's* (1961), *Charade* (1963), and *My Fair Lady* (1964).

12. Hepburn said that she didn't think she was right for the part of Holly Golightly in *Breakfast at Tiffany's*.

13. In 2006, the black dress Audrey Hepburn wore in *Breakfast at Tiffany's* sold at auction for $800,000.

14. Hepburn turned down the lead role in *The Diary of Anne Frank* (1959).

15. She ranks third among all actresses (behind Katharine Hepburn and Bette Davis) in the American Film Institute's list of Greatest American Female Screen Legends.

16. She's also on *Empire* magazine's list of Top 100 Movie Stars of All Time (#50) and 100 Sexiest Stars in Film History (#8).

17. Hepburn has been named one of *People* magazine's 50 Most Beautiful People in the World.

18. *Harper's* and *Queen* named her the most fascinating woman of modern time.

19. Hepburn often downplayed her own fame.

20. "I never think of myself as an icon," she once said. "I probably hold the distinction of being one movie star who, by all laws of logic, should never have made it."

21. Hepburn was fluent in five languages: Dutch (Flemish), English, French, Italian, and Spanish.

22. Hepburn was always self-conscious about her size 10 feet.

23. You've probably heard Marilyn Monroe's birthday song to President John F. Kennedy, but Audrey Hepburn reportedly sang one, too—in 1963, on JFK's final birthday.

24. A breed of tulip is named after Audrey Hepburn.

25. Hepburn presented the Best Picture Oscar at the Academy Awards four times, a record among actresses.

26. Audrey and Katharine Hepburn—who are not related—are the only two winners of a Best Actress Oscar to share a last name.

27. Hepburn died on January 20, 1993, from complications stemming from cancer of the appendix, which spread to her colon.

28. She was posthumously awarded an Emmy and a Grammy, joining the short list of EGOT winners in 1994.

29. Hepburn also received an additional Oscar after her death: the Jean Hersholt Humanitarian Award for her work as a goodwill ambassador to UNICEF.

30. She played that role from 1988 until her death.

35 Presidential Nicknames

1. James Monroe—Last Cocked Hat

2. John Quincy Adams—Old Man Eloquent

3. Andrew Jackson—The Hero of New Orleans; Old Hickory

4. Martin Van Buren—The Little Magician; Martin Van Ruin

5. William Henry Harrison—General Mum; Tippecanoe

6. John Tyler—His Accidency

7. James K. Polk—Napoleon of the Stump; Young Hickory

8. Zachary Taylor—Old Rough and Ready

9. James Buchanan—The Bachelor President; Old Buck

10. Abraham Lincoln—Honest Abe; The Rail-Splitter

11. Andrew Johnson—King Andy; Sir Veto

12. Ulysses S. Grant—Useless; Unconditional Surrender

13. Rutherford B. Hayes—Rutherfraud Hayes; His Fraudulency

14. James Garfield—The Preacher; The Teacher President

15. Grover Cleveland—Uncle Jumbo; His Obstinacy

16. Benjamin Harrison—Little Ben; White House Iceberg

17. William McKinley—Wobbly Willie; Idol of Ohio

18. Theodore Roosevelt—The Trust Buster

19. Woodrow Wilson—The Schoolmaster

20. Warren Harding—Wobbly Warren

21. Calvin Coolidge—Silent Cal

22. Herbert Hoover—The Great Engineer

23. Harry Truman—The Haberdasher; Give 'Em Hell Harry

24. Dwight D. Eisenhower—Ike

25. John F. Kennedy—King of Camelot

26. Lyndon B. Johnson—Big Daddy

27. Richard M. Nixon—Tricky Dick

28. Gerald Ford—The Accidental President; Mr. Nice Guy

29. Jimmy Carter—The Peanut Farmer

30. Ronald Reagan—Dutch; The Gipper; The Great Communicator

32. George H. W. Bush—Poppy; 41; Senior

33. Bill Clinton—Bubba; Slick Willie; The Comeback Kid

34. George W. Bush—Junior; W; Dubya

35. Donald J. Trump—The Donald

57 Hawaiian King Kamehameha I Facts

1. His given name at birth was Pai'ea.

2. It is said that Pai'ea was born on Hawaii's Big Island shortly after an appearance of Halley's Comet, which occurred in 1758.

3. Legend had it that a bright light in the skies would herald the birth of a great unifier.

4. Depictions of him in later life support this time frame.

5. As he got older, Pai'ea earned the name "Kamehameha" (meaning "lonely one") because of his solitary, stern disposition.

6. In 1779, he met his first Europeans when Captain Cook and his men arrived in Hawaii.

7. At first, the Hawaiians thought Cook was a representative of Lono, the harvest god.

8. They grew skeptical when Cook's ship floated back into the bay with storm damage; the battered ship hardly seemed like a godly vessel.

9. Cook's arrival may have prompted Kamehameha to believe that Hawaii's islands should be strengthened through unification.

10. Kamehameha gained a reputation as a feared warrior.

11. During a raid in 1782, his foot became stuck in a crevice, and while he was trapped, a couple of fishermen broke a paddle over his head.

12. Fortunately for Hawaii, the fishermen didn't cut Kamehameha's throat before fleeing.

13. Their small show of mercy would later inspire a crucial policy decision from Kamehameha.

14. Kamehameha worked his way up the Big Island's royal chain of command.

15. The path to kingship was treacherous, but the warrior was a savvy political operative in the deadly waters of royal favor.

16. By 1790, he was well on his way to consolidating the Big Island under his rule as *ali'i nui*, or supreme chief.

17. In 1794, Kamehameha remembered the fishermen who knocked him on the head and ordered that they be brought before him.

18. Instead of punishing them, though, Kamehameha put the blame on himself for having assaulted noncombatants.

19. In a gesture of apology, he granted them land and set them free.

20. Also, a new *kapu*, or law, was enacted that forbade harming civilians.

21. It was called *mamalahoe kanawai*, "the law of the splintered paddle."

22. It decreed, "Leave the elderly, women, and children in peace."

23. By 1795, with the Big Island firmly under control, Kamehameha assembled a large army and thousands of war canoes to invade the rest of the Hawaiian islands.

24. He overtook Maui and Molokai without much trouble, and soon his forces landed on Oahu's beaches.

25. Unfortunately, Oahu's defenders had European cannons to fire down on the invading forces from a mountain ridge.

26. This held up Kamehameha until he sent a couple of battalions to flank the artillery.

27. Soon, Oahu fell to Kamehameha as well.

28. After the capture of Oahu, Kamehameha ran into some snags.

29. He was getting ready to invade Kauai and Niihau when a rebellion on the Big Island forced him to return and shore up his power base.

30. In 1803, he attempted another invasion, which was cut short by an epidemic among his troops.

31. Kamehameha also became ill but managed to survive.

32. When he finally came for Kauai in 1810, Kamehameha had European schooners equipped with cannons to supplement his war canoes.

33. The ali'i nui of Kauai took one look at that armada and surrendered.

34. Kamehameha was now king of all of the Hawaiian islands with the individual rulers of each island acting as his vassals.

35. The great king ended the tradition of human sacrifice but remained a devout follower of traditional Hawaiian religion.

36. Christianity didn't impress him, though he let its followers live in peace.

37. Regardless of religion, he would execute anyone who violated kapu.

38. A figure closely associated with Kamehameha I is his favorite wife, Queen Ka'ahumanu.

39. Traditional kapu was fairly harsh for women, with numerous gender-biased rules and restrictions.

40. Ka'ahumanu could be called an early feminist.

41. She used her strong influence to end most of the kapu that treated women unfairly, thereby expanding Hawaiian women's rights.

42. Kamehameha's law of the splintered paddle has remained in force and even today is part of Hawaii's state constitution.

43. In 1816, Kamehameha introduced another lasting tradition, Hawaii's flag.

44. His banner, with the Union Jack in the upper left and one horizontal stripe for each of the eight major islands, is now the state flag.

45. The king loved to fish. When he didn't have more important things to do, he could be found angling off the Kailua coast.

46. He even scheduled affairs of state around the prime fishing seasons.

47. King Kamehameha I died in 1819.

48. During his illness, tradition dictated that human sacrifice was necessary to save the king, but Kamehameha refused to bend his kapu, even to save his own life.

49. After he died, a close friend hid his body, and his burial site remains unknown to this day.

50. This was done intentionally, as the body of a king was believed to contain *mana*, or spiritual force.

51. This mana could be stolen if someone were to possess the king's remains.

52. Kamehameha's dynasty lasted through eight rulers between 1810 and 1893.

53. Unfortunately for Hawaii, this was also an unhappy time of exploitation and loss of independence.

54. The dynasty ended in 1893, when Americans overthrew Queen Lili'uokalani and took the first steps toward creating the Republic of Hawaii.

55. From the start, this republic served the interests of U.S. businesses in Hawaii rather than those of Hawaii's native people.

56. In 1898, Hawaii was incorporated as a U.S. territory.

57. Every year in Hawaii on June 11, people celebrate King Kamehameha Day in honor of the visionary leader and hero.

23 Notable People with a Twin

1. Kofi Annan   twin sister, Efua

2. Isabella Rossellini   twin sister, Isotta

3. Kiefer Sutherland   twin sister, Rachel

4. Scarlett Johansson   twin brother, Hunter

5. Alanis Morissette   twin brother, Wade

6. Mario Andretti   twin brother, Aldo

7. Vin Diesel   twin brother, Paul

8. Ashton Kutcher   twin brother, Michael

9. Billy Dee Williams   twin sister, Loretta

10. José Canseco   twin brother, Ozzie

11. Aaron Carter   twin sister, Angel

12. John Elway   twin sister, Jana

13. Jerry Falwell   twin brother, Gene

14. Deidre Hall   twin sister, Andrea

15. Pier Angeli   twin sister, Marisa Pavan

16. Montgomery Clift twin sister, Roberta

17. Maurice Gibb twin brother, Robin

18. Ann Landers twin sister, Abigail Van Buren (Dear Abby)

19. Mary-Kate Olsen twin sister, Ashley

20. Jim Thorpe twin brother, Charlie

21. Rami Malek twin brother, Sami

22. Tiki Barber twin brother, Ronde

23. Linda Hamilton twin sister, Leslie

30 Facts About Nefertiti

1. Along with Cleopatra, Nefertiti is one of the most famous queens of ancient Egypt.

2. She's often referred to as "The Most Beautiful Woman in the World."

3. Her legendary beauty is largely due to the 1912 discovery of a painted limestone bust of Nefertiti depicting her stunning features: smooth skin, full lips, and a graceful swan-like neck.

4. The likeness, now housed in Berlin's Altes Museum, has become a widely recognized symbol of ancient Egypt.

5. The bust is one of the most important artistic works of the premodern world.

6. It wasn't until the bust surfaced in the early 20th century that scholars began sorting out information about Nefertiti's life.

7. But the bust, like almost everything about the famous queen, is steeped in controversy.

8. Her name means "the beautiful one is come."

9. Some think Neferiti was a foreign princess, not of Egyptian blood.

10. Others believe she was born into Egyptian royalty, that she was the niece or daughter of a high government official named Ay, who later became pharaoh.

11. Basically, no one knows her origins for sure.

12. When "the beautiful one" was 15, she married Amenhotep IV, who later became king of Egypt.

13. Thus, Neferiti was eventually promoted to queen.

14. Neferiti became queen during the 18th Dynasty, though the exact dates are unknown.

15. Nefertiti appears in many reliefs of the period, accompanying her husband in various ceremonies—a testament to her political power.

16. An indisputable fact about both Nefertiti and Amenhotep IV is that they were responsible for bringing monotheism to ancient Egypt.

17. Rather than worship the vast pantheon of Egyptian gods—including the supreme god, Amen-Ra—the couple devoted themselves to exclusively worshipping the sun god Aten.

18. As a sign of this commitment, Amenhotep IV changed his name to Akhenaten.

19. Similarly, Nefertiti changed her name to Neferneferuaten-Nefertiti, meaning, "The Aten is radiant of radiance [because] the beautiful one is come."

20. Around 14 years into Akhenaten's reign, Nefertiti seems to disappear.

21. From this point on there are no more images of her, nor any historical records.

22. Some think Nefertiti was banished from the kingdom after a conflict in the royal family.

23. Some think she died of the plague.

24. Perhaps most interestingly, some think Nefertiti disguised herself as a man, changed her name to Smenkhkare, and went on to rule Egypt alongside her husband.

25. During a June 2003 expedition in Egypt's Valley of the Kings, an English archeologist named Joann Fletcher unearthed a mummy that she suspected to be Nefertiti.

26. Despite the fact that this mummy probably is a member of the royal family from the 18th Dynasty, most Egyptologists think there is not sufficient evidence to prove that it is Nefertiti, or even female.

27. In 2009, Swiss art historian Henri Sierlin published a book suggesting that the famous Nefertiti bust is a copy.

28. Sierlin claimed that the sculpture was made by an artist named Gerard Marks on the request of Ludwig Borchardt, the German archeologist responsible for discovering the bust in 1912.

29. Despite the mysteries surrounding Nefertiti, there's no question that she was revered in her time.

30. At the temples of Karnak are inscribed the words: "Heiress, Great of Favours, Possessed of Charm, Exuding Happiness...Great King's Wife, Whom He Loves, Lady of Two Lands, Nefertiti."

39 Vlad the Impaler Facts

1. The origins of Dracula, the blood-sucking vampire of Bram Stoker's novel (and countless movies, comics, and costumes), are commonly traced to a 15th-century Romanian noble.

2. His brutal and bloody exploits made him both a national folk hero and an object of horror.

3. The late 1300s and early 1400s were a dramatic time in the area now known as Romania.

4. Vlad III Tepes (1431–76) was born into the ruling family of Wallachia, a principality precariously balanced between the Ottoman (Turkish, Muslim) and Holy Roman (Germanic, Catholic) empires.

5. Wallachia was an elective monarchy, with political backstabbing between the royal family and the *boyars*, the land-owning nobles.

6. His father, Vlad II, became known as Dracul ("Dragon") due to his initiation into the knightly Order of the Dragon.

7. As a part of Dracul's attempt to maintain Wallachia's independence, he sent Vlad III and his brother Radu as hostages to the Turks.

8. Dracul was assassinated in 1447 and a rival branch of the family took over.

9. The Turks helped Vlad III recapture the throne briefly as a puppet ruler before he was driven into exile in Moldova.

10. Eight years later, Vlad III, known as Dracula, or "son of Dracul," regained the throne with Hungary's backing and ruled for six years.

11. Dracula faced several obstacles to independence and control.

12. The Ottoman Empire and Hungary were the largest external threats.

13. There was also internal resistance from the boyars (regional nobles), who kept Wallachia destabilized in their own interests, as well as from the powerful Saxon merchants of Transylvania, who resisted economic control.

14. Lawlessness and disorder were also prevalent.

15. Dracula's successes against these obstacles are the source of the admiration that still pervades local folklore.

16. Locally, he is portrayed as a strong, cunning, and courageous leader who enforced order, suppressed disloyalty, and defended Wallachia.

17. Wallachia remained independent and Christian.

18. Not only that, but, as the story goes, Dracula could leave a golden cup by a spring for anyone to get a cool drink without it being stolen—no one was above the law.

19. His methods, however, are the source of his portrayal elsewhere as a sadistic despot who merited an alternate meaning of Dracula, namely "devil."

20. He eliminated poverty and hunger by feasting the poor and sick and, after dinner, burning them alive in the hall.

21. Dracula eliminated disloyal boyars by inviting them to an Easter feast, after which he put them in chains and worked them to death building his fortress at Poenari.

22. Ambassadors who refused to doff their caps had them nailed to their heads.

23. Dishonest merchants and bands of Roma suffered similar fates.

24. Vlad's favorite method of enforcement was impaling (a gruesome and purposefully drawn-out form of execution in which victims were pierced by a long wooden stake and hoisted aloft).

25. He also found creative ways of flaying, boiling, and hacking people to death.

26. When the Turks cornered him at Tirgoviste in 1461, he created a "forest" of 20,000 impaled captive men, women, and children, which so horrified the Turks that they withdrew.

27. They called him *kaziklu bey* ("the Impaler Prince"), and though Vlad himself did not use it, the epithet Tepes ("Impaler") stuck.

28. The Turks then backed brother Radu and besieged Vlad in Poenari.

29. Dracula escaped to Hungary with the help of a secret tunnel and local villagers, but was imprisoned there.

30. Dracula gradually ingratiated himself and, upon Radu's death, took back the throne in 1476.

31. Just two months later, Dracula was killed in a forest battle against the Turks near Bucharest.

32. The manner of his death is unknown.

33. Some say that he was mistakenly killed by his own army.

34. Others say that he was killed and decapitated by the Turks.

35. One thing is certain: Unlike the hundreds of thousands he killed, Vlad the Impaler was never one of the impalees.

36. Tradition says that his head was put on display by the Turks and his body buried in the Snagov Monastery, but no grave has been found.

37. Bram Stoker, while researching for a vampire story about "Count Wampyr," discovered in a library book that "Dracula" could mean "devil" and changed his character's name.

38. Connecting Stoker's Count to Vlad III has been popular in the press and in the tourist industry.

39. However, Stoker's character probably has little basis in Vlad Dracula.

46 Facts About the Bloody Countess

1. Elizabeth Bathory (born Erzsébet Báthory in 1560) was the daughter of one of the oldest, most influential bloodlines in Hungary.

2. Her wedding in 1575 to Ferenc Nadasdy warranted written approval and an expensive gift from the Holy Roman Emperor.

3. There were rumors that a streak of insanity ran in Elizabeth's family.

4. Some suggest that she may have been related to Vlad the Impaler.

5. However, nobles of the time were given wide latitude when it came to eccentric behavior.

6. Ferenc would go on to become one of the greatest Hungarian military heroes of the age.

7. Ferenc was a battle-hardened man, but even so, his own wife made him nervous.

8. He knew that she treated the servants even more harshly than he did, and he had no reservations when it came to punishing the help.

9. He was known to place flaming oil-covered wicks between the toes of lazy servants.

10. Ferenc saw evidence that Elizabeth's punishments exceeded his brutality when he discovered a servant who had been covered with honey and tied to a tree to be ravaged by ants as punishment for stealing food.

11. But Ferenc spent a great deal of time away at war, and someone had to manage his castle.

12. He turned a deaf ear to complaints about Elizabeth's activities.

13. With her husband's lengthy absences and eventual death, Elizabeth found that she had virtually no restrictions on her behavior.

14. A series of lovers of both sexes occupied some of her time.

15. She also dabbled in black magic, though this was not uncommon in an age when paganism and Christianity were contending for supremacy.

16. She spent hours gazing into a wraparound mirror of her own design, crafted to hold her upright so that she would not tire as she examined her own reflection.

17. The exacting fashion of the day required Elizabeth, always a vain woman, to constantly worry over the angle of her collar or the style of her hair.

18. She had a small army of servants constantly by her side to help maintain her appearance.

19. Her servants were often required to attend to their mistress in the nude as an expression of subservience.

20. If they failed in their duties, Elizabeth would strike out, pummeling them into the ground.

21. On one notable occasion, a servant pulled too hard when combing Elizabeth's hair.

22. Elizabeth struck the offender in the face hard enough to cause the girl's blood to spray and cover the countess.

23. Initially furious, Elizabeth discovered she liked the sensation, believing her skin was softer, smoother, and more translucent afterward.

24. The incident led to the legends, unconfirmed, that Elizabeth Bathory took to bathing in the blood of virgins to maintain her youthful appearance.

25. One rumor has her inviting 60 peasant girls for a banquet, only to lock them in a room and slaughter them one at a time, letting their blood run over her body.

26. Though that incident may be apocryphal, it is certain that the countess began torturing girls without restraint.

27. Aided by two trustworthy servants who recruited a never-ending

supply of hopeful girls from poor families, she would beat her victims with a club until they were scarcely recognizable.

28. When her arms grew tired, she had her two assistants continue the punishment as she watched.

29. She had a spiked iron cage specially built and would place a girl within it, shaking the cage as the individual was impaled over and over on the spikes.

30. She drove pins into lips and breasts, held flames to pubic regions, and once pulled a victim's mouth open so forcefully that the girl's cheeks split.

31. Perhaps most chillingly, allegations of vampirism and cannibalism arose when Elizabeth began biting her victims, tearing off the flesh with her bare teeth.

32. On one occasion, too sick to rise from her bed, the countess demanded that a peasant girl be brought to her.

33. She roused herself long enough to bite chunks from the girl's face, shoulders, and nipples.

34. Elizabeth's chambers had to be covered with fresh cinders daily to prevent the countess from slipping on the bloody floor.

35. Eventually, even the cloak of nobility couldn't hide Elizabeth's atrocities.

36. The situation was compounded by the fact that she got sloppy, killing in such numbers that the local clergy refused to perform any more burials.

37. Thereafter, she would throw bodies to the wolves in full view of local villagers, who naturally complained to the authorities.

38. The final straw was when Elizabeth began to prey on the minor aristocracy as well as the peasants.

39. The disappearance of people of higher birth could not be tolerated.

40. The king decided that something had to be done, and in January 1611, a trial was held.

41. Elizabeth was not allowed to testify, but her assistants were compelled to—condemning themselves to death in the process.

42. They provided eyewitness accounts of Elizabeth's terrible practices.

43. Especially damning was a list, in Elizabeth's own handwriting, describing more than 600 people she had tortured to death.

44. Elizabeth Bathory was convicted of perpetrating "horrifying cruelties" and sentenced to be walled up alive in her own castle.

45. She survived for nearly four years but was finally discovered dead on August 21, 1614, by one of her guards.

46. The countess was unrepentant to the end.

10 Famous People Who Died in the Bathroom

1. Elvis Presley
2. Lenny Bruce
3. Elagabalus
4. Robert Pastorelli
5. Orville Redenbacher

6. Claude François
7. Judy Garland
8. Albert Dekker
9. Jim Morrison
10. Coolio

25 Benjamin Franklin Facts

1. Franklin was an unsurpassed diplomat, writer, and inventor.

2. He may also have been the most eccentric, and funniest, of the founding fathers.

3. Each morning, stout Franklin would step naked through his rooms, opening up the windows to let in the invigorating fresh air.

4. If the weather was mild, Franklin would step outside and peruse the morning's newspaper outside his digs, a gentle breeze lapping at his bare body.

5. The famous printer and scientist urged others to try his scanty approach to apparel, but he was less encouraging to strangers who crowded around his property.

6. Interlopers pressing against the iron fence of one domicile were shocked, literally, by an electric charge sent coursing through the metal by the discoverer of electricity.

7. Franklin was evidently comfortable with his body and his bodily functions: he penned an essay dubbed "Fart Proudly."

8. In in the essay he proposed to "Discover some Drug wholesome & not disagreeable, to be mixed with our common Food, or Sauces, that shall render the natural Discharges of Wind from our Bodies, not only inoffensive, but agreeable as Perfumes."

9. With regard to another key physical function, Franklin urged a friend, who was having trouble landing a young wife, to take an elderly woman as his mistress.

10. He counseled: "as in the dark all Cats are grey, the Pleasure of corporal Enjoyment with an Old Woman is at least equal, and frequently superior."

11. Franklin also had distinctive religious views, often skeptical, sometimes more traditional.

12. He anonymously co-authored an *Abridgment of the Book of Common Prayer*, which shortened funeral services to six minutes to better "preserve the health and lives of the living."

13. In the run-up to the American Revolution, Franklin's Committee of Safety—a sort of state provisional government—was deadlocked over whether Episcopal priests should pray for King George.

14. Franklin told the committee: "The Episcopal clergy, to my certain knowledge, have been constantly praying, these twenty years, that 'God would give the King and his Council wisdom,' and we all know that not the least notice has ever been taken of that prayer."

15. The prayers were canceled.

16. Yet, Franklin believed religion had a salutary effect on society and men's morals.

17. In a letter to Thomas Paine, a decided agnostic, he wrote: "If men are so wicked with religion, what would they be if without it?"

18. Asked to design the Great Seal of the United States, Franklin submitted a sketch of Moses and the Israelites drowning the pharaoh's army in the Red Sea, with the motto: "Rebellion Against Tyrants is Obedience to God."

19. Despite all of his revolutionary ideas, Franklin only had two years of formal education.

20. Later in life Franklin embraced the idea of abolishing slavery.

21. He became president of the first abolition organization, the Philadelphia Abolition Society.

22. Two months before his death, Franklin was the first person to petition Congress to abolish slavery.

23. When Franklin died on April 17, 1790, the French National Assembly declared a 3-day period of mourning.

24. The eccentric, sharp-witted founder composed his own epitaph.

25. The epitaph read: "The Body of B. Franklin Printer; Like the Cover of an old Book, Its Contents torn out, And stript of its Lettering and Gilding, Lies here, Food for Worms. But the Work shall not be wholly lost: For it will, as he believ'd, appear once more, In a new & more perfect Edition, Corrected and Amended By the Author."

23 Ivan the Terrible Facts

1. The first all-powerful Russian ruler, Tsar Ivan the Terrible, was terrible indeed.

2. The Terrible One had an unhealthy dose of paranoia.

3. His erratic behavior is thought to have been largely due to his upbringing, which was tumultuous.

4. Ivan was an orphan: His father died just three years after his son's birth, and his mother was poisoned a few years later.

5. As a child prince in Moscow, Ivan was under the thumb of boyars, or Russia's nobles.

6. Feuding noble families such as the Shuiskis would break into young Ivan's palace, robbing, murdering, and even skinning alive the boy's advisors.

7. Ivan took out his frustrations on animals, poking out their eyes or tossing them off the palace roof.

8. In 1543, at age 13, Ivan took some personal revenge, and had Andrei Shuiski thrown to the dogs—literally.

9. After other vile acts, he'd sometimes publicly repent—by banging his head violently on the ground.

10. When his beloved wife Anastasia died in 1560 (Ivan beat his head on her coffin), the boyars refused allegiance to his young son Dmitri.

11. Ivan then established the Oprichniki, a group of hand-picked thugs.

12. After his forces sacked the city of Novgorod in 1570, he had its "archbishop sewn up in a bearskin and then hunted to death by a pack of hounds."

13. Women and children fared no better; they were tied to sleds and sent into the freezing Volkhov River.

14. Over time, Ivan had the lover of his fourth wife impaled.

15. He had his seventh wife drowned.

16. Ivan the Terrible had eight wives in total, though only four were officially recognized by the church.

17. Perhaps afflicted by encephalitis, and likely by syphilis, his behavior grew ever stranger.

18. He beat up his son's wife, who then miscarried, and later beat his son Ivan to death with a royal scepter.

19. Ivan again beat his own head on his son's coffin after his death.

20. But despite Ivan's violent life, his death was rather quiet.

21. He had a stroke while playing chess with one of his close associates in 1584.

22. Ivan the Terrible may well have been mad as a hatter, and by the same cause that drove 19th-century hat-makers insane— mercury poisoning.

23. When his body was exhumed in the 1960s, his bones were found to have toxic levels of the metal.

50 Victoria Woodhull Facts

1. Known for her passionate speeches and fearless attitude, Victoria Woodhull was a trailblazer for women's rights.

2. But some say she was about 100 years before her time.

3. Woodhull advocated revolutionary ideas, including gender equality and women's right to vote.

4. "Women are the equals of men before the law and are equal in all their rights," she said.

5. America, however, wasn't ready to accept her radical ideas.

6. Victoria Woodhull was born in 1838 in Homer, Ohio, the seventh child of Annie and Buck Claflin.

7. Her deeply spiritual mother often took little Victoria along to revival camps where people would speak in tongues.

8. Her mother also dabbled in clairvoyance.

9. Victoria and her younger sister Tennessee believed they had a gift for it as well.

10. With so many chores to do at home, Victoria only attended school sporadically and was primarily self-educated.

11. Soon after the family left Homer, 28-year-old doctor Canning Woodhull asked 15-year-old Victoria for her hand in marriage.

12. But the marriage was no paradise for Victoria—she soon realized her husband was an alcoholic.

13. She experienced more heartbreak when her son, Byron, was born with a mental disability.

14. While she remained married to Canning, Victoria spent the next few years touring as a clairvoyant with her sister Tennessee.

15. At that time, it was difficult for a woman to pursue divorce, but Victoria finally succeeded in divorcing her husband in 1864.

16. Two years later she married Colonel James Blood, a Civil War veteran who believed in free love.

17. In 1866, Victoria and James moved to New York City.

18. Spiritualism was then in vogue, and Victoria and Tennessee established a salon where they acted as clairvoyants and discussed social and political hypocrisies with their clientele.

19. Among their first customers was Cornelius Vanderbilt, the wealthiest man in America.

20. A close relationship sprang up between Vanderbilt and the two young women.

21. He advised them on business matters and gave them stock tips.

22. When the stock market crashed in September 1869, Woodhull made a bundle buying instead of selling during the ensuing panic.

23. That winter, she and Tennessee opened their own brokerage firm.

24. They were the first female stockbrokers in American history.

25. The sisters did so well that, two years after arriving in New York, Woodhull told a newspaper she had made $700,000.

26. On April 2, 1870, Victoria Woodhull announced that she was running for president.

27. In conjunction with her presidential bid, Woodhull and her sister started a newspaper, *Woodhull & Claflin's Weekly*, which highlighted women's issues including voting and labor rights.

28. It was another breakthrough for the two since they were the first women to ever publish a weekly newspaper.

29. This was followed by another milestone: On January 11, 1871, Woodhull became the first woman ever to speak before a congressional committee.

30. As she spoke before the House Judiciary Committee, she asked that Congress change its stance on whether women could vote.

31. She was not advocating a new constitutional amendment granting women the right to vote.

32. Instead, she reasoned, women already had that right.

33. The Fourteenth Amendment says that, "All persons born or naturalized in the United States...are citizens of the United States."

34. Since voting is part of the definition of being a citizen, Woodhull argued, women, in fact, already possessed the right to vote.

35. Woodhull, a persuasive speaker, actually swayed some congressmen to her point of view.

36. But the committee chairman remained hostile to the idea of women's rights and made sure the issue never came to a floor vote.

37. She had better luck with the suffragists: In May 1872, before 668 delegates from 22 states, Woodhull was chosen as the presidential candidate of the Equal Rights Party.

38. Woodhull was the first woman ever chosen by a political party to run for president.

39. But her presidential bid soon foundered.

40. Woodhull was on record as an advocate of free love, which opponents argued was an attack on the institution of marriage.

41. For Woodhull it had more to do with the right to have a relationship with anyone she wanted.

42. Rather than debate her publicly, her opponents made personal attacks.

43. That same year, Woodhull caused an uproar when her newspaper ran an exposé about the infidelities of Reverend Henry Ward Beecher.

44. Woodhull and her sister were thrown in jail and accused of publishing libel and promoting obscenity.

45. They would spend election night of 1872 behind bars as Ulysses Grant defeated Horace Greeley for the presidency.

46. Woodhull was eventually cleared of the charges against her (the claims against Beecher were proven true), but hefty legal bills and a downturn in the stock market left her embittered and impoverished.

47. She moved to England in 1877, shortly after divorcing Col. Blood.

48. By the turn of the century she had become wealthy once more, this time by marriage to a British banker.

49. Fascinated by technology, she joined the Ladies Automobile Club, where her passion for automobiles led her to one last milestone.

50. In her sixties, she and her daughter Zula became the first women to drive through the English countryside.

36 Facts About Henry VIII

1. Henry VIII (1491–1547) is best known for his six marriages and instigating the English Reformation, in which the Church of England broke from the Catholic Church and papal authority.

2. In defiance of the pope, Henry divorced his first wife Catherine of Aragon to marry one of Catherine's ladies-in-waiting, the charming and sophisticated (and pregnant!) Anne Boleyn.

3. To marry Anne in 1533, Henry had to contravene the Catholic Church's prohibition against divorce and face excommunication.

4. The Church of England—a.k.a. the Anglican faith—was the phoenix that ultimately rose from the ashes.

5. Although some accounts of Henry VIII's break with Rome portray the split as "King Marries Temptress, Starts Own Church," the historical reality is slightly more complex.

6. On one hand, Henry was undeniably enraptured with Anne.

7. Seventeen of his love letters to her—nine of them in French— can be perused today at the Vatican Library in Rome.

8. Henry tells his mistress in one missive that "for more than a year, [I've been] struck with the dart of love."

9. Elsewhere he implores her to give "body and heart to me, who will be, and has been, your most loyal servant."

10. On the other hand, Henry's extramarital love life wasn't happening in a historical vacuum.

11. In 1529, what historians now call "the Reformation Parliament" clapped economic and political shackles upon the operation of the Catholic Church in England.

12. They regulated excessive fees the church levied for burials, reduced clergy's ability to make money on the side, and eliminated their de facto above-the-law status by subjecting church officials to the same secular courts as English commoners.

13. Two reformist members of Henry's court in the early 1530s compounded what momentum Parliament had established.

14. The first was Thomas Cranmer, made Archbishop of Canterbury, the highest ecclesiastical position in the nation, in 1533.

15. The second was Thomas Cromwell, Henry VIII's chief advisor and principal minister.

16. The two Thomases provided the political and theological muscle to establish an independent church body and, with it, put the squeeze on the Catholic Church in England.

17. In 1533, Cranmer ensured swift approval by his bishops of the annulment of King Henry's first marriage.

18. Cromwell's Act in Restraint of Appeals effectively set forth the English monarch as ruler of the British Empire, rendering the papacy irrelevant by deriving a monarch's authority directly from God.

19. The pope, thanks to Cromwell's legislative strong-arm, suddenly had no claim over true English subjects.

20. After three months of threats, Pope Clement VII excommunicated Henry VIII from the Catholic Church in September 1533.

21. Though true theological reform of the fledgling Anglican Church into a bona fide Protestant sect would have to wait, the Church of England had been born.

22. Anne Boleyn lasted all of three years as her king's wife.

23. Eventually Henry, in concert with Cromwell, rounded up four courtiers and servants to Boleyn who were all variously charged with luring Boleyn away from the marriage bed.

24. Boleyn was swiftly tried for adultery and beheaded in May 1536.

25. Cromwell lasted only four more years before he too became a victim to the capricious whims of his monarch.

26. Cromwell eventually faced a rigged jury, and was found guilty of treason.

27. Cromwell's "treason" is unclear, though the king's wrath is thought to have been caused by his disastrous fourth marriage to Anne of Cleves, which Cromwell proposed.

28. At the king's behest, Cromwell faced the brutality of an inexperienced executioner.

29. Three unsuccessful ax blows drew out Cromwell's pain before a fourth finally severed his head.

30. After his death, Cromwell's head was then boiled and placed on a pike on London Bridge.

31. Heads were placed on spikes for all to see; they served as a warning for anyone thinking of challenging the crown.

32. Cranmer outlived his king and his king's son, only to face the furies of the brutally anti-Protestant queen "Bloody" Mary Tudor (reigned 1553–58).

33. Yet perhaps it was Anne Boleyn herself who bequeathed the

greatest legacy in the form of her daughter, who eventually became Queen Elizabeth I.

34. Queen Elizabeth I took the reins of power after Queen Mary's death in 1558.

35. The frail young princess, not expected to survive any longer than her siblings, instead enjoyed a prosperous 45 years on the throne as perhaps England's greatest monarch.

36. Under Elizabeth, the Church of England established itself as the ecclesiastical force that underpinned the rise of a true global empire.

38 Marilyn Monroe Facts

1. Marilyn Monroe was born in Los Angeles on June 1, 1926.

2. The name on her birth certificate was Norma Jean Mortenson, but she was baptized Norma Jean Baker, with her mother's last name, because the identity of her father was considered undetermined.

3. Monroe's mother suffered from bipolar disorder and spent much of her life in mental institutions.

4. As a result of her mother's illness, Marilyn spent two years of her childhood at the Los Angeles Orphans Home.

5. She also lived in many foster homes as a child.

6. Marilyn Monroe was married three times.

7. In 1942, she married 21-year-old Jim Dougherty.

8. The two had only dated for six months before their wedding, but Marilyn was anxious to escape her unhappy childhood, and her foster parents encouraged her to get married.

9. Monroe and Dougherty divorced in 1946 when Marilyn began to focus on her modeling career.

10. In 1954, she married Joe DiMaggio after two years of dating.

11. They divorced nine months later.

12. Monroe married playwright Arthur Miller in 1956.

13. They stayed together until 1961.

14. Monroe signed her first acting contract in August 1946.

15. This first deal with 20th Century Fox paid her $125 a week.

16. By the time she was a teenager, Marilyn Monroe had very dark blonde hair, which looked brunette in black-and-white photos.

17. While working as a model in her late teens, she gradually lightened her hair.

18. The change in hair color got Monroe more modeling jobs.

19. Monroe has been featured in numerous tributes.

20. Elton John's song "Candle in the Wind" was written in her honor.

21. Her image also graced a commemorative U.S. postage stamp.

22. Monroe won the Golden Globe for Best Motion Picture Actress for her role in *Some Like It Hot* (1959).

23. She was also named the female World Film Favorite at the Golden Globes in 1954 and again in 1962, but she was never nominated for an Oscar.

24. Contrary to her on-screen persona, Marilyn Monroe loved to read Tolstoy and other classic writers.

25. In fact, while she was under contract at Fox, she was enrolled at UCLA, studying literature.

26. *Playboy* magazine has shown plenty of love for MM over the years.

27. The men's magazine first featured Monroe in 1953 as "Sweetheart of the Month" in its debut issue.

28. Hugh Hefner reportedly paid $500 for the rights to her nude photo—a shot taken seven years prior for which Monroe herself had only received $50.

29. Fed up with being typecast as the dumb blonde, Monroe defied 20th Century Fox at the height of her popularity and moved to New York City to study with legendary teacher Lee Strasberg at the Actors Studio.

30. Though she used the name Marilyn Monroe as early as 1946, she didn't legally take on the moniker until a decade later.

31. Monroe was her mother's maiden name.

32. Hollywood folklore says that during the production of *Some Like It Hot* (1959) Monroe's notorious habit of being late or not showing up escalated and that when she did show up, she often needed dozens of takes to get even the simplest of lines correct.

33. Few realize that part of the reason for her behavior was due to a miscarriage that occurred during the film's shooting schedule.

34. Monroe was battling a profound depression, and was relying on prescription drugs to help her sleep and to wake her up.

35. Marilyn Monroe completed 29 films during her career.

36. She was working on her last—*Something's Got to Give*—when she died on August 5, 1962.

37. The film was never released.

38. Marilyn Monroe is buried at the Corridor of Memories, marker #24, at Westwood Memorial Park in Los Angeles.

18 People in Line for the Presidency

1. Vice President

2. Speaker of the House of Representatives

3. President Pro Tempore of the Senate

4. Secretary of State

5. Secretary of the Treasury

6. Secretary of Defense

7. Attorney General

8. Secretary of the Interior

9. Secretary of Agriculture

10. Secretary of Commerce

11. Secretary of Labor

12. Secretary of Health and Human Services

13. Secretary of Housing and Urban Development

14. Secretary of Transportation

15. Secretary of Energy

16. Secretary of Education

17. Secretary of Veterans Affairs

18. Secretary of Homeland Security

60 Facts About Killer Queens

Boudicca

1. Boudicca stood six feet tall, sported a hip-length mane of fiery red hair, and had a vengeful streak a mile wide.

2. She was queen of the Celtic Iceni people of eastern Britain.

3. In A.D. 61, she led a furious uprising against the occupying Romans that nearly chased Nero's legions from the island.

4. Boudicca didn't always hate the Romans.

5. The Iceni kingdom, led by her husband Prasutagus, was once a Roman ally.

6. But Prasutagus lived a life of conspicuous wealth on borrowed Roman money.

7. When it came time to pay, he was forced to bequeath half his kingdom to the Romans; the other half was left for his daughters.

8. The Romans, however, got greedy.

9. On Prasutagus's death, they moved to seize all Iceni lands as payment for the dead king's debt.

10. When the widow Boudicca challenged the Romans, they publicly flogged her and raped her daughters.

11. While most of the Roman army in Britain was busy annihilating the druids in the west, Boudicca led the Iceni and other aggrieved Celtic peoples on a bloody rebellion that reverberated back to Rome.

12. Boudicca's warriors annihilated the vaunted Roman Ninth Legion and laid waste to the Roman cities of Camulodunum (Colchester), Londinium (London), and Verulamium (St. Albans).

13. Boudicca's vengeance knew no bounds and was exacted on both Romans and fellow Britons who supported them.

14. Upwards of 80,000 people fell victim to her wrath.

15. The Romans were floored by the ferocity of Boudicca's attack.

16. Nero actually considered withdrawing his army from Britain.

17. But the Romans regrouped, and later a seasoned force of 1,200 legionnaires trounced Boudicca's 100,000-strong rebel army in a decisive battle.

18. The defeated Boudicca chose suicide by poison over capture.

19. Today, a statue of the great Boudicca can be found near Westminster Pier in London, testament to the veneration the British still hold for Boudicca as their first heroine.

Empress Jingo

20. Our next killer queen is Empress Jingo, who led the Japanese in the conquest of Korea in the early third century.

21. In 1881, she became the first woman to be featured on a Japanese banknote—no small feat given the chauvinism of imperial Japan.

22. More than 1,700 years after her rule, Empress Jingo is still revered as a goddess in Japan.

23. Jingo's success as a warrior queen may be attributed to the irresistible sway she held over both the ancient deities and mortal men.

24. As regent ruler of imperial Japan following the death of her emperor husband Chuai, Jingo was determined to make Korea her own.

25. According to Japanese lore, she beguiled Ryujin, the Japanese dragon god of the sea, to lend her his magical Tide Jewels.

26. She then used the Tide Jewels to create favorable tides that destroyed the Korean fleet and safely guided the Japanese fleet to the Korean peninsula.

27. From there, she commanded and cajoled her armies to an illustrious campaign of conquest that secured her exalted status within Japanese cultural history.

28. Jingo purportedly had amazing powers of persuasion over the human reproductive cycle as well.

29. Pregnant with Chuai's son at the time of the invasion, Jingo remained in Korea for the duration of the campaign, which by all accounts lasted well beyond the length of a normal term.

30. Legend has it, however, that she delayed giving birth until after the conquest so that her son and heir, Ojin, could be born in Japan.

31. Once home, Jingo cemented her power by using brute force to convince several rivals to the throne to concede to her rule, which would ultimately last for more than 60 years.

Zenobia

32. Like Boudicca before her, our third warrior queen Zenobia made her name by leading an army against the mighty Romans.

33. Unlike the Celtic queen, however, Zenobia would experience a much different fate for her actions.

34. Zenobia and her husband Odenathus ruled the prosperous Syrian city of Palmyra.

35. Though technically subordinate to Odenathus, she certainly didn't take a backseat to him.

36. She established herself as a warrior queen by riding at her husband's side into battle against the Persians—often overshadowing her more reserved mate.

37. Zenobia became the undisputed ruler of Palmyra in 267

following the assassination of Odenathus (which some attribute to Zenobia herself).

38. As an ostensible ally of Rome, Zenobia launched a campaign of conquest in the Middle East.

39. Within three years she expanded her realm to Syria, Egypt, and much of Asia Minor.

40. Flushed with success, Zenobia declared Palmyra's independence from Rome.

41. But in 272, the Romans struck back.

42. Zenobia was up for the fight, but her forces were overextended.

43. The Romans easily recaptured Zenobia's outlying territories before laying siege to Palmyra itself.

44. After its fall, Palmyra was destroyed, and Zenobia was captured.

45. She was taken to Rome and paraded in golden chains before Emperor Aurelian.

46. But even in defeat, Zenobia triumphed.

47. The striking beauty with the defiant stride struck a chord with Aurelian, who later pardoned her and allowed her to live a life of luxury on an estate outside Rome.

Aethelflaed

48. Aethelflaed, the "Noble Beauty" of the Anglo-Saxons, is our fourth and final killer queen.

49. At the beginning of the 10th century, the Anglo kingdom of Wessex and Saxon kingdom of Mercia in southern England were under siege by the Danish Vikings.

50. The cocksure Vikings were confident of victory, but they hadn't counted on the rise of the Mercian queen, Aethelflaed, who would earn her warrior reputation by leading armies in victory over the Vikings and emerging as one of Britain's most powerful rulers.

51. Aethelflaed's father, Alfred the Great, was king of Wessex.

52. Aethelflaed, at age 15, married the Mercian nobleman, Ethelred, thus forming a strategic alliance of the two kingdoms against the Vikings.

53. Her first fight against the Vikings occurred on her wedding day when the Norsemen tried to kill her to prevent the union.

54. Aethelflaed took up the sword and fought alongside her guards while holed up in an old trench, eventually driving the Vikings away.

55. From then on, battling Vikings became old hat for Aethelflaed.

56. When her husband died in 911, she assumed sole rule of Mercia and began taking the fight to the Vikings.

57. Perhaps remembering her wedding-day experience in the trench, she built formidable fortifications to defend Mercia.

58. She also used exceptional diplomatic skills to form alliances against the Vikings.

59. By the time of her death in 918, she had led armies in several victories over the Vikings, had them begging for peace, and had extended her power in Britain.

60. Aethelflaed made her name as a Viking killer, but her most important legacy was her success in sustaining the union of the Angles and the Saxons, which would later germinate into the English nation.

45 Julius Caesar Facts

1. Caesar's birth in 100 B.C. (sometimes listed as 102 B.C.) coincided with great civil strife in Rome.

2. Although his parents' status as nobles gave him advantages, Caesar's childhood was spent in a politically volatile Rome marked by personal hatreds and conniving.

3. As an adult, he learned to be wary in his dealings with other powerful people.

4. By the time Caesar was 20, a patrician named Sulla had been the Roman dictator for about 20 years.

5. Although Caesar and Sulla were friends, Sulla later became enraged when Caesar refused to divorce his wife who was the daughter of a man Sulla loathed (and murdered).

6. In order to save his neck, Caesar promptly left Rome for Asia.

7. When Sulla died in 78 B.C., Caesar returned to Rome and took up the practice of law.

8. Caesar had everything necessary for success: the best education, impressive oratorical skills, and outstanding writing abilities.

9. He also spent huge sums of money, most of which he had to borrow.

10. The money went to bribes and sumptuous parties for the influential and bought Caesar access to power.

11. Leading politicians rewarded Caesar with a series of increasingly important political positions in Spain and Rome.

12. Caesar's time in Spain was especially useful, as he used his position there to become very wealthy.

13. In 59 B.C., Caesar, by now a general, made a successful bid for power in concert with two others: Marcus Licinius Crassus and Pompey.

14. Marcus Licinius Crassus was the richest man in Rome.

15. Pompey, another ambitious general was known, to his immodest pleasure, as Pompey the Great.

16. These three Type-A personalities ruled Rome as the First Triumvirate, with Caesar becoming first among equals as consul.

17. Caesar had always been popular among the common people and with Rome's soldiers.

18. He aimed to cement that loyalty with reforms that would benefit them.

19. Soon, Caesar was made governor of Gaul and spent the next

11 years conquering all of what is now France, with a couple of profitable trips to Britain for good measure.

20. While on campaign, he wrote an account of his actions, called *Commentaries*, which is among the finest of all military literature.

21. To leave Rome, even for military glory, was always risky for any of the empire's leaders.

22. While Caesar was abroad, Crassus was killed in battle.

23. This void encouraged Pompey, who made it clear that Caesar was no longer welcome in Rome.

24. Caesar and his army responded by crossing the Rubicon River in 49 B.C. to seize control of the city.

25. Within a year of the civil war that followed, Caesar defeated Pompey.

26. He also began a torrid affair with Egypt's Queen Cleopatra.

27. After a few other actions against Rome's enemies, Caesar was acclaimed by all of Rome as a great hero.

28. In turn, he pardoned all who had opposed him.

29. Mindful of the fleeting nature of popularity, Caesar continued to promote a series of important reforms.

30. Some of the land that had been held by wealthy families was distributed to common people desperate to make a living.

31. As one might expect, this didn't go over well with the wealthy.

32. Tax reforms Caesar insisted upon forced the rich to pay their fair share—this innovation didn't win Caesar many new friends among the powerful.

33. Retired soldiers were settled on land provided by the government.

34. Because this land was in Rome's outlying territories, it became populated with a happy, well-trained cadre of veterans meant to be Rome's first line of defense, if needed.

35. Unemployed citizens were also given the opportunity to settle in these areas, where jobs were much more plentiful.

36. This reduced the number of poor people in Rome and decreased the crime rate.

37. As he had done earlier, Caesar made residents of the provinces, such as people living in Spain, citizens of Rome.

38. Many years later, some of the Roman emperors actually came from Spain.

39. In a clever move, Caesar instituted a massive public works program that provided both jobs and a sense of pride among the citizens of Rome.

40. All these reforms notwithstanding, Caesar's enemies feared he would leverage his great popularity to destroy the Roman Republic and institute in its place an empire ruled by one man.

41. So, in a moment of violence that turned the wheel of history, Caesar was assassinated by people he trusted on March 15, 44 B.C.—the Ides of March, for those of you who remember your Shakespeare.

42. The civil war that followed was ultimately won by Caesar's nephew, Octavian, who changed his name to Caesar Augustus.

43. He replaced the republic and instituted in its place an empire ruled by one man!

44. Augustus was the first of a long succession of emperors who ruled virtually independent of the Roman Senate.

45. It was the rulers who followed Julius Caesar, not Caesar himself, who proved the undoing of the system Caesar's enemies so cherished.

16 Facts About Suleiman the Magnificent

1. Suleiman the Magnificent was a warrior-scholar who lived up to his billing.

2. A Turkish sultan who reigned from 1520–66, Suleiman led the Ottoman Empire to its greatest heights.

3. Born in 1494, he was only 26 years old when he became emperor.

4. Not only was Suleiman a brilliant military strategist, he was also a great legislator, a fair ruler, and a devotee of the arts.

5. During his rule, he expanded the country's military empire and brought cultural and architectural projects to new heights.

6. For all this and more, Suleiman is considered one of the finest leaders of 16th-century Europe.

7. His reign is often referred to as the "Golden Age."

8. Under Suleiman's leadership, his forces conquered Mesopotamia (now Iraq), fending off the Safavid's Iran.

9. The Ottomans would then successfully occupy Iraq until the First World War.

10. Suleiman annexed or made allies of the Barbary Pirate states of North Africa, who remained a thorn in Europe's underbelly until the 1800s.

11. He also led an army that went deep into Europe itself, crushing the Hungarian King Louis II at the great Battle of Mohács in 1526, which led to the Siege of Vienna.

12. An accomplished poet, Suleiman was gracious in victory, saying of the young Louis: "It was not my wish that he should be thus cut off while he scarcely tasted the sweets of life and royalty."

13. To his favorite wife Hurrem, he wrote: "My springtime, my merry faced love, my daytime, my sweetheart, laughing leaf...My woman of the beautiful hair, my love of the slanted brow, my love of eyes full of mischief..."

14. While Shari'ah, or sacred law, ruled his far-flung land's religious life, Suleiman reformed the Ottomans' civil law code.

15. In fact, the Ottomans called him Kanuni, or "The Lawgiver."

16. The final form of Suleiman's legal code would remain in place for more than 300 years.

34 Catherine de Medici Facts

1. Speculation still swirls around the life of Italian queen Catherine de Medici (1519–89).

2. Catherine was the wife of Henri II, but she spent her entire married life overlooked in favor of Henri's lifelong mistress, Diane de Poiters, the Duchesse de Valentinois.

3. Even at her own coronation, Catherine had to tolerate excessive attention to Diane's daughters.

4. Diane herself was publicly honored as well.

5. Henri and Catherine didn't have children for the first ten years of their marriage.

6. At the insistence of Diane, the couple finally consummated their marriage and eventually had ten children.

7. Catherine supposedly saw a vision of Henri's death several days before it happened.

8. She begged him not to joust at the tournament celebrating their 13-year-old daughter Elisabeth's marriage to Philip II of Spain.

9. He did so anyway and died from an infection in his eye caused by a splinter from a broken lance.

10. Catherine wore mourning clothes for the rest of her life.

11. She also took a broken lance as her emblem, bearing the motto "From this come my tears and my pain."

12. Among Catherine's ten children were three kings of France (Francois II, Charles IX, and Henri III, respectively) and two queens: Margaret, called Margot, married Henri of Navarre, and Elisabeth became Queen of Spain.

13. Francois II, Catherine's eldest son, married Mary, Queen of Scots.

14. Despite her reputation as a persecutor of Protestants, Catherine actually tried to compromise between the Catholic and Protestant factions.

15. She made concessions to Protestants, allowing them to worship their own way in private, but war broke out nonetheless.

16. Though she is often blamed for the St. Bartholomew's Day Massacre, it is not known what role Catherine played in the disaster.

17. Some historians believe she intended it only to be a culling of Protestant nobles who had been leaders against her in the religious wars, but the situation got out of control.

18. As was common at the time, Catherine married at age 14.

19. The match was arranged by her uncle, Pope Clement VII, and Henri's father, King Francois I.

20. The daughter of Lorenzo the "Magnificent" and a French princess, Catherine was orphaned as an infant.

21. Her mother died 15 days after Catherine's birth, and her father died six days later.

22. The Medicis were overthrown in Florence when Catherine was eight years old.

23. She was taken hostage and moved from convent to convent around the city.

24. Catherine was often threatened with death or with life in a brothel to ruin her value as a bride.

25. The Medici family, though very wealthy, was a merchant family rather than nobility.

26. Catherine was reviled by many of the French because she was thought to be a commoner and, therefore, in their eyes, unfit to be queen of France.

27. Catherine had a great influence on fashion.

28. As a new bride of short stature, she was desperate to make a good show at her introduction to the French court.

29. She allegedly wore specially made high-heeled shoes to her grand entrance, impressing the fashion-obsessed nobles and sparking the style that remains popular today.

30. Once out from under Henri's and Diane's influence, Catherine spread her wings and began to exercise control over her children. In fact, she dominated them.

31. She ruled France as regent during the minority of Charles IX (and continued to rule during his adulthood, though as the king's "advisor" rather than regent).

32. Henri III relied heavily on her throughout his reign.

33. Tragically, Catherine's large brood turned out to be sickly, and she outlived all but two of her ten children.

34. Three had died in infancy, and only Henri and Margot were still alive when she died.

71 Presidential Peculiarities

1. President William Taft had a new bathtub installed in the White House that could hold four grown men.

2. Taft's 300-pound frame wouldn't fit in the original presidential tub.

3. President John Quincy Adams started each summer day with an early-morning skinny-dip in the Potomac River.

4. Grover Cleveland was the first and only president to marry in the White House itself.

5. He wed Frances Folsom in 1886.

6. With 15 children from two marriages, President John Tyler was the most prolific chief executive.

7. Martin Van Buren was characterized as a "dandy" and was known to have a weakness for rich foods, fine wine, and clothing.

8. The only president never to have won a national election was Gerald R. Ford, who took office after President Richard Nixon resigned.

9. William Henry Harrison was the first president to die in office.

10. During his lengthy inaugural speech (more than two hours long!) he contracted a cold that quickly developed into pneumonia.

11. Gerald Ford played football for the University of Michigan from 1931 to 1934.

12. He was offered tryouts by both the Detroit Lions and Green Bay Packers.

13. President Jimmy Carter could speed-read 2,000 words per minute.

14. Franklin Delano Roosevelt was related to 11 other presidents, either directly or through marriage.

15. There hasn't been a bearded president since Benjamin Harrison, who left office in 1893.

16. James Madison was the smallest president, weighing only 100 pounds and standing just 5'4".

17. When George Washington was elected, America's first president had only one tooth.

18. He wore dentures made from hippopotamus and elephant ivory, not wood as commonly thought.

19. Edwin Booth, the brother of Abraham Lincoln's assassin John Wilkes Booth, once saved the life of President Lincoln's son Robert.

20. He kept the boy from falling off a train platform.

21. After a long feud between Thomas Jefferson and John Adams, the two former presidents finally called a truce and developed a friendship that lasted the rest of their lives.

22. Both men died on July 4, 1826—the 50th anniversary of the adoption of the Declaration of Independence.

23. Lyndon B. Johnson and his wife, Lady Bird, were married with a $2.50 wedding ring bought at Sears the day after he proposed to her.

24. Abraham Lincoln was the only president to receive a patent.

25. It was for a device designed to lift boats over shoals.

26. Andrew Johnson never received any formal schooling; he credited his wife with teaching him to read and write.

27. Ronald Reagan was the only president to have worn a Nazi uniform.

28. He donned it as an actor in the 1942 film *Desperate Journey*.

29. Richard Nixon's mother wanted her son to become a Quaker missionary.

30. Nixon's true desire was to be an FBI agent.

31. He applied to the Bureau but wasn't accepted.

32. Woodrow Wilson played golf for exercise, even in winter.

33. He had his golf balls painted red so he could see them in the snow.

34. In his earlier days, Grover Cleveland served as a New York sheriff and carried out at least two hangings, refusing to delegate the unpleasant task to others.

35. After being diagnosed with mouth cancer in 1893, Cleveland had a secret operation on a yacht to remove part of his upper jaw.

36. He was later fitted with a rubber prosthesis to fill the hole.

37. George Washington never shook hands with visitors, choosing to bow instead.

38. Two decades after leaving the White House, John Tyler joined the Confederacy and became the only president named a sworn enemy of the United States.

39. During his presidency, Franklin Pierce allegedly ran over a woman with his horse.

40. He was arrested, but the case was dropped due to lack of evidence.

41. Warren G. Harding suffered his first nervous breakdown at age 24 and spent time in a sanitarium run by J. H. Kellogg of breakfast cereal fame.

42. Having never held an elected office before becoming president, Herbert Hoover was the first self-made millionaire to reside in the White House after earning his fortune in the mining industry.

43. After his election, Chester A. Arthur sold more than two dozen wagons full of White House furniture.

44. Some of the items dated back to the John Adams administration.

45. Arthur commissioned Louis Comfort Tiffany as designer for his White House makeover.

46. Ulysses S. Grant was once cast in the role of Desdemona in an all-soldier production of *Othello* during the Mexican-American War.

47. Benjamin Harrison was the first president to use electricity in the White House.

48. But after getting an electrical shock, he refused to touch light switches, and he and his family would often leave the lights on all night.

49. In 1791, Andrew Jackson married Rachel Donelson, believing that her former husband had applied for a divorce.

50. A few years later, they were informed that no divorce had ever been sought, and the couple was being charged with adultery.

51. Once the situation was remedied, Donelson and Jackson quietly remarried.

52. Though it was after his term in office, Theodore Roosevelt was the first president to ride in an airplane.

53. In 1910, he flew for three minutes and twenty seconds in a plane built by the Wright Brothers.

54. The pilot of that flight, Arch Hoxsey, crashed and died two months later.

55. James Polk was the first president ever to have his photograph taken.

56. In 1910, William H. Taft was the first president to throw the first baseball of a season and the first president to own a car while in office.

57. In 1930, Taft's funeral was also the first presidential funeral to be broadcast on the radio.

58. Harry Truman suffered from bad eyesight, which kept him from attending West Point.

59. When World War I broke out, he passed his vision test by secretly memorizing the eye chart.

60. John F. Kennedy was the first Catholic president.

61. He was also the first president to have been a Boy Scout.

62. At age 19, George H. W. Bush became the youngest pilot in the U.S. Navy's history.

63. He went on to fly 58 combat missions during World War II and was shot down in 1944.

64. Early in his presidency, Bill Clinton suffered from a series of allergies including cat dander, beef, milk, mold spores, and weed and grass pollens.

65. He took injections to control the allergies, which allowed him to partake in a favorite food at the time—cheeseburgers.

66. George W. Bush had some of the highest and lowest approval ratings ever recorded for a president.

67. In October 2001, he held an 88 percent positive rating—the highest in the Harris Poll's 40-year history.

68. However, he also had one of the lowest positive ratings, 28 percent in April 2007.

69. Only Jimmy Carter (22 percent in 1980) and Richard Nixon (26 percent in 1974) ever scored lower.

70. Franklin Delano Roosevelt was an avid poker player.

71. During one of his Fireside Chats, he absentmindedly shuffled some poker chips, rendering portions of his speech inaudible.

41 Facts About Alfred Hitchcock

1. Born in London in 1899, Alfred Joseph Hitchcock became the most recognizable director in Hollywood history, not only for his commanding presence but also in regard to his cinematic style.

2. Hitchcock's films were largely suspense-filled, psychological

thrillers with gallows humor that attracted audiences with their subject matter and dark visual imagery.

3. He had a gift for transforming familiar characters and ordinary-looking locations into frightening stories about the moral failings that lurk in all of us and the evil hidden in the everyday world.

4. The son of a greengrocer in a working-class London neighborhood, Alfred was the only child of strict parents.

5. When Hitch left school, he pursued a career as a draftsman, but became increasingly interested in movies.

6. In 1920, he got a job doing title cards (the intertitles that contain the dialogue and text between the scenes of a silent film) for a Hollywood studio called Famous Players–Lasky, which had an office in London.

7. It was the beginning of his illustrious film career.

8. Hitch put in time doing art direction work and quickly climbed the ranks at the studio, becoming an assistant director in 1922.

9. After some uncredited work, he got his first directorial assignment with *The Pleasure Garden* (1925).

10. Sadly, none of the initial projects embarked upon by the fledgling director took off.

11. It wasn't until 1927 that he had a hit—*The Lodger: A Story of the London Fog* was a commercial success and put Hitch on the map.

12. Two years later, he directed Britain's first "talkie," a thriller called *Blackmail*.

13. The story of a woman who suffers pangs of guilt for killing a would-be rapist, *Blackmail* foreshadowed Hitch's mature style.

14. In 1934, Hitchcock garnered international attention with *The Man Who Knew Too Much*, followed by *The 39 Steps* the next year.

15. His approach to the thriller was established by this time, and he became adept at using a plot device he liked to call the MacGuffin.

16. The MacGuffin was the thing in the story that the characters are

concerned with (important papers, secret microfilm, uranium) but the audience doesn't really care about because they are wrapped up in the suspense and the motivations of the characters.

17. By the end of the 1930s, the rotund English director had made quite the name for himself as a significant auteur. It was time to take on Hollywood.

18. In 1939, mega-producer David O. Selznick offered Hitchcock a deal.

19. Selznick had just made cinematic history producing the record-breaking, Oscar-winning epic *Gone with the Wind*.

20. It gave the producer even more power, so when he offered Hitch a seven-year contract, it was clear that Selznick would have a significant amount of control over the director's work, which was par for the course in Hollywood at the time.

21. It was not an arrangement that sat well with Hitch, but he agreed to the terms; he finally had Hollywood money with which to make his movies.

22. The first picture of the partnership was *Rebecca* (1940), a Gothic melodrama that starred Sir Laurence Olivier and Joan Fontaine.

23. The movie was a critical and commercial hit, and when Oscar time rolled around, *Rebecca* won Best Picture.

24. Hitch didn't win for Best Director, but it secured his reputation in Hollywood.

25. Throughout the 1940s, Hitch worked tirelessly, making movies that used familiar filmmaking techniques and typical conventions.

26. Yet with his exquisite craftsmanship, the films were like works of art.

27. He also toyed with audience expectations in regard to the casting, using leading man Cary Grant in a sinister role in *Suspicion* (1941), for example.

28. Classic Hitch titles such as *Lifeboat* (1944) and—the director's personal favorite—*Shadow of a Doubt* (1943) were made during this time, complete with his expressive use of light and shadows, carefully worked-out compositions, and oblique angles.

29. But if Hitch had a golden age, it was the 1950s.

30. The list of films he made reads like a "best of" list: *Dial M for Murder* (1954), *Rear Window* (1954), *To Catch a Thief* (1955), a remake of *The Man Who Knew Too Much* (1956), and *Vertigo* (1958).

31. Hitch worked multiple times with Jimmy Stewart and Cary Grant as well as leading lady Grace Kelly.

32. His favorite female archetype was the cool, sophisticated blonde, whom he often used as the lead character and even the protagonist.

33. These films featured his key themes: the presence of evil in the everyday world, the deceptive nature of appearances, and the idea that we are all guilty of something.

34. His most famous plotline—the story of the falsely accused man— was epitomized by his film *North by Northwest* (1959).

35. By the 1950s, color film was the norm.

36. Hitchcock used it to his advantage when it suited him, but he still had a fondness for the expressive nature of black and white, evidenced by one of his major masterpieces: *Psycho* (1960).

37. Considered his most famous films, *Psycho* and *The Birds* (1963) mixed suspense with anxiety-inducing soundtracks by legendary composer Bernard Herrmann.

38. *The Birds* chronicles an infestation of avian creatures that are terrorizing a California town, while *Psycho* offers a warning about the darkness that exists in all of us through the character of Norman Bates.

39. As Norman says, "We all go a little mad sometimes...haven't you?"

40. Hitch made two more significant works, *Marnie* (1964) and *Frenzy* (1972), but when his health started to decline, his output began to suffer.

41. In 1976, the undisputed "Master of Suspense" made his last film, *Family Plot*, before dying of kidney failure in 1980.

11 Hitchcock Quotations

1. "What is drama but life with the dull bits cut out?"

2. "There is no terror in a bang, only in the anticipation of it."

3. "Our evil and our good are getting closer together today."

4. "People always think villains are extraordinary, but in my experience they are usually rather ordinary and boring."

5. "I never look through the camera; I think only of that white screen that has to be filled up the way you fill up a canvas. That's why I draw rough setups for the cameraman."

6. "As far as I'm concerned, the film has been made on paper, that's the most important and fascinating stage . . . I wish I didn't have to go into a studio."

7. When a parent complained to the director that his daughter refused to take a shower after seeing *Psycho* (1960), Hitchcock replied, "Have her dry cleaned."

8. "In films, murders are always very clean. I show how difficult it is and what a messy thing it is to kill a man."

9. "Television has brought back murder into the home—where it belongs."

10. "I never said all actors are cattle, what I said was all actors should be treated like cattle."

11. "When an actor comes to me and wants to discuss his character, I say, 'It's in the script.' If he says, 'But what's my motivation?', I say, 'Your salary.'"

67 Famous People Who Died Young

1. Jessica Dubroff (7); Pilot; Plane crash;1996

2. Heather O'Rourke (12); Child actor; Bowel obstruction; 1988

3. Anne Frank (15); Dutch-Jewish author; Typhus in concentration camp; 1945

4. Ritchie Valens (17); Rock 'n' roll singer; Plane crash;1959

5. Eddie Cochran (21); Rockabilly musician; Auto accident; 1960

6. Sid Vicious (21); Punk musician; Heroin overdose; 1979

7. Aaliyah (22); R&B singer; Plane crash; 2001

8. Buddy Holly (22); Rock 'n' roll singer; Plane crash; 1959

9. Freddie Prinze (22); Comedian/actor; Suicide; 1977

10. River Phoenix (23); Actor; Drug overdose; 1993

11. Selena (23); Mexican-American singer; Homicide; 1995

12. James Dean (24); Actor; Auto accident; 1955

13. The Notorious B.I.G. (24); Rapper; Homicide; 1997

14. Tupac Shakur (25); Rapper; Homicide; 1996

15. Otis Redding (26); Soul singer; Plane crash; 1967

16. Brian Jones (27); Rock guitarist; Drug-related drowning; 1969

17. Janis Joplin (27); Rock/soul singer; Heroin overdose; 1970

18. Jim Morrison (27); Rock singer; Heart attack, possibly due to drug overdose; 1971

19. Amy Winehouse (27); Singer; Accidental alcohol poisoning; 2011

20. Jimi Hendrix (27); Rock guitarist/singer; Asphyxiation from sleeping pill overdose; 1970

21. Kurt Cobain (27); Grunge rock singer/guitarist; Gunshot and lethal dose of heroin, presumed suicide; 1994

22. Reggie Lewis (27); Basketball player; Heart attack; 1993

23. Brandon Lee (28); Actor; Accidental shooting on the set of *The Crow*; 1993

24. Shannon Hoon (28); Rock singer; Drug overdose; 1995

25. Heath Ledger (28); Actor; Accidental medication overdose; 2008

26. Hank Williams (29); Country musician; Heart attack, possibly due to an accidental overdose of morphine and alcohol; 1953

27. Andy Gibb (30); Singer; Heart failure due to cocaine abuse; 1988

28. Jim Croce (30); Singer/songwriter; Plane crash; 1973

29. Patsy Cline (30); Country music singer; Plane crash; 1963

30. Sylvia Plath (30); Poet and author; Suicide; 1963

31. Brian Epstein (32); Beatles manager; Drug overdose; 1967

32. Bruce Lee (32); Martial arts actor; Possible allergic reaction; 1973

33. Cass Elliot (32); Singer; Heart failure; 1974

34. Karen Carpenter (32); Singer and musician; Cardiac arrest from anorexia nervosa; 1983

35. Brittany Murphy (32); Actor; Pneumonia and lack of iron, though toxic mold may have been the culprit; 2009

36. Keith Moon (32); Rock drummer; Overdose of medication; 1978

37. Carole Lombard (33); Actor; Plane crash; 1942

38. Chris Farley (33); Comedian/actor; Cocaine and heroin overdose; 1997

39. Darryl Kile (33); Baseball pitcher; Coronary heart disease; 2002

40. Jesus Christ (33); Founder of Christianity; Crucifixion; A.D. 30

41. John Belushi (33); Comedian/actor; Cocaine and heroin overdose; 1982

42. Sam Cooke (33); Soul musician; Homicide; 1964

43. Charlie Parker (34); Jazz saxophonist; Pneumonia and ulcer, brought on by drug abuse; 1955

44. Dana Plato (34); Actor; Prescription drug overdose; 1999

45. Jayne Mansfield (34); Actor; Auto accident; 1967

46. Andy Kaufman (35); Comedian/actor; Lung cancer; 1984

47. Josh Gibson (35); Negro League baseball player; Stroke; 1947

48. Stevie Ray Vaughan (35); Blues guitarist; Helicopter crash; 1990

49. Bob Marley (36); Reggae musician; Melanoma that metastasized into lung and brain cancer; 1981

50. Diana, Princess of Wales (36); British royal; Auto accident; 1997

51. Marilyn Monroe (36); Actor; Barbiturate overdose; 1962

52. Bobby Darin (37); Singer/actor; Complications during heart surgery; 1973

53. Lou Gehrig (37); Baseball player; Amyotrophic lateral sclerosis (ALS); 1941

54. Michael Hutchence (37); Rock singer; Suicide by hanging; 1997

55. Sal Mineo (37); Actor; Homicide; 1976

56. Florence Griffith Joyner (38); Olympian/sprinter; Epileptic seizure; 1998

57. George Gershwin (38); Composer; Brain tumor; 1937

58. Harry Chapin (38); Singer/songwriter; Auto accident; 1981

59. John F. Kennedy Jr. (38); Journalist/publisher; Plane crash; 1999

60. Roberto Clemente (38); Baseball player; Plane crash; 1972

61. Sam Kinison (38); Comedian; Auto accident caused by drunk driver; 1992

62. Dennis Wilson (39); Rock 'n' roll drummer; Drowning due to intoxication; 1983

63. Malcolm X (39); Militant civil rights leader; Assassination; 1965

64. Martin Luther King Jr. (39); Civil rights activist/minister; Assassination; 1968

65. Anna Nicole Smith (40); Model/actor; Accidental prescription drug overdose; 2007

66. Kobe Bryant (41); Basketball player; Helicopter crash; 2020

67. Chadwick Boseman (43); Actor; Colon cancer; 2020

34 Facts About Henry VIII's Wives

1. In total, Henry VIII married six times.

2. His first marriage was to Catherine of Aragon.

3. This arrangement was pushed by Catherine and Henry's fathers.

4. Catherine had been married to Henry's older brother Arthur for only a few months. Upon his passing, Henry was betrothed to the widow.

5. In this way, the fathers were guaranteeing an ongoing union between England and Spain.

6. It took Henry nearly a quarter of a century to end this marriage of convenience, but he was finally granted an annulment on the grounds that Catherine had once been married to his brother.

7. The pope refused to grant this annulment, so Henry arranged to receive it from the archbishop of Canterbury.

8. This severely damaged Henry's—and England's—relationship with the Roman Catholic Church.

9. Three years later, Henry closed Catholic monasteries and abbeys.

10. Anne Boleyn, marchioness of Pembroke, was a lady-in-waiting.

11. Anne was an English noble, educated in France, who provided reputable servitude to Queen Catherine.

12. During her period of personal assistance, Anne and Henry began their affair, with Henry proposing marriage roughly six years prior to his annulment from Catherine.

13. Some argue that the pair did not consummate their affair until Henry's annulment was final (although others claim she was pregnant when she married).

14. According to the theory, Anne was not so moralistic that she wouldn't engage with a married man—she had simply seen the fate that befell her sister Mary when she carried on with Henry.

15. When Mary Boleyn finally gave in and consummated her affair with the king, she was rewarded with a pink slip and sent packing.

16. It turns out Mary was the luckier of the two, being simply sent away.

17. Anne, on the other hand, after failing to produce a male heir, was beheaded on trumped-up charges of adultery and witchcraft.

18. Upon Queen Anne's death, he plucked his third wife, Jane Seymour, from Anne's group of attendants.

19. Of course, it has been suggested that Henry's interest in Jane was the true reason he had Anne killed—not the six fingers he accused her of having (which was seen as the mark of the devil).

20. Unfortunately, Jane only lasted a year as Henry's wife.

21. Although she finally brought Henry a legitimate male heir, Edward, Jane succumbed to a fever caused from childbirth complications.

22. Henry agreed to marry Anne of Cleves for her family's advantageous political ties to both the Catholic Church and the Protestant Reformation.

23. With the nuptials already scheduled, Henry was said to have voiced his misgivings upon finally meeting his bride-to-be.

24. While striking in personality, Anne was not as agreeable in appearance as the king hoped: He likened Anne's visage to a horse.

25. Although he followed through with the vows, Anne—dubbed "Flanders Mare" by Henry—was released from her role as queen through an annulment.

26. The tables were turned on the adulterous King Henry by his fifth wife, Catherine Howard, a woman who was known to engage in illicit affairs of her own.

27. Even before marrying the king, Catherine—who was Anne Boleyn's first cousin—was intimately involved with many men.

28. Soon after they wed, the king discovered that his bride was still sowing her oats with other suitors.

29. Unsurprisingly, he had her beheaded as well.

30. Henry's sixth and final wife, Catherine Parr, was famous for her own lengthy list of marriages.

31. With two marriages before Henry and one after, Catherine holds the record for the most-married queen in English history.

32. She is also the only one of Henry's six brides to make it out of her marriage alive and without an annulment.

33. Only four years after their wedding, Henry died at age 55 from obesity-related complications.

34. Catherine escaped the fates of Henry's first five wives and survived as a widow free to marry one last time—to Thomas Seymour, Jane Seymour's brother.

28 Ida B. Wells Facts

1. Ida B. Wells was born in 1862 to enslaved parents who were freed along with the rest of America's slaves in 1865.

2. When Wells was 16, her parents and youngest sibling died during a yellow fever epidemic.

3. In order to keep her family together, Wells took a job as a schoolteacher and raised her younger siblings.

4. Wells moved her family to Memphis, Tennessee.

5. It was here that she experienced the act of racism that launched her career.

6. Wells had bought a first-class train ticket for a "ladies'" car, but the conductor told her to move to the "colored" car to make room for a White man.

7. When Wells refused, the conductor attempted to forcibly move her.

8. As Wells explained in her autobiography, "the moment he caught hold of my arm I fastened my teeth in the back of his hand."

9. It took two more men to drag her off the conductor and off the train.

10. Wells sued the railroad company.

11. She won the case in lower courts, but lost when it was appealed to the Supreme Court of Tennessee.

12. But the case served to instigate the fight for equality.

13. Wells became the co-owner and editor of *Free Speech*, an anti-segregationist newspaper in Memphis.

14. She focused her energies on revealing the horrors of lynching.

15. In her landmark book, *A Red Record: Tabulated Statistics and Alleged Cause of Lynching in the United States*, she showed how horrifyingly common the practice of lynching was.

16. Wells also picked apart one popular excuse used to justify lynching: A Black man's rape of a White woman.

17. "Somebody must show that the Afro-American race is more sinned against than sinning, and it seems to have fallen upon me to do so," Wells said.

18. Wells argued that whenever the rape defense was brought into a lynching case, the truth was that it usually was a voluntary act between a White woman and a Black man.

19. Wells traced the history of this rape defense, and pointed out that White slave owners would often leave for months at a time, leaving their wives under the care of their Black male slaves.

20. In fact, she argued, White-Black sexual liaisons were typically the other way around, with White owners sleeping with or raping Black female slaves.

21. Wells was the first scholar of note to unearth the hypocrisy behind the White man's so-called protection of White women's honor through lynching.

22. "To justify their own barbarism," Wells wrote, "they assume a chivalry they do not possess...no one who reads the record, as it is written in the faces of the million mulattoes in the South, will for a minute conceive that the southern White man had a very chivalrous regard for the honor due the women of his own race, or respect for the womanhood which circumstances placed in his power."

23. Wells concluded that the brutal lynching epidemic was really the result of fear for economic competition, combined with White men's anger at voluntary liaisons between White women and Black men and a large helping of racism.

24. She married fellow activist and writer F. L. Barnett in 1895.

25. Together the couple had four children and worked to help the African American community in Chicago.

26. Wells also was a founding member of the NAACP and the first president of The Negro Fellowship League.

27. Wells's work for civil rights continued until her death in 1931.

28. In 1930, shortly before her death, she ran for the Illinois Senate.

15 Notable People Who Dropped Out of School

1. Thomas Edison
2. Benjamin Franklin
3. Bill Gates
4. Albert Einstein
5. John D. Rockefeller
6. Walt Disney
7. Richard Branson
8. George Burns
9. Colonel Sanders
10. Charles Dickens
11. Elton John
12. Ray Kroc
13. Harry Houdini
14. Ringo Starr
15. Princess Diana

36 Facts About Genghis Khan

1. Genghis Khan was born to a tribal chief in 1162, probably at Dadal Sum, in the Hentii region of what is now Mongolia.

2. At birth, he was called Temujin.

3. Legend says that he came into the world clutching a blood clot in this right hand.

4. At age nine, his father was poisoned to death by an enemy clan.

5. After his father's death, Temujin's clan deserted him, his mother, and his siblings in order to avoid having to feed them.

6. For three years, he and the remainder of his family wandered the land, living from hand to mouth.

7. He killed his own older half-brother in order to take over as head of the family.

8. Temujin was captured and enslaved by the clan that had abandoned him and his family.

9. He was eventually able to escape.

10. In 1178, when he was 16, Temujin married his first wife Borte.

11. Together, they had four sons and an unknown number of daughters.

12. At one point Borte was kidnapped, and in rescuing her Temujin began building his reputation as a warrior and making alliances.

13. After convincing some tribesmen to follow him, he eventually became one of history's most successful political and military leaders.

14. Temujin formally adopted the name Genghis Khan in 1206.

15. He united the nomadic Mongol tribes into a vast sphere of influence.

16. The Mongol Empire lasted from 1206 to 1368.

17. It was the largest contiguous dominion in world history, stretching from the Caspian Sea to the Sea of Japan.

18. At the empire's peak, it encompassed more than 700 tribes and cities.

19. To this day, its size is only surpassed by the British Empire in the 20th century.

20. Genghis Khan gave his people more than just land.

21. He introduced a writing system that is still in use today.

22. Genghis wrote the first laws that governed all Mongols.

23. He regulated hunting to make sure that everybody had food.

24. Genghis also created a judicial system that guaranteed fair trials.

25. His determination to create unity swept old tribal rivalries aside and made everyone feel like a single people, the "Mongols."

26. Today, Genghis Khan is seen as one of the founding fathers of Mongolia.

27. However, he is not so fondly remembered in Asia, the Middle East, and Europe, where he is regarded as a ruthless and bloodthirsty conqueror.

28. It seems that Genghis was the father of more than the Mongol nation.

29. An international team of geneticists determined that one in every 200 men now living is a relative of the great Mongol ruler.

30. More than 16 million men in Central Asia have been identified as carrying the same Y chromosome as Genghis Khan.

31. Outside of his marriage, Genghis himself is thought to have had over hundreds of children over the course of his life.

32. Another key reason is this: Genghis's sons and other male descendants had many children by many women.

33. One son, Tushi, may have had 40 sons of his own.

34. One of Genghis's grandsons, Chinese dynastic ruler Kublai Khan, fathered 22 sons with recognized wives and an unknown number with the scores of women he kept as concubines.

35. Genetically speaking, Genghis continues to "live on" because the male chromosome is passed directly from father to son, with no change other than random mutations (which are typically insignificant).

36. When geneticists identify those mutations, called "markers," they can chart the course of male descendants through centuries.

60 Facts About Walt Disney

1. Walt Disney was born in Chicago on December 5, 1901.

2. He came from humble beginnings.

3. Walt's father was a farmer and carpenter, running the household with an overly firm hand.

4. Walt and his siblings, working the Disney land near Kansas City, often found themselves on the receiving end of a strap as their dad doled out the discipline.

5. As a young boy, Walt took advantage of his infrequent free time by drawing, improvising by using a piece of coal on toilet paper.

6. When the Disneys moved back to Chicago in 1917, Walt attended art classes at the Chicago Academy of Fine Arts.

7. As a youngster, Walt made extra spending money by selling drawings to his neighbors.

8. Armed with forged birth records, Walt joined the American Red Cross Ambulance Corps and entered World War I in 1918, just before it ended.

9. He was sent to France, where he drove an ambulance.

10. He returned to Kansas City, where two of his brothers continued to run the Disney farm.

11. Walt Disney grew his trademark mustache at the age of 25.

12. Although he was rejected as a cartoonist for the *Kansas City Star*, Walt soon began to create animated film ads for movie theaters, working with a young Ub Iwerks, who eventually became an important member of Disney Studios.

13. In 1922, Disney started Laugh-O-Gram Films, producing short cartoons based on fairy tales.

14. But the business closed within a year, and Walt headed to Hollywood, intent on directing feature films.

15. Finding no work as a director, he revisited film animation.

16. With emotional and financial support from his brother Roy, Walt slowly began to make a name for Disney Brothers Studios on the West Coast.

17. The company introduced "Oswald the Lucky Rabbit" in 1927 but lost the popular character the next year to a different company.

18. Disney was left in the position of having to create another cute and clever cartoon animal.

19. According to Disney, the Kansas City office of Laugh-O-Gram Films was rampant with mice.

20. One mouse was a particular favorite of Walt's—this rodent became the inspiration for Disney's next cartoon character.

21. Working with Iwerks, and borrowing copiously from their former meal ticket Oswald, Disney Studios produced *Steamboat Willie* in 1928.

22. The first commercially released Mickey Mouse cartoon, *Steamboat Willie* was also the first Disney cartoon to feature synchronized sound.

23. The cartoon premiered in New York City on November 18, 1928.

24. The animated star introduced additional Disney icons, including girlfriend Minnie Mouse, the always-exasperated Donald Duck, faithful hound Pluto, and dim-but-devoted pal Goofy.

25. Disney provided the voice for both Mickey and Minnie Mouse for nearly 20 years.

26. Although television made him world-famous, Disney experienced terrible stage fright every time he stepped in front of the camera.

27. Disney's cartoons won every Animated Short Subject Academy Award during the 1930s.

28. As Disney Studios moved into animated feature films (which everyone said would never work), Walt began to wield the power he'd gained as one of Hollywood's most prominent producers.

29. Disney's first animated feature film, *Snow White and the Seven Dwarfs*, cost nearly $1.5 million to produce.

30. *Snow White* was a huge gamble for Disney Studios, but went on to financial success and critical acclaim.

31. Following the success of *Snow White and the Seven Dwarfs*, Disney and his brother Roy gifted their parents with a new house close to their studios.

32. A month later, their mother died from asphyxiation caused by a broken furnace in the new home.

33. It was a tragedy from which Walt Disney never recovered.

34. Disney's strict upbringing and harsh bouts of discipline had left him with a suspicious, ultraconservative mindset.

35. Bad language by employees in the presence of women resulted in immediate discharge—no matter the inconvenience.

36. Disney was prone to creating a double standard between himself and his employees.

37. For example, while Walt kept his mustache for most of his life, all other Disney workers were prohibited from wearing any facial hair.

38. While he considered his artists and animators "family," he treated them in the same way Elias Disney had treated his family—unfairly.

39. Promised bonuses turned into layoffs; higher-paid artists resorted to giving their assistants raises out of their own pockets.

40. By 1941, Disney's animators went on strike, supported by the Screen Cartoonists Guild.

41. Walt was convinced, and stated publicly, that the strike was the result of Communist agitators infiltrating Hollywood.

42. Settled after five weeks, the Guild won on all counts, and the "Disney family" became cynically known as the "Mouse Factory."

43. Disney was also suspected of being a Nazi sympathizer.

44. He often attended American Nazi Party meetings before the beginning of World War II.

45. When prominent German filmmaker Leni Riefenstahl tried to screen her films for Hollywood studios, only Disney agreed to meet her.

46. Yet, when World War II began, Disney projected a strictly all-American image and became closely allied with J. Edgar Hoover and the FBI.

47. According to FBI files, Hoover recruited Disney in late 1940 to be an informant, flagging potential Communist sympathizers among Hollywood stars and executives.

48. In September 1947, Disney was called by the House Un-American Activities Committee to testify on Communist influence in the motion picture industry.

49. He fingered several of his former artists as Reds, again blaming much of the 1941 labor strike on their efforts.

50. He also identified the League of Women Voters as a Communist-fronted organization.

51. Later that evening, his wife pointed out that he meant the League of Women Shoppers, a consumer group that had supported the Guild strike.

52. Disney's testimony contributed to the "Hollywood Blacklist," which included anyone in the industry even remotely suspected of Communist affiliation.

53. The list resulted in many damaged or lost careers, as well as a number of suicides in the cinematic community.

54. Included in the turmoil was Charlie Chaplin, whom Disney referred to as "the little Commie."

55. The FBI rewarded Walt Disney for his efforts by naming him "SAC—Special Agent in Charge" in 1954, just as he was about to open his first magical amusement park, Disneyland.

56. Disney and Hoover continued to be pen pals into the 1960s.

57. The FBI made script "suggestions" for *Moon Pilot*, a Disney comedy that initially spoofed the abilities of the Bureau. The bumbling FBI agents in the screenplay became generic government agents before the film's release.

58. Walt Disney won more Academy Awards than any other individual: 32 total.

59. After a lifetime of chain-smoking, Disney developed lung cancer and died in December 1966.

60. His plans for Disney World in Florida had just begun—the park didn't open until 1971.

22 Famous People Who Were Adopted

1. Babe Ruth
2. Bo Diddley
3. Dave Thomas
4. Debbie Harry
5. Malcom X
6. Steve Jobs
7. Scott Hamilton
8. Marilyn Monroe
9. Melissa Gilbert
10. Dr. Ruth Westheimer
11. Harry Caray
12. Faith Hill
13. Jamie Foxx
14. Frances McDormand
15. Ray Liotta
16. Ted Danson
17. Sarah McLachlan
18. Keegan-Michael Key
19. Liz Phair
20. Simone Biles
21. Run DMC
22. Michael Bay

62 Facts About Women of the Wild West

Annie Oakley

1. Annie Oakley was born Phoebe Ann Oakley Moses in Dark County, Ohio, in 1860.

2. She was shooting like a pro by age 12.

3. Germany's Kaiser Wilhelm II trusted her with a gun so much that he let her shoot the ash off his cigarette while he smoked it.

4. Oakley is the only woman of the Wild West to have a Broadway musical loosely based on her life (*Annie Get Your Gun*), which depicts her stint in Buffalo Bill's famous traveling show.

5. When Oakley died in 1926, it was discovered that her entire fortune had been spent on various charities, including women's rights and children's services.

Calamity Jane

6. Like Annie Oakley, Calamity Jane was a sharpshooter by the time she was a young woman.

7. Born Martha Jane Canary in Missouri around 1856, Jane was said to be a whiskey-drinking, "don't-mess-with-me" kind of gal.

8. She married a man named Burk at age 33. During her brief stint as a married woman, Calamity Jane gave birth to a daughter, who was raised in a convent.

9. Calamity Jane allegedly penned an autobiography, *The Life and Adventures of Calamity Jane*, in 1896. Some sources speculate that the book was actually written by a ghostwriter.

10. When Jane died in 1903, she asked to be buried next to Wild Bill Hickock.

11. Rumor has it that Hickock was the only man she ever loved.

Belle Starr

12. Belle Starr was born Myra Maybelle Shirley in 1848.

13. Frank and Jesse James' gang hid out at her family's farm when she was a kid, and from then on she was hooked on the outlaw life.

14. Later, when her husband Jim Reed shot a man, the two went on the run, robbing banks and counterfeiting.

15. Starr was known to wear feathers in her hair, buckskins, and a pistol on each hip.

16. She was shot in the back while riding her horse in 1889—it's still unclear whether her death was an accident or murder.

Charley Parkhurst

17. Our next western woman—Charley Parkhurst—actually lived out most of her life successfully disguised as a man.

18. Born in 1812, Parkhurst lived well into her 60s, in spite of being a hard-drinking, tobacco-chewing, fearless, one-eyed brute.

19. She drove stages for Wells Fargo and the California Stage Company, not an easy or particularly safe career.

20. Using her secret identity, Parkhurst was a registered voter and may have been the first American woman to cast a ballot.

21. Her true identity wasn't revealed until her death in 1879, much to the surprise of her friends.

Josephine Sarah Marcus

22. Next up is Josephine Sarah Marcus, a smolderingly good-looking actor born in 1861.

23. Marcus came to Tombstone, Arizona, while touring with a theater group.

24. She stuck around to marry sheriff John Behan.

25. But when Wyatt Earp showed up, her marriage went cold, and she and Earp reportedly fell in love.

26. Marcus was supposedly the reason behind the famous gunfight at the OK Corral—a 30-second flurry of gunfire involving Doc Holliday, the Clayton Brothers, and the Earps.

27. She passed away in 1944 and claimed until her dying day that Wyatt Earp was her one and only true love.

Etta Place

28. Like many women of the Wild West, much of Etta Place's life is shrouded in mystery and legend.

29. Evidence seems to indicate that Place was born around 1878 and became a prostitute at Fanny Porter's bordello in San Antonio, Texas.

30. When the Wild Bunch came through, Place went with them to rob banks.

31. She wasn't with the boys when they were killed in South America in 1909, and some believe she became a cattle rustler.

Laura Bullion

32. More commonly referred to as "Rose of the Wild Bunch," our next woman of the West is Laura Bullion.

33. This outlaw was born in 1876 in Knickerbocker, Texas, and learned the trade by observing her bank-robbing father.

34. Eventually hooking up with Butch Cassidy and his Wild Bunch, Bullion fenced money for the group and became romantically involved with several members.

35. Most of those men died by the gun, but "The Thorny Rose" gave up her life of crime after serving time in prison.

36. She died a respectable seamstress in Memphis, Tennessee, in 1961.

Lillian Smith

37. Lillian Smith famously considered Annie Oakley her nemesis.

38. Born in 1871, Smith joined Buffalo Bill's show at age 15 and was notorious for bragging about her superior skills, wearing flashy clothes, and cursing like a sailor.

39. When the show went to England in 1887, Smith shot poorly and was ridiculed while Oakley rose to the occasion.

40. This crushing blow put Smith behind Oakley in the history books, and she died in 1930, a relatively obscure relic of the Old West.

Pearl de Vere

41. Pearl de Vere is next—one of the most famous madams in history.

42. This red-haired siren was born in Indiana around 1860 and made her way to Colorado during the Silver Panic of 1893.

43. De Vere told her family she was a dress designer, but in fact rose to fame as the head of The Old Homestead, a luxurious brothel in Cripple Creek, Colorado.

44. The price of a night's stay could cost patrons more than $200—at a time when most hotels charged around $3 a night!

45. The building was reportedly equipped with an intercom system and boasted fine carpets and chandeliers.

46. An overdose of morphine killed Pearl de Vere in 1897, but it's unclear whether it was accidental or not.

Ellen Liddy Watson

47. Also known as "Cattle Kate," this lady of the West made a name for herself in the late 1800s when she was in her mid-20s.

48. Ellen Liddy Watson worked as a cook in the Rawlins House hotel and there she met her true love, James Averell.

49. The two were hanged in 1889 by vigilantes who claimed Averell and Watson were cattle rustlers.

50. But it is now believed that their murder was unjustified, the result of an abuse of power by land and cattle owners.

Pearl Hart

51. Pearl Hart was born in Canada around 1870.

52. By the time she was 17, she was married to a gambler and on a train to America.

53. She especially liked life in the West, and, at 22, tried to leave her husband to pursue opportunities there.

54. Her husband followed her and won her back, but Hart was already living it up with cigarettes, liquor, and even morphine.

55. After her husband left to fight in the Spanish-American War, Hart met a man named Joe Boot.

56. They robbed stagecoaches for a while before she was caught and jailed.

57. Hart is famous for saying, "I shall not consent to be tried under a law in which my sex had no voice in making."

58. She was eventually released, but the rest of her life is unknown.

Rose Dunn

59. In a family of outlaws, it was only a matter of time before "the Rose of Cimarron" started working in the business, too.

60. Rose Dunn met Doolin Gang member George Newcomb and joined him as he and his crew robbed stagecoaches and banks.

61. During a particularly nasty gunfight, Dunn risked her life to supply Newcomb with a gun and bullets and helped him escape after he was wounded in battle.

62. Dunn died around 1950 in her mid-70s, a respectable citizen married to a local politician.

Heights & Zodiac Signs of 45 U.S. Presidents

1. George Washington: 6'2"; Pisces

2. John Adams: 5'7"; Scorpio

3. Thomas Jefferson: 6'2"; Aries

4. James Madison: 5'4"; Pisces

5. James Monroe: 6'0"; Taurus

6. John Quincy Adams: 5'7"; Cancer

7. Andrew Jackson: 6'1"; Pisces

8. Martin Van Buren: 5'6"; Sagittarius

9. William Henry Harrison: 5'8"; Aquarius

10. John Tyler: 6'0"; Aries

11. James Polk: 5'8"; Scorpio

12. Zachary Taylor: 5'8"; Sagittarius

13. Millard Fillmore: 5'9"; Capricorn

14. Franklin Pierce: 5'10"; Sagittarius

15. James Buchanan: 6'0"; Taurus

16. Abraham Lincoln: 6'4"; Aquarius

17. Andrew Johnson: 5'10"; Capricorn

18. Ulysses S. Grant: 5'8"; Taurus

19. Rutherford B. Hayes: 5'8"; Libra

20. James Garfield: 6'0"; Scorpio

21. Chester Arthur: 6'2"; Libra

22. Grover Cleveland: 5'11"; Pisces

23. Benjamin Harrison: 5'6"; Leo

24. William McKinley: 5'7"; Aquarius

25. Theodore Roosevelt: 5'10"; Scorpio

26. William Howard Taft: 6'0"; Virgo

27. Woodrow Wilson: 5'11"; Capricorn

28. Warren Harding: 6'0"; Scorpio

29. Calvin Coolidge: 5'10"; Cancer

30. Herbert Hoover: 5'11"; Leo

31. Franklin Delano Roosevelt: 6'2"; Aquarius

32. Harry Truman: 5'9"; Taurus

33. Dwight Eisenhower: 5'10"; Libra

34. John F. Kennedy: 6'0"; Gemini

35. Lyndon B. Johnson: 6'3"; Virgo

36. Richard M. Nixon: 5'11"; Capricorn

37. Gerald Ford: 6'0"; Cancer

38. Jimmy Carter: 5'9"; Libra

39. Ronald Reagan: 6'1"; Aquarius

40. George H. W. Bush: 6'2"; Gemini

41. Bill Clinton: 6'2"; Leo

42. George W. Bush: 5'11"; Cancer

43. Barack Obama: 6'2"; Leo

44. Donald Trump: 6'3"; Gemini

45. Joe Biden: 6'0"; Scorpio

29 Buster Keaton Facts

1. Buster Keaton was born on October 4, 1895 in Piqua, Kansas.

2. Keaton spent his entire childhood performing with his mother and father in the family vaudeville act in which he was constantly heaved, hurled, and hoisted across the stage.

3. During this era, he learned many of the physical skills he later employed in his silent films, including mimicry, stunt work and falls, and costuming.

4. He never attended school, but he learned to read while on the road.

5. Keaton's first film was *The Butcher Boy* (1917), in which he costarred with Roscoe "Fatty" Arbuckle, one of the silver screen's first superstars.

6. Keaton learned about filmmaking while working on Arbuckle's shorts, which were made for producer Joseph Schenck.

7. When Arbuckle moved to Paramount, Keaton took over Schenck's Comique Films (later called Buster Keaton Productions).

8. Keaton perfected his comic persona and learned how to direct comedy for cinema on the shorts he made for Schenck.

9. Buster Keaton, Charlie Chaplin, and Harold Lloyd were the three top comedians of Hollywood's silent era.

10. In 1923, Keaton made his first feature-length comedy, *The Three Ages*, a spoof of D. W. Griffith's *Intolerance*.

11. Keaton's comedy included large-scale stunts and gags, often involving vehicles and buildings, which he performed with remarkable grace and agility.

12. His stunts were not only remarkable but dangerous.

13. While working on *Sherlock Jr.*, he broke his neck when gallons of water spilled on top of him as part of a gag.

14. He got up and walked away as scripted in order to finish the scene correctly, broken neck and all.

15. Known as "The Great Stone Face," Buster Keaton is instantly recognizable by his trademark porkpie hat.

16. After his Civil War opus *The General* flopped at the box office in 1927, Keaton sold his studio and the rights to his films to MGM.

17. Chaplin and Lloyd continued to enjoy great success and immeasurable wealth, while Keaton endured years of financial problems as well as creative failures because of MGM's inability to use him to his best advantage.

18. In a case of art imitating life, Keaton appeared as a down-on-his-luck silent screen star who plays bridge with Gloria Swanson in Billy Wilder's *Sunset Boulevard* (1950).

19. Although he was reportedly paid only $1,000 and mutters just one line in his brief cameo, Keaton's performance was a brilliant character study of lost dreams and shattered illusions.

20. Keaton played an aging comedian in Chaplin's *Limelight* in 1952.

21. It was the only time in their careers that the two great stars appeared together.

22. In 1959, Buster Keaton was finally appreciated by Hollywood when he was awarded an honorary Oscar.

23. In 1965, Keaton starred in a silent short called *The Railrodder* for the National Film Board of Canada.

24. Wearing his porkpie hat, he rode the rails and traveled from one end of Canada to the other on a motorized handcar, performing gags similar to those that made him a box-office star in the 1920s.

25. *The Railrodder* was the last silent movie of his illustrious career.

26. Although it was completed before *The Railrodder*, the last film